Plant molecular biology

a practical approach

TITLES PUBLISHED IN
THE
PRACTICAL APPROACH
SERIES

Series editors:
Dr D Rickwood
Department of Biology, University of Essex
Wivenhoe Park, Colchester, Essex CO4 3SQ, UK
Dr B D Hames
Department of Biochemistry, University of Leeds
Leeds LS2 9JT, UK

Affinity chromatography

Animal cell culture

Antibodies

Biochemical toxicology

Biological membranes

Carbohydrate analysis

Centrifugation (2nd Edition)

DNA cloning

Drosophila

Electron microscopy
in molecular biology

Gel electrophoresis of nucleic acids

Gel electrophoresis of proteins

Genome analysis

HPLC of small molecules

Human cytogenetics

Human genetic diseases

Immobilised cells and enzymes

Iodinated density gradient media

Lymphocytes

Lymphokines and interferons

Mammalian development

Microcomputers in biology

Mitochondria

Mutagenicity testing

Neurochemistry

Nucleic acid and
protein sequence analysis

Nucleic acid hybridisation

Oligonucleotide synthesis

Photosynthesis:
energy transduction

Plant cell culture

Plant molecular biology

Plasmids

Prostaglandins
and related substances

Spectrophotometry
and spectrofluorimetry

Steroid hormones

Teratocarcinomas
and embryonic stem cells

Transcription and translation

Virology

Yeast

Plant molecular biology

a practical approach

Edited by
C H Shaw

Department of Biological Sciences, University of Durham, Science Laboratories, South Road, Durham DH1 3LE, UK

IRL PRESS

OXFORD · WASHINGTON DC

IRL Press
Eynsham
Oxford
England

British Library Cataloguing in Publication Data

Plant molecular biology.
  1. Plants. Molecular biology
  1. Shaw,C.H. (Charles H.)   II. Series
  581.8′8

ISBN 1 85221 057 5 (hardbound)
ISBN 1 85221 056 7 (softbound)

Library of Congress Cataloging in Publication Data
Plant molecular biology.

  (The Practical approach series)
  Includes index.
  1. Plant molecular biology—Technique. 2. Plant genetic
engineering—Technique. I. Shaw.C.H.
(Charles H.), 1955–   II. Series.
QK728.P54   1988   581.8′8   88–23072

ISBN 1 85221 057 5 (hardbound)
ISBN 1 85221 056 7 (softbound)

Typeset by Élite Typesetting and printed by Information Printing Ltd,
Oxford, England.

Preface

The advent of recombinant DNA technology has resulted in an explosion in plant molecular biology. An unfortunate side-effect has been the increase in the myths associated with this area, in particular the difficulties of working with plant material. In truth, as long as scientists know what they are doing, plants pose no greater hazards as experimental subjects than any other organisms. This is the purpose for which this book was designed: to provide the newcomer and the practitioner with clear and sensible protocols to enable them to perform meaningful experiments in plant molecular biology. In that respect, it does not aim to supplant the many excellent practical texts dealing with molecular biology and recombinant DNA methodology, but merely to supplement them, providing plant-specific protocols.

This book was not completed without the assistance of a large cast of characters. I would like to thank David Hames for the opportunity to edit this volume and for valuable guidance in the execution of that task, the authors for agreeing to the onerous task of preparing their excellent contributions, and the staff at IRL Press for advice, and acting as occasional heavies. I wish to thank particularly the staff of Ward 16S, The Royal Victoria Infirmary, Newcastle-upon-Tyne, for helping to keep my family whole. I would especially like to extend heartwarm thanks to my research group and family, for their forebearance during a difficult period.

Most importantly, I hope that this book will be read and used by the scientific community, for it will be judged a success only if it is useful. I have learnt a lot by editing *Plant Molecular Biology — A Practical Approach*, and I hope that others will gain from reading it.

Charles H.Shaw

for Sam: fight the good fight

Contributors

L.K.Barnett
Department of Biological Sciences, University of Warwick, Coventry CV4 7AL, UK

A.Brennicke
Lehrstuhl Spezielle Botanik, Universität Tübingen, Auf der Morgenstelle 1, D-7400 Tübingen 1, FRG

R.H.A.Coutts
Department of Pure and Applied Biology, Imperial College of Science and Technology, Prince Consort Road, London SE7 2BB, UK

K.H.Cox
Department of Biology, UCLA, Los Angeles, CA 90024-1606, USA

V.A.Dzelzkalns
The Biological Laboratories, Harvard University, 16 Divinity Avenue, Cambridge, MA, USA

J.Erickson
Département de Biologie Moléculaire, Université de Genève, Sciences II, 30 Quai Ernest Ansermet, CH-1211 Genève 4, Switzerland

R.B.Goldberg
Department of Biology, UCLA, Los Angeles, CA 90024-1606, USA

M.Goldschmidt-Clermont
Département de Biologie Moléculaire, Université de Genève, Sciences II, 30 Quai Ernest Ansermet, CH-1211 Genève 4, Switzerland

C.Hawes
Department of Plant Sciences, University of Oxford, South Parks Road, Oxford OX1 3PF, UK

L.Herrera-Estrella
Centro de Investigacion y Estudios Avanzados del IPN, Universiti Irapuato, Apartado Postal 629, 36500 Irapuato, GTO, Mexico

R.Hiesel
Lehrstuhl Spezielle Botanik, Universität Tübingen, Auf der Morgenstelle 1, D-7400 Tübingen 1, FRG

K.D.Jofuku
Laboratorium voor Genetika, Rijksuniversiteit Gent, B-9000 Gent, Belgium

S.MacDowell
Department of Pure and Applied Biology, Imperial College of Science and Technology, Prince Consort Road, London SE7 2BB, UK

S.Mayfield
Département de Biologie Moléculaire, Université de Genève, Sciences II, 30 Quai Ernest Ansermet, CH-1211 Genève 4, Switzerland

B.J.Mulligan
Department of Botany, University of Nottingham, University Park, Nottingham NG7 2RD, UK

C.Robinson
Department of Biological Sciences, University of Warwick, Coventry CV4 7AL, UK

J.-D.Rochaix
Département de Biologie Moléculaire, Université de Genève, Sciences II, 30 Quai Ernest Ansermet, CH-1211 Genève 4, Switzerland

M.W.Saul
Institute for Plant Sciences, Swiss Federal Institute of Technology, Universitätstrasse 2, CH-1892 Zürich, Switzerland

W.Schobel
Lehrstuhl Spezielle Botanik, Universität Tübingen, Auf der Morgenstelle 1, D-7400 Tübingen 1, FRG

W.Schuster
Lehrstuhl Spezielle Botanik, Universität Tübingen, Auf der Morgenstelle 1, D-7400 Tübingen 1, FRG

N.S.Shepherd
Central Research & Development Department, E.I. du Pont de Nemours & Co. Inc., E328/246, Wilmington, DA 19898, USA

R.D.Shillito
CIBA-GEIGY Corporation, PO Box 12257, Research Triangle Park, NC 27709-2257, USA

J.Simpson
Centro de Investigacion y Estudios Avanzados del IPN, Universiti Irapuato, Apartado Postal 629, 36500 Irapuato, GTO, Mexico

M.Szekeres
Institute of Plant Physiology, Biological Research Centre, Hungarian Academy of Sciences, 6701-Szeged, Hungary

P.J.Wise
Department of Pure and Applied Biology, Imperial College of Science and Technology, Prince Consort Road, London SE7 2BB, UK

B.Wissinger
Lehrstuhl Spezielle Botanik, Universität Tübingen, Auf der Morgenstelle 1, D-7400 Tübingen 1, FRG

Contents

Abbreviations

APH(3′)II	aminoglycoside 3′ phosphotransferase, NPT
BAP	6-benzylaminopurine
BCTV	beet curly top virus
BGMV	bean golden mosaic virus
BMe	β-mercaptoethanol
BSA	bovine serum albumin
CAT	chloramphenicol acetyl transferase
ccc	covalently closed circular (supercoiled) DNA
CETAB	hexadecyltrimethylammonium bromide
CIP	calf intestinal alkaline phosphatase
CLV	cassava latent virus
Con A	Concanavalin A
cos	cohesive ends phage λ, site of concatemer cleavage
cms	cytoplasmic male sterility
CNBr	cyanogen bromide
cp	chloroplast
CTAB	hexadecylpyridinium chloride
2,4-D	2,4-dichlorophenoxyacetic acid
DAB	3,3′-diaminobenzidine
DAPI	4-6-diamidino-2-phenylindole
DEPC	diethyl pyrocarbonate
DMSO	dimethylsulphoxide
DNase	deoxyribonuclease
ds	double-stranded
DTT	dithiothreitol
EDTA	ethylene diamine tetraacetic acid
EGTA	ethyleneglycol-bis-(β-aminoethyl ether) $N'N'$-tetraacetic acid
ELISA	enzyme-linked immunosorbent assay
EM	electron microscopy
EtBr	ethidium bromide
FAA	formalin, alcohol, acetic acid
FITC	fluorescein isothiocyanate
HEPES	N-2-hydroxyethylpiperazine-N'-2-ethanesulphonic acid
HP	hygromycin phosphotransferase
IAA	indoleacetic acid
IgG	immunoglobulin G
IPTG	isopropyl-β-D-thiogalactopyranoside
MDV	wheat dwarf virus
MES	2-[N-morpholino]ethanesulphonic acid
m.o.i.	multiplicity of infection
MSV	maize streak virus
mt	mitochondrial
n	nuclear
NAA	naphthaleneacetic acid

nos	nopaline synthase
NPT II	neomycin phosphotransferase
oc	open circular DNA
ocs	octopine synthase
onc	T-DNA oncogenicity functions
PAP	peroxide anti-peroxide complex
PBS	phosphate-buffered saline
PEG	polyethylene glycol
p.f.u.	plaque forming units
PIPES	piperazine-N,N'-bis[2-ethanesulphonic acid]
PMSF	phenylmethylsulphonylfluoride
p.s.i.	pounds per square inch
pt	plastid
PVP	polyvinylpyrrolidone
RFLP	restriction fragment length polymorphism
RhP	rhodamine-conjugated phalloidin
RNase	ribonuclease
rubisco	ribulose bisphosphate carboxylase
SDS	sodium dodecyl sulphate
SGAT	serine glyoxylate amino transferase
ss	single-stranded
SSC	standard saline citrate
T-DNA	DNA of Ti-plasmid transferred to plant
TEMED	$N'N'N'N'$ tetramethyl ethylene diamine
TGMV	tomato golden mosaic virus
Ti-plasmid	tumour-inducing plasmid of *A.tumefaciens*
TNV	tobacco necrosis virus
Tris	tris(hydroxymethyl)aminomethane
vir	Ti-plasmid virulence genes
WGA	wheat germ agglutinin
X-gal	5-bromo-4-chloro-3-indolyl-β-D-galactoside

CHAPTER 1

Analysis of plant gene expression

KATHLEEN H.COX and ROBERT B.GOLDBERG

1. INTRODUCTION

Our knowledge of the cellular processes that control plant gene expression has increased enormously during the last 5 years. This has been due in part to innovative technological advances in molecular biology and the application of these techniques to the analysis of plant gene structure and expression. In this chapter, we describe several qualitative and quantitative gene expression assays that are being used currently in our laboratory. We have concentrated on two types of techniques: those that are specific for plants, or have been optimized for plant systems; and those that are more recent and are not generally available. We have avoided discussion of many techniques pioneered in animal systems that are directly applicable to plant research such as the construction of cDNA libraries (1–3), gel electrophoresis techniques (4) and S1 and primer extension assays (5,6).

This chapter describes procedures for the following techniques.

(i)   RNA isolation. In this section, we have provided procedures for isolating whole cell, nuclear, polysomal and poly(A^+) RNA.

(ii)   Nuclei isolation and run-off transcription. This section describes how to isolate nuclei from a variety of organs. In addition, it provides procedures to label the pre-initiated nuclear RNA to measure the relative transcription rates of individual genes.

(iii)   Synthesis of single-stranded RNA probes and use in RNA blot analysis. This section describes how to synthesize RNA probes from any of the commercially available bacteriophage promoter–RNA polymerase systems. These probes can be used for *in situ* hybridization, RNA titration analysis and RNase mapping. This section also contains detailed protocols for their use in blot analysis.

(iv)   RNA titration analysis. RNA titration is the method of choice for determining the number of copies of a particular RNA sequence. This section provides detailed procedures for this technique, as well as the equations required to analyse the data.

(v)   *In situ* hybridization. This section provides procedures for every aspect of *in situ* hybridization from tissue fixation through data analysis. Steps in the technique that are particularly sensitive to the organ system or tissue source are pointed out to aid in optimizing this procedure for other plant systems.

1

2. RNA ISOLATION

2.1 **Introduction**

A successful RNA purification technique yields undegraded and biologically active RNA that is uncontaminated with other macromolecules. Isolating intact plant RNA has been complicated by high levels of endogenous RNases in plant cells. RNA degradation during isolation has been inhibited by the use of a variety of RNase inhibitors including chelators such as EDTA and EGTA (7), detergents such as sodium dodecyl sulphate (8) and strong denaturants such as guanidinium isothiocyanate (9).

The choice of an extraction procedure will depend primarily on the fraction of RNA that is desired. This section provides procedures for isolating whole cell, nuclear, polysomal and poly(A^+) RNA.

Table 1. Total RNA extraction—guanidinium isothiocyanate.

Stock solutions
Solutions are filtered through 0.45 μm Millipore filters.
1. Guanidinium isothiocyanate extraction buffer
    4 M Guanidinium isothiocyanate
    0.5% Sodium lauroylsarcosinate
    50 mM Tris
    10 mM Na_2EDTA
    5 mM Sodium citrate
    0.1 M β-mercaptoethanol
    Adjust the pH to 7.0, sterile filter and store up to a month at room temperature.
2. 1 M Acetic acid
    Make this solution up in DEPC-treated (see Section 2.2.2) water.
3. Tris/Sarc
    10 mM Tris
    1% Sodium lauroylsarcosinate
    Adjust the pH to 8.0, treat with DEPC (see Section 2.2.2), filter and autoclave.
4. Tris/Sarc/CsCl
    10 mM Tris
    1% Sodium lauroylsarcosinate
    5.7 M CsCl
    Adjust the pH to 8.0.

Procedure
1. Grind the tissue to a fine powder in liquid nitrogen in a Waring blender[a].
2. Transfer the powder and liquid nitrogen to an omni-mixer cup and allow the liquid nitrogen to evaporate[b].
3. Add a 5× excess (v/w) of guanidinium isothiocyanate extraction buffer (0°C) to the powder (50 ml buffer/10 g tissue).
4. Homogenize in an omni-mixer on ice at top speed for 60 sec.
5. Spin the homogenate at 10 000 r.p.m. for 10 min using the SS34 Sorvall rotor at 4°C. This will pellet particulate material.
6. Measure the supernatant. Add 0.025 vols of 1 M acetic acid to lower the pH from 7 to 5. Add 0.75 vols of ethanol. Precipitate overnight at −20°C.
7. Spin the ethanol precipitate at 10 000 r.p.m. for 20 min in the SS34 Sorvall rotor at 4°C.
8. Pour off the supernatant and drain the tube. Wash the pellet with 70% ethanol and dry.

9. Resuspend the pellet in 8 ml of Tris/Sarc. Incubate on ice for 1 h. If there are problems resuspending the pellet, incubate it briefly at 68°C[c].
10. Pellet the insoluble material by spinning at 8000 r.p.m. for 20 min in the SS34 rotor at 4°C.
11. Measure the supernatant and bring it back up to 8 ml by adding Tris/Sarc.
12. Add 8 g of CsCl.
13. Add 1.25 ml of Tris/Sarc/5.7 M CsCl to three 5 ml ultracentrifuge tubes.
14. Overlayer the RNA/CsCl solution. Bring the solution to within ⅛ in of the top of the tube by adding Tris/Sarc.
15. Spin at 35 000 r.p.m. for 20 h in the AH650 Sorvall rotor at 20°C. This spin will pellet the RNA.
16. Remove the CsCl down to the cushion. Carefully and thoroughly wash the sides of the tube with DEPC-treated water[d]. Remove the CsCl cushion.
17. Wash the pellet with 70% ethanol and dry.
18. Resuspend the pellets in DEPC-treated water on ice for 1 h. Heat briefly to 68°C if necessary.
19. Spin out the insoluble material with a brief spin in a microfuge at 4°C.
20. Add ¹⁄₁₀ vol. of 3 M sodium acetate and 2 vols of ethanol.
21. Store at −20°C.

---

[a] Steps 1–4 should be done in a 4°C coldroom.
[b] When the liquid nitrogen is completely evaporated, the tissue will be lighter in colour and will not clump. It is important that the liquid nitrogen evaporate so that ice crystals do not form in the buffer. However, it is even more important that the tissue does not thaw.
[c] It is important at this step that the pellet be totally resuspended or the RNA yield will be low. Tissues that contain a large amount of starch or polysaccharides may result in large pellets which are difficult to resuspend.
[d] This step must be done carefully to prevent contamination of the RNA pellet by the supernatant.

2.2 General guidelines for RNA isolation

2.2.1 *Glassware*

All glassware and plastics should be dipped in water containing 0.02% diethylpyrocarbonate (DEPC). All items that can withstand the heat should be baked overnight at a temperature greater than 200°C or autoclaved. The rest should be dried overnight at room temperature.

2.2.2 *Solutions*

Most of the solutions (see individual notes) should be treated with 0.05% DEPC. DEPC is added to the solution, it is shaken vigorously and autoclaved for 30 min to break down the DEPC. The hot solution should be swirled after autoclaving to remove residual CO_2, a DEPC breakdown product.

2.2.3 *Plant harvesting*

If insects or eggs are present, the plant material should be rinsed with water before harvesting. It is then frozen in liquid nitrogen as soon as it is harvested and should not be allowed to thaw at any subsequent point. Plant material can be stored at −80°C for years without RNA degradation.

2.2.4 *Technique*

Gloves should be worn at all times. All solutions, rotors and glassware should be pre-cooled to 4°C before use. Stock solutions should be stored at −20°C between uses.

2.3 **Whole cell RNA isolation**

This total cellular RNA isolation procedure relies on a high concentration of the chaotropic agent, guanidinium isothiocyanate, in conjunction with the reductant β-mercaptoethanol. Guanidinium isothiocyanate rapidly denatures proteins resulting in dissociation of ribonucleoprotein complexes and inactivation of enzymes including RNases. The procedure shown in *Table 1* is a modification of the procedure of Chirgwin *et al.* (9) that we have used to isolate total RNA from tobacco roots and stems. These RNAs appear to be undegraded as analysed by denaturing gels, and they have been used for RNA blot analysis. We have not determined if this procedure results in RNA of sufficient purity to be used as a template for enzymatic reactions.

The cells are lysed in guanidinium isothiocyanate. After lysis there are two different procedures that can be used to isolate the RNA. One involves a series of precipitations in guanidinium hydrochloride (9) and the other, which is provided in our protocol, uses centrifugation of the RNA through a CsCl cushion to isolate the RNA from the rest of the cellular macromolecules (9).

The advantage in using this technique is that it is fast and results in undegraded RNA. The disadvantage of this technique is that we have observed that, depending on the source of the plant material, the pellet formed after the initial ethanol precipitation can be very hard to resuspend. When this occurs, the RNA yield is low.

2.4 **Nuclear RNA isolation**

Nuclear RNA can be isolated by the same procedure as outlined in Section 3.4. This technique has been used in our laboratory to isolate RNA for *in vitro* transcription and solution hybridization analysis. We have not used it for blot analysis and do not know if these RNAs are full length.

2.5 **Polysomal RNA isolation**

In order to analyse a mRNA population that is translated, as opposed to total cytoplasmic RNA, it is necessary to isolate polysomes and then release the mRNA from those structures. We have found that a modification of the Jackson and Larkins (7) procedure shown below results in high yields of undegraded polysomal mRNA. Stock solutions are given in *Table 2*.

This technique utilizes high pH, detergents, EGTA and β-mercaptoethanol to inhibit RNases during the lysis steps. The mRNA is separated from the cytoplasmic extract containing endogenous RNases by pelleting the RNA through a sucrose cushion. The pellet contains the polysomes along with some chromatin (a DNase step can be included in the protocol if the RNA must be free of DNA) and membrane bound polysaccharides, while RNases remain in the supernatant. For this reason it is critical that steps in which the supernatant is removed (steps xi and xii) be done carefully so the pellet is not contaminated.

The first centrifugation spin (step vi) removes the nuclei from the cellular homogenate. Some of the nuclei lyse before this step, contaminating the subsequent polysomal RNA with nuclear RNA. If the polysomal RNA must be free of nuclear RNA, the EDTA-release procedure described in *Table 3* can be used (10,11). The rationale behind this procedure is that the mRNAs associated with polysomes will be released by the EDTA and be found in a fraction which sediments at less than 80S, while contaminating nuclear RNAs will be unaffected and sediment at greater than 100S. The level of contaminating nuclear RNA is low without the EDTA-release procedure. Therefore for general purposes, we do not include these steps.

The polysome profile (step xvi) can be used to assess the quality of the isolated RNA. Profiles vary slightly from organ to organ. However, on average, polysome preparations from organs other than leaf peak at 5-mers to 6-mers. Distinct peaks can be seen up to 10-mers. Leaf profiles peak at 7-mers to 8-mers with distinct peaks to 12-mers.

This technique has been used with minor modifications to isolate mRNA from various plants and organs. We have modified the extraction buffer slightly for use with different plants (see *Table 2*). mRNA isolated by this procedure has been used in a variety of techniques including blot analysis, S1 analysis, cDNA synthesis and R-loop analysis.

For the polysomal mRNA extraction carry out steps i–v and vii in the coldroom (all solutions are given in *Table 2*).

(i)   Grind the tissue in liquid nitrogen in a Waring blender for 2 min or until the tissue becomes a uniform fine powder (see *Table 5* for tissue:buffer ratios).

(ii)   Allow the liquid nitrogen to completely evaporate without thawing the tissue. Residual liquid nitrogen will cause formation of ice crystals and foaming during the homogenization step.

(iii)   Add the powder to 200 ml of extraction buffer (0°C) in the omni-mixer cup.

(iv)   Immerse the omni-mixer cup in ice and homogenize the tissue for 1.5 min at high speed.

(v)   If the homogenate contains a large amount of debris, filter it through Nitex cloth (12 in × 12 in, 44 micron, monofilament, nylon mesh, Small Parts Inc., Miami, FL). Nitex should be DEPC-treated (Section 2.2.2) and autoclaved before use.

(vi)   Spin the homogenate at 10 000 r.p.m. at 4°C in a Sorvall SS34 rotor for 20 min. This step removes debris, unlysed cells and organelles.

(vii)   Carefully decant the supernatant into a glass beaker with a stir bar on ice. Add 20% detergent solution to a final concentration of 1%. Stir at moderate speed for 30 min. This step solubilizes membrane-bound polysomes.

5

Table 2. Stock solutions for polysomal RNA isolation.

Solutions are filtered through 0.45 μm Millipore filters and stored at −20°C.
1. 2× Salts for extraction buffer (see Section 2.5)

Tobacco extractions	Soybean extractions
0.4 M Tris	0.4 M Tris
0.2 M KCl	0.8 M KCl
0.05 M EGTA	0.05 M EGTA
0.07 M MgCl$_2$·6H$_2$O	0.07 M MgCl$_2$·6H$_2$O

Adjust the pH to 9.0 with KOH, filter, treat with DEPC (see Section 2.2.2) and autoclave.
2. 10× Salts for sucrose cushion and resuspension buffer[a]
    0.4 M Tris
    2.0 M KCl
    0.05 M EGTA
    0.3 M MgCl$_2$·6H$_2$O
    Adjust the pH to 9.0 with KOH, filter, treat with DEPC (see Section 2.2.2) and autoclave.
3. 10× Salts for analytical polysome gradients[a]
    0.4 M Tris
    0.2 M KCl
    0.1 M MgCl$_2$·6H$_2$O
    Adjust the pH to 8.5 with HCl, filter, treat with DEPC (see Section 2.2.2) and autoclave.
4. 2 M Sucrose–DEPC
    Treat with DEPC (see Section 2.2.2) and autoclave.
5. 2 M Sucrose[b]
    Autoclave.
6. Detergents[c]
    20% Brij 35 (Sigma)
    20% Tween 40 (Sigma)
    20% Nonidet P-40 (BRL)
    Autoclave.
7. 2× Salts for proteinase K digestion (2× TES)
    20 mM Tris
     0.2 mM EDTA
     2% Sodium lauroylsarcosinate
    Adjust the pH to 7.6 with HCl, filter, treat with DEPC (see Section 2.2.2) and autoclave.
8. 3 M Sodium acetate
    Adjust the pH to 6.0 with glacial acetic acid, filter, treat with DEPC (see Section 2.2.2) and autoclave.
9. Extraction buffer
    Final concentrations are: 0.2 M Tris/0.4 M (0.1 M) KCl/0.025 M EGTA/0.035 M MgCl$_2$/1% Triton X-100/0.5% sodium deoxycholate/1 mM spermidine–HCl/5 mM (25 mM) β-mercapto-ethanol/0.5 M sucrose
      100 ml H$_2$O
       50 μl DEPC
    Autoclave. Swirl. While still hot, pour 48 ml into a 250 ml Erlenmeyer flask containing:
      1 g deoxycholate (Sigma)
      51 mg spermidine–HCl (Sigma)
       2 ml Triton X-100
    Shake into solution.
    Add:
      100 ml 2× salts for extraction buffer
       50 ml 2 M sucrose

Keep the solution on ice. Just before the extraction add:

    70 µl β-mercaptoethanol (soybean buffer)

    350 µl β-mercaptoethanol (tobacco buffer)

10. Sucrose cushion

Final concentrations are: 0.04 M Tris/5 mM EGTA/0.2 M KCl/0.03 M MgCl$_2$/1.8 M sucrose/ 5 mM (25 mM) β-mercaptoethanol/pH 9

    90 ml 2 M sucrose

    10 ml 10× salts for sucrose cushion

    35 µl (175 µl) β-mercaptoethanol

Cool to 0°C.

11. Resuspension buffer

Final concentrations are: 40 mM Tris/5 mM EGTA/0.2 M KCl/0.03 M MgCl$_2$/5 mM β-mercaptoethanol

    45 ml DEPC-treated H$_2$O

    5 ml 10× salts for sucrose cushion and resuspension

    17.5 µl β-mercaptoethanol

Cool to 0°C.

12. Sucrose gradients

Make 5 ml of sucrose gradients from the following solutions.

%Sucrose	2M Sucrose (No DEPC) (ml)	10× Salts (ml)	DEPC-treated H$_2$O (ml)
60	8.8	1	0.2
45	6.6	1	2.4
30	4.4	1	4.6
15	2.2	1	6.8

Construct the gradients as follows: 0.8 ml 60% sucrose, 1.6 ml 45% sucrose, 1.6 ml 30% sucrose and 0.8 ml 15% sucrose. Let the gradients diffuse for 18 h at 4°C.

[a] As the pH gets close to 9.0, a precipitate will form. It will disappear after stirring for a few minutes.
[b] DEPC interferes with UV scanning.
[c] Before using, heat this solution to 40°C with stirring to get the detergents back into solution.

(viii) Spin the supernatant at 10 000 r.p.m. at 4°C for 30 min in a SS34 Sorvall rotor.

(ix) Carefully decant the supernatant into a beaker on ice. Transfer the supernatant to eight 36 ml polycarbonate tubes containing a 5 ml sucrose cushion on ice. Fill the tubes to just above the shoulder.

(x) Spin at 4°C for 16 h at 33 000 r.p.m. in a Ti865 Sorvall rotor.

(xi) Remove the tubes from the rotor and place them at an angle on ice so that the RNA pellets are covered by the sucrose cushion. The pellets will be small and slightly amber.

(xii) Aspirate off the supernatant down to the cushion. Rinse the sides of the tube with cold DEPC-treated H$_2$O (Section 2.2.2). Aspirate down to the sucrose cushion. Quickly invert the tube so that the pellet is on top and remove the last of the sucrose cushion. Repeat for the remainder of the tubes.

(xiii) Resuspend the pellets in a total of 5 ml of resuspension buffer for 2 h on ice.

Table 3. EDTA-release of polysomal mRNA.

Stock solutions
1. 10× Salts for polysome gradients
   0.4 M Tris
   0.2 M KCl
   0.1 M MgCl$_2$
   Adjust the pH to 8.5 with KOH, treat with DEPC (see Section 2.2.2) and autoclave.
2. 10× Salts for EDTA-release gradients
   0.4 M Tris
   0.2 M KCl
   10 mM EDTA
   Adjust the pH to 8.5 with KOH, treat with DEPC (see Section 2.2.2) and autoclave.
3. 10× Resuspension buffer minus MgCl$_2$
   0.4 M Tris
   2.0 M KCl
   50 mM EGTA
   Adjust the pH to 8.5 with KOH, treat with DEPC (see Section 2.2.2) and autoclave. Add
   5 mM BMe to the 1× buffer shortly before use.
4. TES
   10 mM Tris
   0.1 mM EDTA
   0.1% SDS
   Adjust the pH to 7.6, treat with DEPC (see Section 2.2.2) and autoclave.

Procedure
1. Follow steps i–xv of the polysomal RNA isolation procedure (Section 2.5).
2. Layer up to 8 mg on each preparative sucrose gradient. Use 10× polysome gradient salts.

% Sucrose	2 M Sucrose (ml)	10× Salts (ml)	DEPC-treated H$_2$O (ml)
32.5	23.8	5	21.2
27	39.4	10	50.6
21	23.0	7.5	44.5
16	11.7	5	33.3
10	7.3	5	37.7

Construct the gradient as follows: 5.5 ml of 32.5% sucrose, 11.8 ml of 27% sucrose, 7.8 ml of
21% sucrose, 5.0 ml of 16% sucrose and 4.5 ml of 10% sucrose. Let the gradients diffuse for
18 h at room temperature and cool in the 4°C room before use.
3. Spin at 27 000 r.p.m. for 1.25 h in the Sorvall AH627 rotor at 2°C.
4. Under these conditions, the >100S polysomes occupy 63% of the gradient. Fractionate in the
   4°C room. Collect the >100S polysomes directly into ¹⁄₁₀ vol. of 3 M sodium acetate and 2 vols
   of ethanol.
5. Precipitate overnight at −20°C.
6. Pellet the polysomes by spinning the ethanol precipitate at 25 000 r.p.m. for 30 min in the
   Sorvall AH627 rotor at −10°C. Drain the supernatant and dry the pellet.
7. Resuspend the pellets in 1 ml/pellet of resuspension buffer minus MgCl$_2$ on ice. Determine the
   RNA recovery.
8. Add EDTA to a final concentration of 0.1 M.
9. Layer up to 3 mg RNA on each preparative sucrose gradient using 10× salts for EDTA
   gradients instead of 10× salts for polysome gradients. Spin at 27 000 r.p.m. for 8 h in the
   Sorvall AH627 rotor at 2°C.
10. Under these conditions, the <80S portion occupies 70% of the gradient. Fractionate in the 4°C

room and collect the <80S portion directly into ¹/₁₀ vol. of 3 M sodium acetate and 2 vols of ethanol. Precipitate at −20°C overnight.

11. Pellet the <80S mRNA by spinning the ethanol precipitate at 27 000 r.p.m. for 30 min in the Sorvall AH627 rotor at −10°C. Drain the supernatant and dry the pellet.
12. Resuspend the pellets in 0.5 ml of TES/pellet and assay the RNA concentration.
13. Add an equal volume of 1 mg/ml pre-digested (30 min, 37°C, in TES) proteinase K. Incubate at room temperature for 30 min.
14. Add ¹/₁₀ vol. of 3 M sodium acetate and 2 vols of ethanol. Store at −20°C.

(xiv)   Pellet the remaining insoluble material by spinning at 1000 r.p.m. at 4°C for 10 min in the SS34 Sorvall rotor. Firm up the pellet by briefly increasing the speed to 5000 r.p.m.

(xv)   Remove the supernatant and assay the RNA recovery.

(xvi)   Assay the polysome profile using sucrose gradients. Layer 80 µg of RNA on the gradient. Spin the gradients at 50 000 r.p.m. at 6°C for 45 min in a Sorvall AH650 rotor. Take the gradients up and down as slowly and as gently as possible (A-slow and reograd on Sorvall centrifuges). Analyse the gradients with a ISCO UV monitor.

(xvii)   Remove the ribosomal proteins by adding an equal volume of 1 mg/ml pre-digested proteinase K (pre-digest proteinase K at 37°C for 30 min in 2× TES). Incubate at room temperature for 30 min.

(xviii) Add ¹/₁₀ vol. of 3 M sodium acetate (pH 6.0) and 2 vols of 100% ethanol to precipitate the RNA. Store at −20°C.

2.6 Isolation of poly(A⁺) RNA

Poly(A⁺) RNA can be isolated from total or polysomal RNA by affinity chromatography on oligo-dT−cellulose columns detailed in *Table 4* (12). The RNA is bound to the column in the presence of high salt. The column is then extensively washed with high salt buffer to remove contaminating poly(A⁻) RNA (mostly ribosomal RNA). The poly(A⁺) RNA is eluted in low ionic strength buffer. We find that two passages through the column are required to remove the contaminating poly(A⁻) RNA. This purification method works well for RNA molecules with average to long poly(A) tails (12). *Table 5* shows typical polysomal and poly(A⁺) RNA yields from a variety of tobacco organs and from tobacco seeds.

3. NUCLEI ISOLATION AND RUN-OFF TRANSCRIPTION

Regulation of gene expression occurs at many different levels in eukaryotes ranging from processes that initiate transcription to those that control protein turnover. Insight into the mechanisms by which individual genes are regulated requires techniques that can measure these dynamic processes. One such technique, run-off transcription, can be used to measure the relative transcription rates of genes (13–18). It is most often used to measure the relative transcription rates of multiple genes in the same developmental state, or to measure the transcription rates of the same gene in different developmental states, such as in different organs or at different stages of development.

Table 4. Poly(A$^+$) RNA isolation.

Solutions
1. Elution buffer (EB)
     10 mM Tris–HCl (pH 7.6)
      1 mM EDTA
      0.1% SDS
     Treat with DEPC (see Section 2.2.2) and autoclave.
2. Binding buffer (BB)
      0.5 M NaCl
     10 mM Tris–HCl (pH 7.6)
      1 mM EDTA
     0.1% SDS
     Treat with DEPC (see Section 2.2.2) and autoclave.
3. 0.1 M NaOH
     Dissolve the NaOH in DEPC-treated (see Section 2.2.2) and autoclaved water.

Preparation of the columns
1. Place 0.2 g of oligo-dT cellulose over a thin layer of 80 μm glass beads into a glass-jacketed
     column 1 cm in diameter.
2. Rinse the column with 10 ml of the following solutions: 0.05% DEPC; 0.1 M NaOH, repeat;
     EB, repeat twice; BB, repeat. This step should be repeated each time the column is re-used.

Isolation of poly(A$^+$) RNA
1. Resuspend the RNA in EB on ice to give a final concentration <0.5 mg/ml.
2. Heat at 68°C for 3 min. Cool quickly on ice.
3. Bring to room temperature and add NaCl to a final concentration of 0.5 M.
4. Pass it over the column at a flow rate of 1 ml/min.
5. Collect the flow through on ice and re-pass it over the column. If the SDS comes out of solution,
     warm it slightly in a 30°C waterbath.
6. Wash the column with BB until all of the unbound RNA is removed. Assay this by comparing
     the A_{260} of the flow through with BB.
7. Elute the A$^+$ RNA with EB, collect ten 1 ml aliquots on ice.
8. To remove residual A$^-$ RNA, regenerate the column (see step 2 of Preparation of the columns),
     heat the poly(A$^+$) RNA to 68°C for 3 min, cool quickly on ice and repeat steps 3–7.
9. Precipitate the RNA by adding 1/10 vol. of 3 M sodium acetate and 2.5 vols of ethanol. Store at
     −20°C.

    To measure relative rates of transcription, nuclei are isolated from the plant material of interest, RNA is elongated from pre-initiated complexes in the presence of radiolabelled nucleotide precursors, and the labelled RNA is hybridized to cloned DNA sequences immobilized on blots. The relative transcription rates of individual genes can be measured by quantitating the hybridization level of the labelled RNA to the cloned DNAs.

    The analysis of plant gene regulation by run-off transcription has been optimized by Luthe and Quantrano (17,19). The major modification was the addition of a purification step in which the nuclei are separated from contaminating endogenous RNases on a Percoll gradient.

    This section provides detailed protocols for isolating plant nuclei, and synthesizing and isolating nuclear RNA. It has been used successfully to analyse the relative transcription rates of genes expressed in tobacco petals and leaves, tomato fruit, and soybean embryos, leaves, stems and roots (16).

Table 5. Tobacco polysomal RNA and poly(A$^+$) RNA yields.

Source[a]	Maximum tissue/buffer[b]	Yield[c]	%Poly(A) RNA
Leaf (PM)[d]	1:4	1.11 ± 0.47 (5)[e]	0.9 ± 0.2 (5)
Stem	1:5	0.09 ± 0.03 (16)	1.1 ± 0.2 (6)
Root	1:5	0.15 ± 0.05 (13)	0.6 ± 0.1 (5)
Petal	1:5	0.13 ± 0.04 (14)	1.1 ± 0.2 (4)
Ovary	1:8	1.13 ± 0.37 (5)	0.9 ± 0.3 (4)
Anther	1:10	0.70 ± 0.23 (7)	0.9 ± 0.3 (4)
Seed[f]	1:5	0.88 ± 0.34 (22)	0.6 ± 0.2 (22)

[a] Young leaves, 1–4 cm in length, were harvested from 35–50 cm plants. Stems were harvested from 15–20 cm plants cut 2 cm above the soil line. Leaves were excised at the base of the petiole and the shoot apex was removed. Roots were harvested from 35–50 cm plants grown hydroponically. Petals were harvested from fully opened flowers with no sign of senescence. Pistils and anthers were harvested from 1 cm unopened flowers. Filaments were removed by hand-dissection.
[b] We have found that a high tissue to buffer ratio results in lower yields and smaller polysomes. The units are g tissue/ml buffer.
[c] mg polysomal RNA/g tissue.
[d] RNA yields are higher from leaves harvested during peak light hours (noon–2 pm).
[e] The figures in parentheses represent the number of independent RNA isolations.
[f] These seeds were collected 19 days after flowering, the peak of RNA accumulation during tobacco embryogenesis.

3.1 Harvesting plant material

Plant material is harvested and immediately frozen in liquid nitrogen. It can then be stored at −80°C. We have found that fresh tissue and tissue that has been stored at −80°C for more than 1 year result in equally active nuclei.

3.2 Nuclei isolation

In the isolation procedure described in detail below, solutions for which are given in *Table 6*, the cells are lysed in an iso-osmotic buffer containing the non-ionic detergent, Triton X-100. This buffer has been optimized to minimize nuclear damage during extraction (20) and has been used to produce nuclei that are highly active in RNA synthesis (16,17). After lysis, the nuclei are collected by low speed centrifugation and purified on a discontinuous Percoll gradient, a polyvinylpyrrolidone (PVP)-coated silica suspension. Percoll gradients are used to purify plant nuclei instead of the more common sucrose gradients, because they have been shown to be more effective in separating nuclei from endogenous RNases (19). Cellular debris, lipids, chloroplasts and membrane fragments will be found above the 40% Percoll layer. The nuclei will be found either between the 80% Percoll layer and the 2 M sucrose or will pellet depending on their density. Nuclei from soybean embryos at later stages of development, and from tobacco petals, are found between the 80% Percoll layer and the 2 M sucrose. Nuclei from young soybean embryos, leaf, stem, root and tomato fruit pellet, as do the starch granules. The nuclei are washed several times to remove residual Percoll and resuspended in 50% glycerol. They can be

Table 6. Stock solutions for the isolation of nuclei.

1. Honda buffer
   3.3% Ficoll (Sigma 400)
   6.6% Dextran–T40
   33 mM Tris–HCl (pH 8.5)
   6.6 mM $MgCl_2$
   3.3% Triton X-100
   Heat to 45°C to get the Dextran into solution. The final solution will be a bit turbid. Add
   DEPC to a final concentration of 0.05% (see Section 2.2.2). Autoclave.
2. 200 mM Spermine
   Make this stock in DEPC-treated (Section 2.2.2), autoclaved water. Do not autoclave the
   spermine solution.
3. Percoll (Pharmacia)
   Autoclave. Store Percoll at 4°C, do not freeze.
4. 3 M Sucrose
   Heat to 75°C to get the sucrose into solution. Autoclave.
5. 10× Percoll gradient buffer
   250 mM Tris–HCl (pH 8.5)
   100 mM $MgCl_2$
   Filter, treat with DEPC (see Section 2.2.2) and autoclave.
6. 2 M Sucrose
   Treat with DEPC (see Section 2.2.2) and autoclave.
7. Nuclei wash buffer
   50 mM Tris–HCl (pH 8.5)
   5 mM $MgCl_2$
   20% Glycerol (v/v)
   Treat with DEPC (see Section 2.2.2) and autoclave.
8. Nuclei resuspension buffer
   50 mM Tris–HCl (pH 8.5)
   5 mM $MgCl_2$
   50% Glycerol (v/v)
   Treat with DEPC (see Section 2.2.2) and autoclave.
9. Honda buffer/DEPC/spermine
   Final concentrations are: 2.5% Ficoll/5.0% Dextran–T40/25 mM Tris (pH 8.5)/5 mM
   $MgCl_2$/2.5% Triton X-100/0.44 M sucrose/10 mM β-mercaptoethanol/0.04% DEPC/2 mM
   spermine.
      145 ml Honda buffer
       44 ml 2 M Sucrose
      140 μl β-Mercaptoethanol
     72.5 μl DEPC
     1.89 ml 200 mM Spermine
10. Percoll gradients

% Percoll	Percoll (ml)	3 M Sucrose (ml)	10× Grad buffer (ml)	H_2O (ml)	Volume (ml)
40	8	3	2	7	20
60	12	3	2	3	20
76	16	3	2	0	21
0	0	13	2	6	21

Put the solutions in ice. It is easier to layer the solutions when they are cold. The gradients are
formed in 30 ml Corex tubes. The steps are layered from the bottom up. Each layer is 4.5 ml.
Let the gradients stand at 4°C for at least 30 min before they are used.

stored at −80°C for at least a year. General guidelines for RNA isolation (Section 2.2) should be followed. The glassware and solutions must be RNase-free. Carry out steps i–v in the coldroom.

(i)   Grind the tissue to a fine powder in liquid nitrogen in a Waring blender.

(ii)   Transfer the powder and liquid nitrogen to an omni-mixer cup and allow the liquid nitrogen to evaporate. When the liquid nitrogen is completely evaporated, the tissue will be lighter in colour and will not clump. It is important that the liquid nitrogen evaporates so that the buffer will not freeze. However, it is even more important that the tissue does not thaw.

(iii)   Add a 10× excess (v/w) of Honda buffer/DEPC/spermine to the powder (190 ml of Honda/19 g of tissue).

(iv)   Homogenize in the omni-mixer on ice at top speed for 30 sec. Repeat twice.

(v)   Filter the homogenate through 60–80 μm mesh Nitex [DEPC treat (see Section 2.2.2) and autoclave before use, Small Parts Inc., Miami, FL].

(vi)   Spin the homogenate at 5000 r.p.m. for 5 min using the Sorvall JA20 or SS34 rotor at 4°C. This will pellet the nuclei.

(vii)   Remove the supernatant by vacuum aspiration.

(viii) Using a Pasteur pipette, resuspend the nuclei in 1 ml of Honda buffer/sucrose/β-mercaptoethanol (no DEPC or spermine) per pellet on ice. If the pellet contains a lot of starch, it will be very hard and difficult to resuspend. In this case, disperse the pellet with a glass rod before resuspending with a Pasteur pipette. Pool the nuclei and bring the volume up to 20 ml (total) with Honda buffer/sucrose/β-mercaptoethanol. If the nuclei are contaminated with other compounds, resuspend and repeat the spin.

(ix)   Gently layer about 5–5.5 ml of nuclei on each of four Percoll gradients.

(x)   Spin the Percoll gradients in a Sorvall HB-4 swinging bucket rotor for 30 min at 5000 r.p.m. at 4°C.

(xi)   Remove the upper layers with a vacuum aspirator. If the nuclei pellet, leave about 2 ml in the tube. Resuspend and remove the nuclei with a Pasteur pipette. When isolating nuclei from developmental stages or from organs where starch, polysaccharides or protein bodies are a major problem, a second round of Percoll gradients are carried out.

(xii)   Pool the nuclei. Add 15–20 ml of Honda buffer/sucrose/β-mercapto-ethanol and mix with a Pasteur pipette until there are no aggregates.

(xiii) Spin down the nuclei in a Sorvall HB-4 rotor at 5000 r.p.m., for 5 min at 4°C. Remove the supernatant by vacuum aspiration.

(xiv) Resuspend the nuclei in 2 ml of Honda buffer/sucrose/β-mercaptoethanol on ice with a Pasteur pipette. Bring the volume up to 20 ml with the same solution. Repeat step xiii.

(xv)   Add 2 ml of nuclei wash buffer/10 mM β-mercaptoethanol to the pellet. Resuspend the pellet with a Pasteur pipette on ice. Add 18 ml of nuclei resuspension buffer/10 mM β-mercaptoethanol and mix. Repeat step xiii.

(xvi) Resuspend the nuclei in a minimal volume of nuclei resuspension buffer/10 mM β-mercaptoethanol. We have resuspended the nuclei from 19 g of tissue in 400 μl.

(xvii) Put into a pre-chilled box and store at −80°C. The nuclei appear to be stable at −80°C for a long period of time (>1 year). This is assayed by their ability to incorporate UTP and by the analysis of their transcription products.

3.3 RNA labelling

A detailed procedure for RNA synthesis in isolated nuclei is provided in *Table 7*. RNA is synthesized in 100 mM $(NH_4)_2SO_4$ and 4 mM $MgCl_2$,

Table 7. *In vitro* transcription.

Do not DEPC treat any of these solutions.

1. 10× Salts solution
    1 M $(NH_4)_2SO_4$ (2 M stock—filtered, autoclaved.)
    40 mM $MgCl_2$ (1 M stock—filtered, autoclaved.)
    3 μM Phosphocreatine (3 mM stock—made in autoclaved H_2O, pH 7.0)
2. 5 mM ATP, GTP, CTP in autoclaved H_2O. Adjust the pH to 7.0 with NaOH.
3. 20 μg/ml α-Amanitin in autoclaved H_2O.
4. 2.5 mg/ml Creatine phosphokinase in autoclaved H_2O.
5. Stop buffer
    0.2% SDS
    5% Na_4PPi
    2 mM UTP
    0.5 M Phosphate buffer

Test synthesis
1. Thaw nuclei on ice and mix gently to disperse the nuclei.
2. Add (in the following order) and incubate for 20 min at 30°C.

Nuclei[a]	10 μl
10× salts	2 μl
H_2O[b]	2 μl
0.25 mg/ml Creatine phosphokinase (1:10 dilution of stock)	2 μl
ATP, GTP, CTP stock	2 μl
[^{32}P]UTP (3000 Ci/mM, 10 mCi/ml)	2 μl

3. Stop the reaction by adding 100 μl of Stop buffer. Use Stop buffer only when testing the nuclei. For the preparative synthesis, go directly to step 1 of the RNA isolation (*Table 8*).
4. Vortex the reaction at a high speed for a few seconds to decrease the viscosity of the reaction so that an accurate volume can be sampled.
5. Assay incorporated counts by TCA precipitation or DE81 filter analysis.

Preparative synthesis
1. Scale up and carry out preparative synthesis at 30°C for 20 min. Follow steps 1 and 2 of test synthesis and then continue with step 1 of the RNA isolation (*Table 8*).

---

[a] Use a yellow pipetman tip with the end cut off to transfer the nuclei.
[b] 20 μg/ml α-Amanitin can be used in a separate reaction instead of H_2O (final concentration 2 μg/ml α-amanitin) to determine what percent of the synthesized RNA is made from polymerase II.

monovalent and divalent cation concentrations in which RNA polymerase has maximum activity (17). The concentration of unlabelled nucleotides is 500 μM. The reaction is incubated at 30°C for 20 min. We have observed that the rate of synthesis is linear during this time.

Synthesis rates will vary depending on the source of the nuclei. Because we have observed as much as a 10-fold variability in the amount of RNA synthesized from nuclei from various sources, we suggest that a test reaction be done first. After this the reaction can be scaled up by increasing the reaction volume and number of nuclei, or by increasing the [^{32}P]UTP in the reaction. We synthesize approximately 1×10^8 d.p.m. of incorporated counts and use at least 1×10^7 d.p.m. per blot (75–150 cm^2).

To determine how much of the synthesis is due to RNA polymerase II activity, a separate aliquot of nuclei is incubated in the presence of 2 μg/ml α-amanitin (*Table 7*). At this concentration, most of the polymerase II activity is inhibited while polymerases I and III are unaffected. We find that 40–80% of the RNA synthesis is inhibited at this α-amanitin concentration.

3.4 **RNA isolation**

After the RNA has been synthesized, the DNA and proteins are removed by enzymatic digestion and phenol extraction (*Table 8*). The RNA is purified from other cellular components by centrifugation through a CsCl cushion. Unincorporated nucleotides and CsCl are removed by chromatography on a 25 ml column of Sephadex G-150. The RNA is then separated from contaminating polysaccharides by extraction with hexadecylpyridinium chloride (CTAB) (21) (see *Table 9*). This purification step reduces non-specific background binding when the RNA is used as a hybridization probe. The rationale behind this extraction is that the quaternary ammonium cation in CTAB forms a complex with the anionic nucleic acid under low salt conditions. In the two-phase water–butanol system the complex partitions to the alcohol phase and the polysaccharides and other contaminants remain behind in the water phase. When the salt concentration is raised, the complex dissociates and the nucleic acid can be recovered in the water layer. Chloroform is then used to remove residual CTAB.

3.5 **Hybridization**

The labelled RNA can be used to determine the relative rates of gene transcription by one of the following two methods. The RNA can be hybridized to cloned DNA sequences dotted onto a nitrocellulose filter (15) or to DNA separated by electrophoresis on an agarose gel and blotted onto nitrocellulose (18). The resulting hybridization can be quantitated either by autoradiography followed by densitometric analysis or by cutting out the dots and counting them in a liquid scintillation counter.

Table 8. RNA isolation.

Stock solutions
1. 10× TES
    100 mM Tris–HCl (pH 7.6)
    50 mM Na$_2$EDTA
    10% SDS
    Make the SDS in sterile, autoclaved water and add to Tris and EDTA that have been autoclaved.
2. TESar
    10 mM Tris–HCl (pH 7.6)
    1 mM Na$_2$EDTA
    1% Sodium lauroylsarcosinate
    Filter and autoclave.
3. Phenol
    Double-distilled phenol is equilibrated with 50 mM Tris–HCl (pH 8.0), 0.1 mM EDTA.
4. Phenol/Sevag
    Equal volumes of phenol and Sevag are mixed together. Sevag is chloroform and isoamyl alcohol (24:1).

Procedure
1. Add RNase-free DNase to a final concentration of 20 μg/ml. Incubate for 10 min at 30°C.
2. Add ¹⁄₁₀ vol. of 10× TES.
3. Add pre-digested Proteinase K (pre-digest, 30 min, 37°C in TESar) to a final concentration of 100 μg/ml. Incubate for 30 min at 42°C.
4. Add 50 μg of tRNA. The tRNA should be phenol/Sevag and Sevag extracted and ethanol precipitated before use.
5. Extract with an equal volume of phenol/Sevag at room temperature. Re-extract with a small volume of TES. Repeat until no interface can be detected after spinning.
6. Add ¹⁄₁₀ vol. of 3 M sodium acetate (pH 6.0) and 2.5 vols of ethanol. Precipate overnight at −20°C.
7. Pellet the RNA for 30 min at 4°C in the microfuge and dry.
8. Resuspend the RNA in 400 μl of TESar on ice.
9. Add 400 μl of 1 g/ml CsCl in TESar (filtered and autoclaved).
10. Put 1.2 ml of 5.7 M CsCl/TESar into a 5 ml Sorvall AH650 tube. Gently layer the 0.8 ml of RNA/CsCl onto the 5.7 M CsCl cushion. Fill the tube with 0.5 g/ml CsCl/TESar.
11. Spin at 35 000 r.p.m. for 46 h at 20°C in a AH650 Sorvall rotor.
12. Carefully remove all of the gradient except for the last 0.2 ml. Assay this aliquot for incorporated counts. If that fraction has >20% of the total incorporated counts in the pellet, add that aliquot to the resuspended pellet.
13. Resuspend the pellet in 200 μl of TESar. Rinse the tube with an additional 200 μl of TESar. Pool the aliquots.
14. Run the RNA over a 25 ml Sephadex G-150 column equilibrated against TESar[a].
15. Collect the RNA and add ¹⁄₁₀ vol. of 3 M sodium acetate (pH 6.0) and 2.5 vols of ethanol.
16. Precipitate overnight at −20°C.

[a] We use a column that is 1.2 cm in diameter and 26 cm long. We collect 35 drop fractions and the volume of the pooled fractions is ~7 ml.

Table 9. CTAB purification of RNA.

Solutions

CTAB/butanol and CTAB/aqueous

(i) In a separatory funnel, shake 75 ml of 1-butanol and 75 ml of H$_2$O. Allow the phases to separate. Collect the phases.

(ii) Add 1.84 g of CTAB to 50 ml of the butanol saturated with water. Add 50 ml of water saturated with butanol. Shake in a separatory funnel. Allow the phases to separate overnight.

(iii) Butanol/CTAB (Bu/CTAB) is the upper layer and aqueous/CTAB (Aq/CTAB) is the lower layer. Store separately.

Procedure

1. Pellet the RNA by spinning at 22 000 r.p.m. for 1 h at 4°C and dry.
2. Resuspend the pellet in 200 μl of TESar (*Table 8*).
3. Add 200 μl of Aq/CTAB and 200 μl of Bu/CTAB.
4. Vortex for at least 2 min.
5. Separate the phases by spinning in a microfuge for 2 min.
6. Transfer the upper (butanol) phase to a microfuge tube. This phase will contain the nucleic acids.
7. Re-extract the aqueous phase with 200 μl of Bu/CTAB. Pool the butanol layers.
8. Add 150 μl of 0.2 M NaCl. Vortex for 30 sec. Separate the phases by spinning in a microfuge for 2 min.
9. Transfer the lower (aqueous) phase to a microfuge tube. This now contains the RNA.
10. Re-extract the butanol phase with 150 μl of 0.2 M NaCl. Pool the aqueous layers.
11. Add 300 μl of chloroform, dropwise, to the pooled aqueous phases. Incubate on ice for 15 min.
12. Separate the phases by spinning in the microfuge for 2 sec.
13. Remove the lower (chloroform) phase which now contains the CTAB.
14. Ethanol precipitate the aqueous layer by adding ⅒ vol. of 3 M sodium acetate (pH 6.0) and 2.5 vols of ethanol.
15. Precipitate overnight at −20°C.

4. SYNTHESIS OF SINGLE-STRANDED RNA PROBES FOR RNA BLOT ANALYSIS

4.1 **Introduction**

Single-stranded, high specific activity RNA sequences are very efficient hybridization probes. They can be used in a variety of analyses including *in situ* hybridization (Section 6), blot analysis (Section 4) and RNA titration analysis (Section 5). RNA probes have been shown to be as much as ten times more sensitive than nick translated probes of the same specific activity in blot analysis (22) and *in situ* hybridization (23). This section will describe in detail synthesis of RNA probes, as well as provide suggestions for their use. A more detailed discussion of the uses of these probes can be found in other sources (22,24).

4.2 **Template preparation**

The synthesis of single-stranded RNA probes was made possible by the construction of vectors containing bacteriophage promoters (22), the most common of which are SP6, T7 and T4, and the commercial availability of their respective RNA polymerases. Although many of these vectors are now available, we recommend the use of a vector system similiar to pGEM Blue

(Promega Biotec). This vector contains both the SP6 and T7 RNA polymerase promoters flanking a multiple cloning region. After insertion of the cloned DNA into the multiple cloning site, both the coding and non-coding strands can be synthesized from a single recombinant clone using the appropriate polymerase. In addition, it also contains the lac α-peptide which complements the lacZ M15 gene to produce β-galactosidase and can be used to identify clones containing inserts by colour selection.

We generally use DNA that has been purified on CsCl gradients for *in vitro* transcription templates. However, others have shown that intact RNA can be synthesized using DNA templates prepared by mini-lysate procedures (25).

After isolation of the template, the clone is linearized downstream of the insert by digestion with the appropriate enzyme (*Table 10*). The synthesis of 'run-off' transcripts from truncated templates eliminates labelled vector from subsequent hybridization reactions. This increases the sensitivity of the blot by lowering non-specific background binding. Restriction enzymes that produce 3' overhangs should be avoided (25). There is evidence that non-specific initiation can occur at these sites giving rise to RNA from the opposite strand and the vector. If a restriction enzyme that produces a 3' overhang must be used, the end should be converted to blunt form using T4 DNA polymerase or the Klenow fragment of DNA polymerase I. After linearization, the DNA is phenol extracted and precipitated.

4.3 *In vitro* transcription

In this system, transcription begins at the bacteriophage promoter, proceeds through the cloned insert and terminates when the enzyme reaches the end of the template. The detailed transcription procedure, provided in *Table 11*, is a modification of the procedures of Melton *et al.* (22). They have determined optimal transcription reaction conditions, and several aspects of their work will be discussed below.

4.3.1 *Nucleotide concentration*

Measurements of the amount and length of the transcribed RNA have been made at various ribonucleotide concentrations. All four nucleotides saturate at 250 μM. However, nucleotide concentrations of this magnitude are not practical

Table 10. Preparation of template DNA.

1. Digest the DNA with the appropriate restriction enzyme.
2. Add 1 vol. of 1 mg/ml pre-digested [30 min, 37°C, in 20 mM Tris–HCl (pH 7.0), 0.2 mM EDTA and 2% sodium lauroylsarcosinate] Proteinase K. Incubate at 37°C for 30 min.
3. Extract with an equal volume of phenol/Sevag (*Table 8*). Extract with an equal volume of Sevag. Double-distilled phenol is equilibrated with 50 mM Tris–HCl (pH 8.0), 0.1 mM EDTA.
4. Add ¹⁄₁₀ vol. of 3 M sodium acetate (pH 6.0) and 2 vols of ethanol. Precipitate overnight at −20°C.
5. Spin down the precipitate in a microfuge at 4°C. Wash the pellet with 70% ethanol and spin. Dry the pellet.
6. Resuspend the DNA in TE (10 mM Tris, pH 7.0, 0.1 mM EDTA) at 1 μg/μl.

for synthesizing high specific activity probes. We routinely use a concentration of 250–500 μM for the three unlabelled nucleotides and 12–25 μM for the labelled nucleotide. At these concentrations, it is still possible to incorporate 50–80% of the labelled nucleotides and to get a large fraction of full length probe.

4.3.2 *Time course*

The reaction rate is constant for about 1 h and drops off slowly after that.

4.3.3 *Temperature optimum*

A temperature curve for RNA synthesis shows a sharp temperature optimum at 40°C. Synthesis occurs at approximately 10% of this rate at 30°C or 50°C.

Table 11. RNA transcription.

Solutions
1. 5× Transcription buffer
    200 mM Tris–HCl (pH 7.5)
    30 mM MgCl$_2$
    10 mM Spermidine
    50 mM NaCl
    DEPC treat (Section 2.2.2) and autoclave all the components except the spermidine. Filter sterilize the spermidine stock.
2. 100 mM Dithiothreitol (DTT) in DEPC-treated H$_2$O (Section 2.2.2).
3. 10 mM NTP stocks (10 mM ATP, 10 mM UTP, 10 mM GTP) in DEPC-treated H$_2$O (Section 2.2.2) and adjust the pH to 7.0.

Procedure
1. Combine the following[a]:

	Final concentration
5× Transcription buffer	1×
100 mM DTT	10 mM
RNasin (25 U/μl)	1 U/μl
10 mM ATP, UTP, GTP	250–500 μM each
Template	0.1 μg/μl
[^{32}P]CTP[b], (400 Ci/mmol, 10 mCi/ml)	12–25 μM
SP6 or T7 Polymerase (10 U/μl)	0.2 U/μl

Final volume: 50 μl. Incubate for 45 min at 40°C.
2. Add RNase-free DNase (Promega Biotec) to a final concentration of 1 U/μg DNA. Incubate for 15 min at 37°C.
3. Add 10 μg of yeast tRNA carrier.
4. Add 1 vol. of 1 mg/ml pre-digested [30 min, 37°C, in 20 mM Tris–HCl (pH 7.0), 0.2 mM EDTA, 2% sodium lauroylsarcosinate] Proteinase K. Incubate for 30 min at room temperature.
5. Extract with an equal volume of phenol/Sevag (*Table 8*). Extract with an equal volume of Sevag.
6. Pass the RNA over a Sephadex G-50 column to remove unincorporated nucleotides[c].
7. Add ¹⁄₁₀ vol. of 3 M sodium acetate and 2.5 vols of ethanol. Precipitate overnight at −20°C.

---

[a]Combine the components in the order shown. The mixture should be kept at room temperature. Because of the spermidine, the DNA can sometimes precipitate if incubated on ice.
[b]The reaction can be run in the absence of unlabelled CTP if maximum specific activity is desired. For a 50 μl reaction, 250 μCi of 400 Ci/mmol [^{32}P]CTP is needed. This is ~12 μM. The yield of full length transcripts is reduced as the concentration of the nucleotides falls below 12 μM.
[c]This step is not required for preparation of *in situ* probes.

4.3.4 *Length of RNA transcripts*

Full length transcripts are made from virtually any size DNA template, although varying amounts of transcripts shorter than full length are observed. In addition, these RNA polymerases are able to transcribe continuous stretches of poly(C), poly(G) and poly(A) of about 20 bases without termination.

4.3.5 *Specificity*

These phage polymerases are totally specific for their own promoters and will not initiate transcription on other prokaryotic or eukaryotic promoters.

4.4 **Hybridization**

Many procedures have been published for RNA blot analysis using RNA probes (22,26,27). Two of these are described in *Table 12*. In addition, procedures developed for double-stranded DNA probes can also be used for single-stranded RNA probes with the following modifications.

(i)   Hybridization and wash criteria must be raised. RNA–RNA hybrids have a higher thermal stability than DNA–RNA hybrids. The melting temperature (T_m) in solution of a RNA–RNA hybrid of average GC content in 50% formamide and 0.3 M NaCl is approximately 80°C. Optimum hybridization rates occur at $T_m - 25$. Therefore, RNA blot hybridizations using RNA probes are generally incubated at 55–60°C. Some researchers have even found higher temperatures (65–70°C) preferable because they result in less non-specific background binding (26). Wash temperatures should also be raised for the same reason.

(ii)   Hybridization buffers must be prepared RNase-free. This includes DEPC-treating the buffers and glassware, and using RNase-free methods for experimental manipulations (see RNA extraction, Section 2.2).

(iii) Yeast RNA should be included as an RNA competitor in the pre-hybridization and hybridization mix.

(iv)   An RNase A wash can be included to further lower the non-specific background binding of probe (27).

5. RNA TITRATION ANALYSIS

5.1 **Introduction**

RNA titration analysis is the most accurate and the most sensitive method to measure the absolute number of copies of a particular RNA transcript in an organism, organ, tissue or cell. Briefly, excess labelled RNA homologous to the sequence of interest (tracer) is hybridized to various amounts of cellular RNA. After the hybridizations have reached completion, unhybridized tracer is digested with RNase and the tracer–cellular RNA hybrids are assayed. Under tracer-excess conditions, a plot of the amount of hybridized tracer versus the amount of cellular RNA for each reaction yields a straight line, the slope of which represents the percent of the cellular RNA that is homologous to the

Table 12. Hybridization conditions.

METHOD 1

Hybridization solution
50% Formamide
  5× SSC (0.75 M NaCl, 75 mM sodium citrate, pH 7.0)
  0.1 M Sodium phosphate (pH 6.5)
  1× Denhardt's (10× Denhardt's in the pre-hybridization solution) 10× Denhardt's is 0.2% bovine
  serum albumin (BSA), 0.2% Ficoll and 0.2% PVP
    0.1% SDS
250 µg/ml tRNA
  100 µg/ml Sheared single-stranded salmon sperm DNA
  10% Dextran sulphate (no Dextran sulphate in the pre-hybridization solution)

Procedure
1. Pre-hybridize the filter in hybridization buffer without Dextran sulphate and with 10×
    Denhardt's overnight at 55°C.
2. Denature the probe at 90°C for 10 min.
3. Hybridize with 3×10^6 c.p.m. of probe per ml. The total volume is 0.3 ml of hybridization
    buffer per cm² of blot. Incubate for 20 h at 55°C.
4. Wash with 50% formamide, 5× SSC, 0.1 M sodium phosphate and 0.1% SDS for 30 min at
    55°C. Repeat five times.

METHOD 2

Hybridization solution
  50% Formamide
    0.1 M Pipes (pH 6.8)
    0.5 M NaCl
    0.2% SDS
  10× Denhardt's (0.2% BSA, 0.2% Ficoll, 0.2% PVP)
250 µg/ml Yeast RNA (Sigma type III)

Procedure
1. Pre-hybridize the filter in hybridization buffer for >3 h at 68°C.
2. Hybridize with 1×10^6 d.p.m. of probe per ml of hybridization solution. Use 0.1 ml of
    hybridization solution per cm² of blot. Hybridize for 20 h at 68°C[a].
3. Wash briefly with 2× SSC at room temperature. Repeat. Wash with 0.2× SSC, 0.01% SDS at
    68°C for 20 min. Repeat three times.

[a]Nytran or a similar nylon membrane will withstand a 68°C incubation better than nitrocellulose.

tracer. If the number of cells expressing the gene and the total number of RNA
molecules in those cells are known, the absolute number of molecules of a given
sequence per cell can be calculated. The details of this technique are described
in *Table 13* (28,29).

Correct interpretation of such data requires that the following criteria be met.

(i)   The tracer must be in excess (Section 5.3.1) and the reaction must go to
        completion (Section 5.3.2). It is only under these conditions that the
        percent of the tracer in hybrid is proportional to the amount of input
        cellular RNA and that a linear positive slope will be generated. As

21

Analysis of plant gene expression

Table 13. RNA titration analysis.

Solutions
1.   5× Hybridization mix
    2 M NaCl
   125 mM Pipes
    1 mM EDTA
Adjust the pH to 6.8, filter through 45 μm Millipore filter, DEPC treat (Section 2.2.2) and autoclave.
2.   SET
   0.375 M NaCl
   75 mM Tris
   5 mM EDTA
Adjust the pH to 8.0, filter through 45 μm Millipore filter and autoclave.

Probe synthesis
1. Synthesize a ^{32}P-labelled RNA tracer with a specific activity of 6×10^8 d.p.m./μg using the procedures detailed in Section 4, *Table 10* (steps 1–6) and *Table 11* (steps 1–7).

Hybridization
1. Mix the appropriate amounts of tracer (generally 50–200 pg) and cellular RNA.
2. Add yeast total RNA to a final RNA content of 100 μg. [The yeast RNA should be digested for 30 min at 37°C with 1 mg/ml proteinase K, extracted with an equal volume of double-distilled, equilibrated (50 mM Tris, pH 8.0, 0.1 mM EDTA) phenol:Sevag (*Table 8*) and an equal volume of Sevag, ethanol precipitated and resuspended in DEPC-treated water (Section 2.2.2) before use.]
3. Add 1/10 vol. of 2 M ammonium acetate and 2 vols of ethanol. Precipitate overnight at −20°C.
4. Spin down the precipitate in a microfuge at 4°C. Wash the pellet with 70% ethanol and dry. Resuspend in 6 μl of DEPC-treated water.
5. Add 10 μl of 100% de-ionized formamide.
6. Add 4 μl of 5× hybridization mix.
7. Cover the reaction with 50 μl of mineral oil[a].
8. Heat to 85°C for 5 min.
9. Hybridize at 50°C[b] for the calculated time (see Section 5.3.2).

Hybridization assay
1. Stop the hybridization by placing the reaction on ice.
2. Dilute the samples in 280 μl of SET.
3. Add sufficient RNase A and RNase T1 to get complete digestion of single-stranded RNA[c]. Incubate at 37°C for 1 h.
4. Assay undigested counts using TCA precipitation of DE81 filters.

---

[a] An alternative method for this is to seal the reaction in DEPC-treated, silanized and baked capillary tubes.
[b] 50°C is the optimum hybridization temperature for 500–1000 nt probes of average GC content.
[c] The activity of each RNase lot should be titrated. We find that 100 μg of RNase A and 225 units of RNase T1 is sufficient to digest 100 μg of RNA to completion.

saturation of the tracer molecules is approached, the slope of the curve describing tracer hybridized versus input of cellular RNA approaches zero.

(ii)   The precise specific activity of the tracer must be known. This is possible with the *in vitro* transcription systems (Section 4.3) in which the specific activity of the tracer is dependent only on the specific activity of the nucleotide precursors.

(iii)   The tracer's specificity at the hybridization criteria and the length of the tracer which will hybridize to the RNA transcript must be known.

RNA titration analysis is very sensitive. If the background level of RNase-resistant tracer is low (<0.5% of the total counts), 0.1–0.3 pg of a specific RNA transcript can be measured (29). We have observed that some tracers contain a larger percent of RNase-resistant nucleotides due to foldback regions. In these instances, a preliminary analysis of different restriction fragments in the tracer may be necessary in order to identify and delete these regions.

5.2 Probe synthesis

The tracer must be uniformly labelled and single-stranded. It can be either a single-stranded RNA synthesized using the *in vitro* transcription technique outlined in Section 4 (22), or a single-stranded DNA made from an M13 vector (30). We recommend use of a single-stranded RNA tracer for the following reasons.

(i)   The tracer is completely single-stranded. Complementary sequences in the template are removed by DNase digestion.
(ii)   The hybridization can be done at higher stringency since RNA–RNA duplexes are more stable than DNA–RNA duplexes. This tends to decrease non-specific background signals.
(iii)   Differentiation of unhybridized single-stranded tracer and hybridized double-stranded tracer is highly reproducible because RNA–RNA duplexes are completely resistant to RNase in moderate to high ionic strength buffers (28).

5.3 Hybridization reaction

5.3.1 *Tracer/cellular RNA ratio*

The labelled tracer RNA must be present at greater than 5-fold sequence excess. An estimate of the concentration of the RNA sequence being analysed must be made. This is most easily done by an initial RNA blot analysis where the intensity of the signal for the specific RNA is compared to the signal obtained from hybridization of a probe to an RNA of known prevalence. If this preliminary analysis cannot be done, hybridizations should be carried out with a wide range of cellular RNA concentrations to determine a range in which the amount of tracer in duplex is proportional to the input of cellular RNA.

5.3.2 *Hybridization time*

Since the concentration of unreacted tracer does not change significantly during the hybridization, titration reactions follow pseudo-first-order kinetics. We suggest that each reaction be hybridized to $10\times R_0T_{1/2}$, that is 10 times the length of time required for 50% of the hybrids to form. $R_0T_{1/2}$ can be calculated from equation 1.

$$R_0 T_{1/2} = \ln 2 / K_t \tag{1}$$

R_0 is the concentration of the tracer in moles of nucleotides per litre, T is the incubation time in seconds and K_t is the rate constant. The rate constant for any single-stranded RNA tracer can be calculated by assuming kinetics similar to single-stranded DNA tracers and using the following standard. A sequence with a complexity of 5400 nt hybridizes with a rate constant of 169 M^{-1} sec^{-1} under standard conditions ($T_m - 25$, 0.18 M Na^+, fragment length $= 400-500$ nt) (31). This standard can be inserted into equation 2 to determine the rate constant, K_t, for the tracer.

$$K_t = K_s \times C_s / C_t \tag{2}$$

$$K_t = 169 \ M^{-1} \ sec^{-1} \times 5400 \ nt / C_t$$

C_t equals the complexity of the tracer. In titration experiments where the tracer consists of multiple copies of a single RNA sequence, the complexity is equal to the number of nucleotides in the sequence.

The RNA titration hybridization is not done under standard conditions. These factors should be used in arriving at a final rate constant.

(i)   Under aqueous conditions, 0.4 M NaCl increases the hybridization rate 4-fold (32).
(ii)   50% formamide retards the rate 2-fold (33).
(iii)   The hybridization rate changes as the square root of the probe length (34).

Due to the low complexity of the tracer, $10\times R_0 T_{1/2}$ can generally be reached in 24–48 h.

5.4 Analysis of the data

5.4.1 *Amount of hybridized tracer*

$$\text{Tracer hybridized} = (\text{c.p.m.}^h - \text{c.p.m.}^b)/(\text{sp. act.} \times C) \tag{3}$$

The amount of tracer hybridized in each reaction can be calculated using equation 3. C.p.m.[h] is the counts per minute resistant to RNase digestion after incubation of the tracer with the cellular RNA. C.p.m.[b] is the counts per minute resistant to RNase digestion after incubation of the tracer alone. Sp. act. is the specific activity of the tracer and C is the counting efficiency of the isotope.

5.4.2 *Fraction of the cellular RNA homologous to the tracer*

$$F = S \times L_r / L_t \tag{4}$$

The fraction of the cellular RNA that is homologous to the tracer can be calculated using equation 4. S is the slope of the line generated by plotting the amount of hybridized tracer (Section 5.4.1) versus the amount of cellular RNA in each reaction. L_r is the length of the RNA sequence being assayed and L_t is the length of the tracer in hybrid. Obtaining a non-linear plot or a slope of zero suggests that the tracer RNA was not in sufficient excess and that the hybridizations should be redone with a higher input of tracer molecules.

5.4.3 *Number of molecules of homologous RNA per cell*

$$\text{molecules/cell} = (F \times R \times 6 \times 10^{23})/(E \times 339 \times L_r) \qquad (5)$$

The absolute number of molecules per cell of RNA homologous to the tracer can be calculated using equation 5. F is the fraction of the cellular RNA that is homologous to the tracer (Section 5.4.2). R is the amount of RNA per cell (see below). 6×10^{23} is the number of nucleotides per mole. E is the fraction of the total number of cells assayed that express the gene (see below). The average molecular weight of a ribonucleotide is 339 and L_r is the length in nucleotides of the RNA sequence being assayed. In order to calculate the number of molecules of a particular sequence per cell, it is necessary to know the total RNA content per cell. This is most easily done by determining the RNA content of a mass of cells using the phloroglucinol reaction (35) and determining the number of cells by measuring the amount of DNA using the diphenylamine procedure (36). From these measurements, the average RNA content per cell can be calculated. It is important to remember that this is the average and that different cell types may have different RNA contents. A second value required for this calculation is the percent of cells included in the assay that are expressing the gene. This can be measured by *in situ* hybridization techniques discussed in Section 6.

6. IN SITU HYBRIDIZATION

6.1 **Introduction**

In situ hybridization is used to analyse spatial patterns of RNA accumulation at the cellular and subcellular levels. In this technique, labelled RNA probes are hybridized *in situ* to homologous RNA sequences in cytological preparations. Use of an autoradiographic emulsion allows the direct identification of cells that contain target RNA sequences.

In situ hybridization has already proved to be a powerful technique for analysing gene expression in animals, particularly during embryogenesis. Recently this technique has been modified for plant systems where we believe it will have similar utility. It can be used to assay gene expression at stages of plant development when standard molecular techniques cannot be used, such as, early stages of embryogenesis or organogenesis when the embryo or organ is very small and cannot be dissected from other structures. This hybridization technique can also be used to identify RNA localization patterns when there are no available methods for biochemical isolation of individual cell types.

This section will discuss various aspects of *in situ* hybridization and provide procedures that have been used successfully in our laboratory and others to analyse RNA localization patterns in all tobacco organs, *Lemna* roots and fronds, maize leaves and tobacco (37), soybean and *Brassica* embryos. Other techniques such as that described in Chapter 5 have been published, which use alternative fixatives, sectioning methods and pre-hybridization treatments (38–40). Depending upon the plant, the organ, and perhaps the developmental stage to be analysed, one procedure may be superior to others.

RNase-free methods (see RNA extraction, Section 2.2) should be used for all experimental manipulations until the post-hybridization washes.

6.2 Fixation, dehydration, clearing and embedding

Optimal fixation of plant material for *in situ* hybridization has two goals. The first is good preservation of tissue morphology and second is retention of RNA molecules at their *in vivo* positions. This must be accomplished without overfixation of the tissue which limits the accessibility of the target RNA to the probe. There are a considerable number of standard fixation methods for plant tissue. They can be grouped into two classes, those that fix by precipitating molecules and those that fix by crosslinking, such as the aldehydes. Early work has conclusively shown that crosslinking fixatives are more effective at retaining RNA molecules during *in situ* hybridization (41,42). Two fixation procedures have been provided in *Table 14*. The glutaraldehyde fixation method has been used successfully by this laboratory and others for fixation of tobacco leaves, stems, roots, petals, ovaries and anthers, *Lemna* roots and fronds, maize leaves and soybean roots and embryos. This fixation procedure was not successful for fixing late stage embryos in tobacco seeds (37). For this plant material, we found that the more standard plant fixative FAA (formalin, alcohol, acetic acid) was superior (43). This is probably due to better penetration of the tobacco seed coat by the smaller formalin molecules. Direct comparisons of *in situ* signals resulting from hybridization of root tips fixed with FAA or glutaraldehyde show that there is no detectable difference in the hybridization efficiency.

We have used ethanol and xylene for dehydration and clearing steps. Tertiary butyl alcohol is recommended by some sources because it is a gentler clearing agent and results in tissues that are less hardened. Tertiary butyl alcohol is compatible with this technique. It is important to remember that regardless of which chemicals are used, changes in concentration should be gradual. Drastic changes cause shrinking of cytoplasm and cellular distortion.

Plant organs that contain cells with large vacuoles should be dehydrated under vacuum to facilitate infiltration of the solutions, otherwise the tissue will not be properly embedded. All changes in pressure should be gradual to avoid tearing.

Finally, all tissues and organs are different in their biochemical composition. Fixation, dehydration, clearing and embedding techniques may have to be optimized. In changing these procedures it is important to remember that some of these steps are interrelated. For example, the extent of fixation as well as the choice of a fixative will have an effect on the optimum extent of protein digestion during pre-hybridization and the optimum probe length for hybridization. Changes in any portion of the technique may have to be coordinated with changes at other steps.

Table 14. Fixation, dehydration, clearing and embedding.

Glutaraldehyde fixation
1. Cut the tissue and immediately place it in a vial containing 10 ml of 1% glutaraldehyde in 0.05 M sodium cacodylate trihydrate buffer[1] (pH 7.0) (Ted Pella, Inc., Tustin, CA).
2. Fix for 3 h at room temperature swirling occasionally. De-gas (in a dessicator) for 10 min at 1 h and 2 h. When de-gassing, bring the vacuum up and down slowly.
3. Remove the fixative from the vials and wash the tissue with 10 ml of 0.05 M sodium cacodylate trihydrate for 30 min. Swirl occasionally. Repeat.
4. Remove the buffer and add 10 ml of 5% ethanol (in water). Incubate at room temperature for 30 min under constant vacuum. Repeat this for the following ethanol solutions: 10%, 15%, 20%, 25%, 30%, 40%, 50%, 60%, 70%, 80%, 95%, 100%. Leave the tissue overnight in 100% ethanol, no vacuum.
5. Replace the 100% ethanol with fresh 100% ethanol. Incubate at room temperature for 1 h, no vacuum. Repeat.
6. Remove the 100% ethanol and add 10 ml of 25% xylene (in ethanol). Incubate at room temperature for 30 min, no vacuum. Repeat this for the following solutions: 50% xylene (in ethanol) and 75% xylene (in ethanol), no vacuum. Remove the 75% xylene solution and replace it with 100% xylene. Incubate at room temperature for 1 h, no vacuum. Repeat twice with 100% xylene.
7. Without removing the 100% xylene, add 15 chips of Paraplast Plus (Fisher). Leave it overnight at room temperature.
8. The paraplast will partially go into solution. Incubate the tissue at 42°C to solubilize the remaining paraplast. Add five more chips and incubate at 42°C for 3–4 h. Repeat.
9. Pour off the paraplast/xylene solution. Add 10 ml melted paraplast plus. Incubate at 57–62°C for at least 8 h[b]. Repeat at least six times.

FAA fixation
FAA
50% Ethanol
 5% Acetic acid
10% Formalin (3.7% formaldehyde)

1. Carry out step 1 above, except that FAA is used as the fixative instead of glutaraldehyde/cacodylate.
2. Carry out step 2 above.
3. Remove the fixative from the vial and wash the tissue with 10 ml of 50% ethanol (in water) for 30 min under constant vacuum. Repeat.
4. Remove the 50% ethanol solution and add 10 ml of 60% ethanol. Incubate at room temperature for 30 min under constant vacuum. Repeat this for the following ethanol solutions: 70%, 80%, 95%, 100%. Leave overnight in fresh 100% ethanol, no vacuum.
5. Continue with step 5 as above.

[a]The tissue to fluid ratio should never be greater than 1:10. The tissue pieces should be small, 1 mm × 1 mm–1 mm × 10 mm, to facilitate quick fixation.
[b]Do not allow the temperature to exceed 62°C at any time. Temperatures in excess of 62°C will cause breakdown of the polymer structure of the paraffin.

6.3 Pre-hybridization treatments

The pre-hybridization treatments (see *Table 16*) have two aims, to increase the accessibility of the target RNA and to lower non-specific background binding. Efficient hybridization of target RNA after crosslinking fixation requires some protease digestion. Proteinase K and pronase have both been used for this

Table 15. Slide preparation.

1. Incubate the slides in chromerge overnight or longer.
2. Rinse in running distilled water for >1 h.
3. Bake at 200°C to dry.
4. Immerse in 100 μg/ml poly-L-lysine, 10 mM Tris–HCl (pH 8.0) for 10 min at room temperature[a].
5. Loosely cover to keep dust off the slides and allow to air dry.

Sectioning
1. Cut 5–10 μm thick sections and float them on a 45°C waterbath to spread.
2. Mount them on poly-L-lysine-coated slides. Incubate at 40°C overnight.

[a] Gelatin and albumin have been tested as alternative subbing agents for retention of sea urchin embryos. Poly-L-lysine was vastly superior. These alternative subbing agents have not been tested for retention of plant material.

Table 16. Pre-hybridization treatments.

1. Incubate the slides in xylene for 10 min with stirring in a staining dish to remove the paraplast. Repeat.
2. Hydrate the sections by passing the slides sequentially through the following solutions, 100%, 100%, 95%, 85%, 70%, 50% and 30% ethanol and two changes of water. Dip the slides about 20 times in each solution.
3. Incubate the slides in 1% BSA in 10 mM Tris–HCl (pH 8.0) at room temperature for 10 min. This will block the positive charges on the slide from the poly-L-lysine and reduce background.
4. Remove the BSA by washing twice with water in a staining dish.
5. Incubate the slides in 1 μg/ml Proteinase K in 100 mM Tris–HCl (pH 7.5), 50 mM Na$_2$EDTA in a Coplin jar for 30 min at 37°C. Pre-warm the Tris/EDTA solution before use.
6. Remove the Proteinase K by washing twice with water in staining dishes.
7. Equilibrate the slides in 0.1 M triethanolamine (pH 8.0). Add undiluted acetic anhydride to a dry staining dish, add the slides, then add 0.1 M triethanolamine (pH 8.0) to the dish to give a final acetic anhydride concentration of 0.25% (v/v), mix. Dip the slides several times. Incubate at room temperature for 10 min. This reaction acetylates any remaining positive charges in the tissue or on the slides, further reducing background[a].
8. Wash the slides briefly in a staining dish with 2× SSC (0.3 M NaCl, 0.03 M sodium citrate, pH 7.0).
9. Dehydrate the sections by passing them through 30%, 50%, 70%, 85%, 95%, 100% and 100% ethanol as in step 2[b].
10. Dry under vacuum.

[a] Tissue sections are sometimes lost from the slides at this step. This can be minimized by dipping the slides gently.
[b] All but the 100% ethanol solutions, which are contaminated with xylene, can be re-used.

(23,42). The extent of deproteinization may have to be optimized for individual tissues. The goal in this case is to maximize the hybridization signal without losing morphological structures or causing the sections to fall off the slides during the procedure. We have determined, however, that the published proteinase K digestion procedure for sea urchin embryos (23) also results in maximum hybridization signals when tobacco root tips are analysed. We conclude from this and other studies that the choice of fixative may be more important in setting the proteinase K conditions than the tissue type. The other pre-treatments, BSA and acetic anhydride, are used to lower non-specific background binding to both the sections and the slides.

6.4 **Hybridization**

Discussions of probes, hybridization conditions, sensitivity and hybridization kinetics can be found in other sources (23,44,45). The following simply highlights the important features for constructing experiments.

6.4.1 *Choice of probe*

Previous work has demonstrated that single-stranded probes result in higher signals than double-stranded probes (23). We favour single-stranded RNA probes over DNA probes for the following reasons.

(i)   Post-hybridization RNase digestion allows removal of unbound probe without loss of signal.
(ii)   The higher thermal stability of RNA–RNA duplexes permits higher wash temperatures, resulting in lower non-specific probe background.

The simplest way to generate single-stranded RNA probes is using the SP6-T7 *in vitro* transcription system which has been discussed previously (Section 4).

Probes can be labelled with ^{3}H, ^{32}P or ^{35}S. ^{3}H results in the highest resolution as well as the lowest background levels at equal specific activity. However, ^{3}H has a very low autoradiographic efficiency (46) and requires extremely long incubation periods to get high grain density for all but the most abundant RNAs. ^{32}P-labelled probes can be very high in specific activity; however, the resolution is usually insufficient to localize expression on a cellular level (45). We favour ^{35}S-labelled probes (see *Table 17*). Using ^{35}S, one can obtain specific activities 10-fold higher than ^{3}H, and the autoradiographic efficiency is about five times higher. Resolution using these probes is sufficient to assay RNA accumulation in individual cells. The hybridization procedure shown in *Table 18* has been modified for use with ^{35}S-labelled probes.

After synthesis the probe is partially hydrolysed to 100–200 nt. Several laboratories have shown that short probes result in higher signals (23,47).

6.4.2 *Probe concentration*

Choosing the correct probe concentration is important to maximize sensitivity. It has been shown previously that non-specific probe binding increases approximately linearly as a function of probe concentration (23). Therefore, maximum signal/noise ratios will be obtained at a probe concentration close to target sequence saturation. Saturation concentrations have been determined for several probes hybridized to sea urchin embryos (48). As predicted, the probe concentration required for saturation is proportional to its sequence complexity. About 0.3 µg/ml probe per kb of probe complexity is required. Similar saturation analyses have not been made in any plant systems.

Table 17. Probe preparation.

We synthesize ^{35}S-labelled RNA probes with specific activities ranging over $1–5 \times 10^8$ d.p.m./µg. Procedures for synthesizing these probes can be found in Section 4, *Table 10* (steps 1–6) and *Table 11* (steps 1–5,7).

Alkaline hydrolysis of probes
1. Resuspend the probe in 50 µl of H_2O.
2. Add 30 µl of 0.2 M Na_2CO_3 and 20 µl of 0.2 M $NaHCO_3$.
3. Incubate at 60°C for the calculated time.
   $$t = (L_o - L_f)/(KL_oL_f)$$
   L_o = starting length (kb)
   L_f = final length (kb) (we suggest 0.15 kb)
   K = 0.11
   Calculated time is in minutes.
4. Stop the reaction by adding 3 µl of 3 M sodium acetate (pH 6.0) and 5 µl of 10% glacial acetic acid.
5. Add $^1\!/_{10}$ vol. of 3 M sodium acetate and 2.5 vols of ethanol. Ethanol precipitate at −20°C overnight.
6. Size the probes on a glyoxyl gel (49).

6.4.3 Hybridization criteria

The optimum temperature for *in situ* hybridization using homologous RNA probes of average GC content (50%) in the buffer listed in *Table 18* is 45–50°C (23). This is 25°C below the T_m for duplexes formed *in situ*.

Hybridizations have been done using heterologous probes (23). The decrease in thermal stability for hybrids formed *in situ* was shown to be the same as for hybrids formed in solution, that is 1% sequence divergence lowers the T_m by 1°C.

6.5 Post-hybridization treatments

6.5.1 Post-hybridization washes

We use a combination of RNase digestion and low salt–high temperature washes to lower non-specific background binding (*Table 19*). RNase digestion has been shown to be essential for identifying hybridization patterns over background. If the signal/noise ratio is low, it may be possible to raise the stringency of the final wash.

The original wash procedure shown in *Table 19* has been modified (Robert Angerer, personal communication). That wash procedure is also shown in *Table 19*. The wash temperature is only about 10°C below the *in situ* T_m for a probe of average GC content. One should be cautious when using this procedure with AT-rich probes.

6.5.2 Autoradiography

We carry out all steps involving the emulsion in absolute darkness (*Table 20*). However, it has been reported that the 'Duplex Super Safelight' consisting of

Table 18. Hybridization.

Final concentrations in the hybridization mix are: 50% formamide, 300 mM NaCl, 10 mM Tris–HCl (pH 7.5), 1 mM Na$_2$EDTA, 1× Denhardt's (0.02% BSA, 0.02% Ficoll and 0.02% PVP), 10% Dextran sulphate, 100 mM DTT, 25 U/ml RNasin, 500 μg/ml poly(A) and 150 μg/ml tRNA.

1. Combine the probe[a], tRNA, poly(A) and one half of the DTT in water. Incubate at 80°C for 5 min. Add the rest of the hybridization components.
2. Add 100 μl of hybridization mix to each slide.
3. Cover the sections with a 22 × 60 mm silanized, baked (>200°C, 2 h) coverslip[b].
4. Immerse the slides in pre-warmed mineral oil and incubate at the appropriate temperature for 16 h.

[a]The final probe concentration should be 0.2–0.5 μg/ml per kb of probe complexity.
[b]Saturate the sections with the hybridization mix before applying the coverslip or bubbles will form in the tissue. The easiest method to apply the coverslip is to hold it at an angle with a pair of electron microscope forceps and slowly lower the coverslip onto the slide. Avoid bubbles.

Table 19. Post-hybridization treatments.

Method 1
1. Drain the excess oil from the slides. Pass the slides through three changes of chloroform in Coplin jars taking care that the coverslips remain on and that the slides are free from oil. Let the chloroform evaporate from the slides.
2. Incubate the slides in 4× SSC/5 mM DTT at room temperature for about 5 min or until the coverslips fall off (1× SSC is 150 mM NaCl, 15 mM sodium citrate). Pass the slides through two more changes of 4× SSC/5 mM DTT. This removes the bulk excess probe. Store in fresh 4× SSC/5 mM DTT until all of the slides have been processed.
3. Incubate the slides in 50 μg/ml RNase A in 0.5 M NaCl, 10 mM Tris–HCl (pH 7.5), 1 mM Na$_2$EDTA for 30 min at 37°C. Pre-warm the buffer to 37°C before beginning the incubation.
4. Wash in RNase buffer/5 mM DTT at 37°C for 20 min. Repeat three more times.
5. Place the slides vertically in a test tube rack and wash with stirring in 5 litres of 2× SSC/1 mM DTT for 30 min at room temperature.
6. Wash the slides with stirring in 5 litres of 0.1× SSC/1 mM DTT for 1 h at the appropriate temperature. A wash temperature of 45–50°C is roughly equivalent to hybridization criteria of 45°C, 50% formamide, 300 mM Na^{+a}.
7. Dehydrate the tissue by passing it through the following filtered solutions (0.45 μm Millipore filter) in staining dishes: 30%, 50%, 70%, 85%, 95% ethanol each containing 100 mM ammonium acetate and 100%, 100% ethanol. Dip the slides 20 times each. Dry under vacuum at room temperature.

Method 2
1. Clean the slides as described in step 1 and 2 above.
2. Dehydrate the tissue by passing it through graded ethanols containing 300 mM ammonium acetate: 30%, 50%, 70%, 85% and 95%. Pass the slides through two changes of 100% ethanol.
3. Incubate the slides in hybridization buffer [50% formamide, 0.3 M NaCl, 20 mM Tris–HCl (pH 7.0), 1 mM Na$_2$EDTA (pH 8.0), 10 mM DTT] at 55–65°C for 10 min.
4. Transfer the slides to 0°C 2× SSC/10 mM DTT and continue as described above from step 3.

[a]We routinely wash the slides at 57°C. This high stringency wash reduces non-specific background without significantly lowering the signal.

31

Table 20. Autoradiography.

The slides are dipped in NTB-2 Liquid Track Emulsion (Kodak) that has been diluted 1:1 with 600 mM ammonium acetate and aliquoted. The aliquots contain 10 ml of emulsion and are stored in light-tight containers at 4°C. We carry out all darkroom work in absolute darkness.

1. Melt an aliquot of emulsion in a suitable dipping container in a 45°C waterbath. This takes about an hour.
2. After the emulsion has warmed to 45°C, dip the slides by immersing once as smoothly as possible.
3. Holding the slides vertically, blot the bottom edge on a paper towel. Place the slides vertically in a test tube rack to dry. Allow the slides to dry for 1 h at room temperature.
4. Transfer the slides to a black slide box containing a tube of desiccant. The slides are exposed at 4°C in a light-tight box.

Developing
1. Develop the slides for 2.5 min in D-19 developer (Kodak) at 15°C. Dip the slides up and down five times then let the slides sit for a total of 2.5 min[a].
2. Stop in 2% acetic acid at 15°C for 30 sec dipping continuously.
3. Fix in Kodak Fixer (not Rapid Fix) at 15°C. Dip up and down five times. Let the slides sit in fixer a total of 5 min.
4. Rinse in distilled water.
5. Rinse in cold running water for 15 min.

Staining
1. Stain the slides in 0.1% toluidine blue (water) for ~1 min. Do test slides for each tissue because the amount of time necessary for optimal staining varies.
2. Rinse quickly with water.
3. Dehydrate the tissue by passing the slides quickly through the following solutions: 25%, 50%, 75%, 100% and 100% ethanol.
4. Dip in xylene.
5. Add a few drops of permount. Wipe the coverslip while pressing down to remove excess xylene/permount. Let the slides dry overnight.

[a]Longer developing times and higher temperatures preferentially produce grains in the emulsion background.

FDY filters in the top slots and FDW filters in the bottom slots did not increase background emulsion grains after a 3 h exposure (45).

6.5.3 *Exposure time*

We have analysed RNA sequences present at 10–0.1% of the poly(A^+) RNA. Using ^{35}S-labelled probes of the specific activity discussed previously, high grain densities (e.g. see *Figure 1*) were obtained after 1–21 days of exposure. Of course exposure time must be determined empirically for each new probe. Grain density depends on the target RNA concentration per cell which cannot be predicted from alternative molecular analyses.

6.5.4 *Staining*

There are many staining procedures for plant tissue. We have used toluidine blue because it provides good differentiation between cytoplasm and nucleus,

A

B

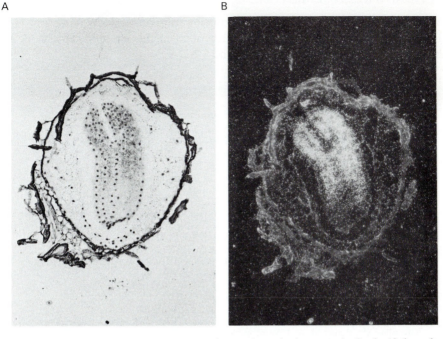

Figure 1. Localization of β-conglycinin mRNA in transformed tobacco seeds. Seeds, 19 days after pollination, were fixed in FAA, embedded in paraffin, and cut into 10 μm sections. A ^{35}S-labelled anti-sense RNA probe (1.5×10^8 d.p.m./μg) was hybridized to plant sections for 14 h at hybridization criteria of 42°C, 0.3 M Na$^+$ and 50% formamide. Wash criteria were 57°C and 0.02 M Na$^+$. The slides were developed after 3.5 days of exposure. Seed sections were stained with 0.5% toluidine blue. (**A**) Longitudinal section of a tobacco seed. The photograph was taken using bright-field microscopy. (**B**) Hybridization with a ^{35}S-labelled anti-sense probe. The photograph was taken using darkfield microscopy. The hybridization signal can be seen over embryo cotyledon cells and upper axis cells. [Reproduced from Barker *et al.* (37).]

and yet stains uniformly enough not to interfere with visualization of silver grains under dark field illumination.

6.6 Quantitative analysis

It has been shown that 3- to 8-fold differences in RNA concentrations in sea urchin embryos were accurately reflected in relative grain densities using this *in situ* hybridization procedure (23). We use these results to suggest that data derived from this technique may be interpreted quantitatively as well as qualitatively with the following caveats. In the two experiments described above, comparisons were made using a single probe hybridized to RNA in the same or very similar cell types. It remains possible that different cell types may have different RNA retention or hybridization efficiency. Second, in order to compare signals with the greatest degree of accuracy, the sections being examined should be from the same experiment and on the same slide. We have observed slight quantitative differences in signals from experiment to experiment and from slide to slide.

7. ACKNOWLEDGEMENTS

We would like to acknowledge the contributions of Linda Walling and Gary Drews in developing the techniques described in this chapter. This work has been supported by grants from NSF and USDA.

8. REFERENCES

1. Brandis,J., Larocca,D. and Monahan,J. (1987) In *Genetic Engineering*. Setlow,J. and Hollaender,A. (eds), Plenum Press, New York, Vol. 8, p. 299.
2. Kimmel,A.R. and Berger,S.L. (1987) In *Methods in Enzymology*. Berger,S.L. and Kimmel, A.R. (eds), Academic Press, New York, Vol. 152, p. 307.
3. Gubler,U. and Hoffmen,B.J. (1983) *Gene*, **25**, 263.
4. Ogden,R. and Adams,D. (1987) In *Methods in Enzymology*. Berger,S.L. and Kimmel,A.R. (eds), Academic Press, New York, Vol. 152, p. 61.
5. Berk,A.J. and Sharp,P.A. (1977) *Cell*, **12**, 721.
6. Calzone,F.J., Britten,R.J. and Davidson,E.H. (1987) In *Methods in Enzymology*. Berger,S.L. and Kimmel,A.R. (eds), Academic Press, New York, Vol. 152, p. 611.
7. Jackson,A.O. and Larkins,B.A. (1976) *Plant Physiol.*, **57**, 5.
8. Zabel,P., Jongen-Neven,I. and Van Kammen,A. (1976) *J. Virol.*, **17**, 679.
9. Chirgwin,J.M., Przybyla,A.E., MacDonald,R.J. and Rutter,W.J. (1979) *Biochemistry*, **18**, 5294.
10. Penman,S., Vesco,C. and Penman,M. (1968) *J. Mol. Biol.*, **34**, 49.
11. Perry,R.P. and Kelley,D.E. (1968) *J. Mol. Biol.*, **35**, 37.
12. Aviv,H. and Leder,P. (1972) *Proc. Natl. Acad. Sci. USA*, **69**, 1408.
13. McKnight,G.S. and Palmiter,R.D. (1979) *J. Biol. Chem.*, **254**, 9050.
14. Groudine,M., Peretz,M. and Weintraub,H. (1981) *Mol. Cell Biol.*, **1**, 281.
15. Marzluff,W.F. and Huang,C.C. (1984) In *Transcription and Translation—A Practical Approach*. Hames,B.D. and Higgins,S.J. (eds), IRL Press, Oxford, p. 89.
16. Walling,L., Drews,G.N. and Goldberg,R.B. (1986) *Proc. Natl. Acad. Sci. USA*, **83**, 2123.
17. Luthe,D.S. and Quatrano,R.S. (1980) *Plant Physiol.*, **65**, 309.
18. Derman,E., Krauter,K., Walling,L., Weinberger,C., Ray,M. and Darnell,J.E. (1981) *Cell*, **23**, 731.
19. Luthe,D.S. and Quatrano,R.S. (1980) *Plant Physiol.*, **65**, 305.
20. Honda,S.I., Hongladarom,T. and Laties,G.G. (1966) *J. Exp. Bot.*, **17**, 460.
21. Langridge,J., Langridge,P. and Bergquist,P.L. (1980) *Anal. Biochem.*, **103**, 264.
22. Melton,D.A., Krieg,P.A., Rebagliati,M.R., Maniatis,T., Zinn,K. and Green,M.R. (1984) *Nucleic Acids Res.*, **12**, 7035.
23. Cox,K.H., DeLeon,D.V., Angerer,L.M. and Angerer,R.C. (1984) *Dev. Biol.*, **101**, 485.
24. Krieg,P.A., Rebagliati,M.R., Green,M.R. and Melton,D.A. (1985) In *Genetic Engineering*. Setlow,J.K. and Hollaender,A. (eds), Plenum Press, New York, Vol. 7, p. 165.
25. *Promega Biotech Instruction Manual*, 2800 S.Fish Hatchery Road, Madison, WI 53711, USA.
26. Zinn,K., DiMaio,D. and Maniatis,T. (1983) *Cell*, **34**, 865.
27. DeLeon,D.V., Cox,K.H., Angerer,L.M. and Angerer,R.C. (1983) *Dev. Biol.*, **100**, 197.
28. Lee,J.J., Calzone,F.J., Britten,R.J., Angerer,R.C. and Davidson,E.H. (1986) *J. Mol. Biol.*, **188**, 173.
29. Lee,J.J. and Costlow,N.A. (1987) In *Methods in Enzymology*. Berger,S.L. and Kimmel,A.R. (eds), Academic Press, New York, Vol. 152, p. 633.
30. Hu,N.-T. and Messing,J. (1982) *Gene*, **17**, 271.
31. Galau,G.A., Britten,R.J. and Davidson,E.H. (1977) *Proc. Natl. Acad. Sci. USA*, **74**, 1020.
32. Britten,R.J., Graham,D.E., Eden,F.C., Painchaud,D.M. and Davidson,E.H. (1976) In *Methods in Enzymology*. Grossman,L. and Moldave,K. (eds), Academic Press, New York, Vol. 29, p. 363.
33. Hutton,J.R. (1977) *Nucleic Acids Res.*, **4**, 3537.
34. Wetmur,J.G. and Davidson,N. (1968) *J. Mol. Biol.*, **31**, 349.
35. Dische,Z. and Borenfreund,E. (1957) *Biochim. Biophys. Acta*, **23**, 639.
36. Burton,K. (1956) *Biochem. J.*, **62**, 315.
37. Barker,S.J., Harada,J.J. and Goldberg,R.B. (1988) *Proc. Natl. Acad. Sci. USA*, **85**, 458.
38. Harris,N. and Croy,R.R.D. (1986) *Protoplasma*, **130**, 57.

39. Anderson,M.A., Cornish,E.C., Mau,S.-L., Williams,E.G., Hoggart,R., Atkinson,A., Bonig, I., Grego,B., Simpson,R., Roche,P.J., Haley,J.D., Penschow,J.D., Niall,H.D., Tregear, G.W., Coghlan,J.P., Crawford,R.J. and Clarke,A.E. (1986) *Nature*, **321**, 38.
40. Meyerowitz,E.M. (1987) *Plant Mol. Biol. Rep.*, **5**, 242.
41. Angerer,L.M. and Angerer,R.C. (1981) *Nucleic Acids Res.*, **9**, 2819.
42. Hafen,E., Levine,M., Garber,R.L. and Gehring,W.J. (1983) *EMBO J.*, **2**, 617.
43. Avery,G.S. (1933) *Am. J. Bot.*, **20**, 309.
44. Angerer,R.C., Cox,K.H. and Angerer,L.M. (1985) In *Genetic Engineering*. Setlow,J.K. and Hollaender,A. (eds), Plenum Press, New York, Vol. 7, p. 43.
45. Angerer,L.M., Cox,K.H. and Angerer,R.C. (1987) In *Methods in Enzymology*. Berger,S.L. and Kimmel,A.R. (eds), Academic Press, New York, Vol. 152, p. 649.
46. Ada,G.L., Humphrey,J.H., Askonas,B.A., McDevitt,H.D. and Nossal,G.V. (1966) *Exp. Cell Res.*, **41**, 557.
47. Brahic,M. and Haase,A.T. (1978) *Proc. Natl. Acad. Sci. USA*, **75**, 6125.
48. Cox,K.H., Angerer,L.M., Lee,J.J., Davidson,E.H. and Angerer,R.C. (1986) *J. Mol. Biol.*, **188**, 159.
49. Carmichael,G.G. and McMaster,G.K. (1980) In *Methods in Enzymology*. Grossman,L. and Moldave,K. (eds), Academic Press, New York, Vol. 65, p. 380.

CHAPTER 2

Analysis of plant gene structure

K.DIANE JOFUKU and ROBERT B.GOLDBERG

1. INTRODUCTION

There has been an explosive interest in the field of plant molecular biology. Recent technical breakthroughs have facilitated the cloning of plant genes and the subsequent analysis of their expression programmes during plant development. In this chapter, we focus on those techniques that are currently used in our laboratory to study the structure and sequence organization of developmentally regulated plant genes (1–6). We have optimized procedures for plant systems and in many cases, have simplified them so that they are easy to use and require minimal amounts of time without sacrificing quality. We have chosen not to discuss procedures that are well-characterized and of common knowledge and, where appropriate, we will direct the reader to published references.

In brief, this chapter deals with:

(i)   DNA isolation. In this section, we describe a detailed protocol for the isolation of high molecular weight plant nuclear DNA suitable for genomic DNA cloning. This procedure minimizes chloroplast and mitochondrial DNA contamination and is gentle enough to allow for the isolation of nuclear DNA greater than 150 kb in length.

(ii)   DNA cloning. This section provides references that discuss and present detailed protocols for the construction of genomic DNA libraries using bacteriophage λ Charon and bacterial cosmid vectors. For convenience, we have included our modified procedure for creating plant gene banks. Using this procedure, we can construct a plant genomic DNA library containing greater than one million independent recombinant phage and less than 0.1% non-recombinant phage within 2 weeks.

(iii) Analysis of plant gene structure. This section provides several procedures used in our laboratory to characterize plant gene structure. We also discuss techniques recently established in the laboratory that allow for the identification and characterization of plant DNA binding proteins.

2. ISOLATION OF HIGH MOLECULAR WEIGHT PLANT DNA

2.1 Introduction

Intact, high molecular weight (>150 kb) plant nuclear DNA is essential for the construction of genomic DNA libraries. Factors that must be considered when choosing a procedure for isolating nuclear DNA include DNA yield, state of

DNA (e.g. intact, free of substances that can inhibit enzyme reactions), and non-nuclear (e.g. chloroplast and mitochondria) DNA contamination.

The procedure given below is optimized for isolating high molecular weight plant nuclear DNA as previously described by Fischer and Goldberg (1). We have used this procedure successfully to isolate intact DNA from several plant species including corn, soybean, tobacco and tomato. In brief, plant nuclei are isolated free of chloroplasts and mitochondria by disrupting tissues and extracting the crude cellular extracts in the presence of the non-ionic detergent Triton X-100. Triton X-100 specifically acts on organellar membranes, effectively lysing chloroplasts and mitochondria while leaving nuclei intact (1). Nuclei are then lysed and the DNA is purified by CsCl gradient ultracentrifugation. Although a more simple and less time consuming procedure has been reported (7; see Chapter 6) we feel that this procedure provides the highest quality plant nuclear DNA with minimal non-nuclear DNA contamination. All solutions are given in *Table 1*.

(i)   Grind tissue to a fine powder with a Waring blender in liquid nitrogen.

(ii)   Allow the liquid nitrogen to escape. The liquid nitrogen is evaporated when the tissue is lighter in colour and does not form clumps. It is important that no liquid nitrogen is present as the homogenized tissue will freeze when added to the extraction buffer. Be extremely careful, however, that the tissue does not thaw. Thawing of the tissue prior to the addition of extraction buffer will cause cellular lysis and the release of endogenous nuclease activities that can degrade the DNA.

(iii)   Add 5–10 ml of cold 1× H buffer (*Table 1*) per g tissue to a sterile beaker containing a stirbar.

(iv)   Transfer the tissue to a beaker containing cold 1× H buffer. Stir until the powder is in solution.

(v)   Wet Nitex with 1× H buffer; filter the homogenate through Nitex. Nitex is a nylon mesh that can be purchased from Small Parts, Inc., Miami, FL; before use, boil Nitex (28 cm × 28 cm; 88 or 100 μm mesh) in 1 mM Na_2EDTA; cool to 4°C before use.

(vi)   Pour the supernatant into six centrifuge bottles (Sorvall GSA) and pellet the nuclei from solution at 3500 r.p.m. for 20 min at 4°C.

(vii)   Discard the supernatants. The supernatant should be green in colour due to the lysis of chloroplasts; if not, check whether Triton X-100 was added to the 1× H buffer.

(viii) Mush the nuclear pellets with a rounded glass rod; add 25 ml of 1× H buffer to each bottle and swirl until the nuclei are resuspended.

(ix)   Pool the supernatants; spin down the nuclei, 3500 r.p.m., in a GSA or HB-4 rotor (Sorvall) for 20 min at 4°C.

(x)   Repeat steps vii–ix at least twice. At this point, the nuclear pellet should be greyish-white and not too viscous; chloroplast or starch contamination is evidenced by a green or viscous nuclear pellet.

(xi)   Resuspend the nuclei in 1× H buffer such that the final volume is 25 ml. This volume is ideal for isolating DNA from small amounts of

(xii) Transfer to 125 ml Erlenmeyer flask.

(xiii) Add 25 ml of *cold* lysis buffer dropwise, with gentle swirling. Addition of Sarkosyl should lyse the nuclei; the amount of viscosity observed is an indication of the amount and size of the DNA released upon lysis.

(xiv) *Immediately* add 48.6 g of optical grade CsCl. Once the nuclei are lysed, the nucleic acids become susceptible to endogenous nucleases; addition of CsCl will inactivate nucleases. Gently swirl the solution at room temperature until the CsCl is in solution. Excessive mixing can result in shearing of the DNA.

(xv) Transfer the DNA solution to 50 ml Sorvall SS34 nylon tubes; spin at 10 500 r.p.m. for 30 min at 4°C.

(xvi) Carefully decant the supernatant into a sterile glass graduated cylinder. This step helps reduce the amount of protein and starch contaminants prior to CsCl gradient ultracentrifugation; protein should form a pellicle at the top of the solution, and starch and other carbohydrates should pellet.

(xvii) Determine the volume of the supernatant.

(xviii) In *dim* light, add 38.6 μl of 10 mg/ml ethidium bromide per ml of supernatant. Never expose the DNA to light in the presence of ethidium bromide which can cause nicks in the DNA.

(xix) Transfer to polyallomar tubes; overlay with mineral oil, if necessary.

(xx) Spin in a Sorvall T865 fixed-angle rotor at 40 000 r.p.m. for 38 h at 20°C.

(xxi) Locate the DNA band by using *long wave* UV light.

(xxii) Remove the upper portion of the gradient to within 2 cm of the DNA band with a 10 ml pipette.

(xxiii) Remove the DNA band from the top with a 10 ml sterile plastic pipette; place in a polyallomar tube.

(xxiv) Bring to volume with a 1.57 g/ml CsCl solution in 50 mM Tris–HCl (pH 9.5), 20 mM Na_2EDTA.

(xxv) Reband the DNA as in step xx at 40 000 r.p.m. for 30 h at 20°C.

(xxvi) Remove the DNA band as in steps xxi–xxiii.

(xxvii) Extract the DNA with an equal volume of NaCl-saturated isopropanol [in 10 mM Tris (pH 7.5), 1 mM Na_2EDTA] to remove ethidium bromide. Ethidium bromide should pass into the upper, isopropanol phase.

(xxviii) Repeat step xxvii four times. At this point, the DNA should be free of ethidium bromide; if not, the DNA is probably too concentrated and should be diluted with a 1.57 g/ml CsCl solution in 50 mM Tris–HCl (pH 9.5), 20 mM Na_2EDTA. Do not dilute with a low salt buffer as addition of isopropanol will precipitate the DNA out of solution.

(xxix) Dialyse the DNA against several (e.g. >5) changes of 10 mM Tris–HCl (pH 7.5), 10 mM NaCl, 0.1 mM Na_2EDTA. Leave room in the dialysis sack for uptake of H_2O due to the high concentration of salt (e.g. CsCl) in the DNA solution.

(xxx) Determine the DNA concentration by UV absorption. Determine the average length by 0.3% agarose gel electrophoresis using appropriate molecular weight markers. Normally, we use intact T2 phage DNA (150 kb) and λ phage DNA (48.5 kb) as DNA standards.

2.2 Guidelines for DNA isolation

2.2.1 *Considerations when working with plant material*

(i) *Choice of tissue.* Choice of tissue plays a critical role in obtaining high yields of DNA. Theoretically, yield depends on the number of cells, and therefore, the number of nuclei per gram of tissue; the efficiency of nuclear isolation from various tissues; and the recovery of DNA from these crude nuclear preparations. We have isolated nuclear DNA from leaves and stems of several plant species as well as from soybean embryos. In our experience, unexpanded leaves (e.g. third or fourth trifoliate tobacco leaves) provide the highest DNA yields typically ranging from 100 μg to as high as 200 μg/g tissue. Leaves are easily obtained without extensive damage to the plant and extraction of as few as five leaves can produce enough DNA for several experiments. Finally, leaf DNA is virtually free of starch and other carbohydrates that often co-extract with nucleic acids.

(ii) *Contamination problems.* Chloroplasts and mitochondria are the major sources of non-nuclear plant DNA contamination. As mentioned previously, we have alleviated the problem in part by isolating intact nuclei in the presence of Triton X-100, a detergent that specifically lyses chloroplasts and mitochondria. In addition, we found that passing the homogenized plant material (in extraction buffer) through a single layer of Nitex nylon mesh (0.2 μm) eliminated much of the undissociated green plant tissue containing chloroplasts and mitochondria that can then co-purify with nuclei during low speed centrifugation.

In our experience, starch and other carbohydrates do not pose significant problems in isolating nuclear DNA because, in general, they can be separated from the DNA by CsCl gradient centrifugation. Occasionally, however, carbohydrate-associated particulate material adheres to the DNA following centrifugation and is therefore co-extracted with it. This problem can be resolved by removing the particulate from the DNA by low speed centrifugation following dialysis (step xxix).

(iii) *Harvesting plant material.* Tissue should be frozen in liquid nitrogen immediately after harvesting, and then broken up into smaller pieces (e.g. with a sterile plastic 10 ml pipette). Frozen tissue can then be stored at −70°C for an indefinite period (e.g. 1 year or more) prior to use.

2.3 Glassware and solutions

2.3.1 *Glassware*

Glassware should be washed thoroughly, dipped in 0.1% diethylpyrocarbonate (DEPC; Sigma) and then baked at 200°C for at least 2 h. Plastic equipment should be washed and, if possible, sterilized by autoclaving at 121°C for 15 min. Glassware used for plasmid DNA preparations should be kept separate from that used to isolate plant DNA. We have determined that plasmid-contaminated plant nuclear DNA preparations are due to plasmids sticking to laboratory centrifuge tubes and glassware.

2.3.2 *Solutions*

All solutions should be prepared as described in *Table 1* and filtered to remove any particulate matter. It is not necessary to sterilize the solutions by autoclaving; in fact, it has been our experience that autoclaving the extraction

Table 1. Solutions for isolation of high molecular weight plant nuclear DNA[a].

---

1. 10× HB (pH 9.4)
   40 mM Spermidine[b]
   10 mM Spermine[b]
   0.1 M Na$_2$EDTA[c]
   0.1 M Trizma base
   0.8 M KCl
   Adjust the pH to 9.4–9.5 with 10 M NaOH; filter through a Millipore filter (45 μm).
2. 2 M Sucrose
   Add sucrose to H$_2$O and bring to volume. Heat at 45–50°C until sucrose is in solution. Autoclave to sterilize for *only* 15 min[d].
3. 100 mM Phenylmethylsulphonylfluoride (PMSF; Sigma)
   Dissolve PMSF in 100% ethanol. Discard after use.
4. 1× H buffer (extraction buffer)[e]
   1× HB
   0.5 M Sucrose[f]
   1 mM PMSF[g]
   0.5% (v/v) Triton X-100
   0.1% (v/v) β-Mercaptoethanol[h]
5. Lysis buffer
   0.1 M Tris–HCl (pH 9.5)
   0.04 M Na$_2$EDTA
   2% Sodium Sarkosyl
6. NaCl-saturated isopropanol in 10 mM Tris–HCl (pH 7.5), 1 mM Na$_2$EDTA
   Add saturating amounts of NaCl to a solution of 10 mM Tris–HCl (pH 7.5), 1 mM Na$_2$EDTA with stirring at 40°C. Allow to cool. Add an equal volume of isopropanol and mix well.
7. Dialysis buffer
   10 mM Tris–HCl (pH 7.5)
   10 mM NaCl
   0.1 mM Na$_2$EDTA

---

[a] All solutions should be filtered through 0.45 μm membrane (Millipore BA 0.45) and chilled to 4°C before use. Do not autoclave unless otherwise directed.
[b] The polycations spermine and spermidine serve to stabilize the nuclear membrane as well as the isolated chromatin, thereby increasing the yield of nuclei and DNA.
[c] EDTA inhibits nuclease activities released upon cellular lysis.
[d] Autoclaving longer than 15 min caramelizes the sucrose solution.
[e] Cool extraction buffer to 7°C before use.
[f] Sucrose is used in the extraction buffer as an osmoticum to prevent nuclear lysis during isolation.
[g] PMSF is used to inhibit contaminating protease that may destabilize intact nuclei.
[h] β-Mercaptoethanol serves as an antioxidant during extraction. Add to the extraction buffer just before use.

buffer severely affects DNA yield, presumably because high heat inactivates the polycations spermidine and spermine and therefore adversely affects the yield of intact nuclei. For this reason, we autoclave only the 2 M sucrose solution in order to prevent bacterial and fungal growth.

Extraction and lysis buffer pH should be carefully monitored before and during extraction. In our experience, extraction of soybean tissue at neutral pH results in the isolation of DNA less than 50 kb in length, suggesting that soybean nuclei contain endogenous nucleases that are active at pH 7.0 (E.Ralston and R.B.Goldberg, unpublished observations). We alleviated this problem by raising the pH of the extraction and lysis buffers to 9.5 (E.Ralston and R.B.Goldberg, unpublished results). Although we do not know whether other plant species contain different nuclease activities, we have used these same extraction conditions to successfully isolate high molecular weight DNA from various dicotyledonous and monocotyledonous plants.

3. DNA CLONING

3.1 **Introduction**

High quality plant DNA is no more difficult to manipulate than any other kind of DNA. For this reason, we strongly recommend Maniatis *et al*. (8) and Kaiser and Murray (9) as well as Meyerowitz *et al*. (10); Ish-Horowicz *et al*. (11); Grosveld *et al*. (12) for excellent, detailed treatments of DNA cloning using bacteriophage λ replacement and cosmid vectors, respectively. For convenience, however, we have included our procedure for cloning plant nuclear DNA into the λ Charon 35 vector (13) the solutions for which are given in *Table 2*. We will only briefly discuss modifications of published procedures that we feel are important to consider.

3.2 **Preparation of DNA**

3.2.1 *Limited Sau3A digestion of plant DNA*

The standard method for generating DNA fragments of defined lengths for cloning is to digest the target DNA with an appropriate restriction enzyme under conditions that achieve partial digestion. Normally, this would involve varying enzyme concentration in test reactions, and then scaling the reaction up to generate large quantities of DNA fragments for cloning purposes. The DNA is then fractionated by sucrose density gradient centrifugation and size-selected DNA fragments are isolated and dialysed into a low-salt buffer in preparation for ligation to vector DNA.

An alternate method suggested by Maniatis *et al*. (8) for generating size-selected DNA fragments is to vary reaction time rather than enzyme concentration. In our experience, this procedure provides the most reproducible partial enzyme digestion and should be the method of choice.

(i)   Dilute *Sau*3A with storage buffer to a final concentration of 1 U/ml. Store at −20°C until ready to use.

Table 2. Solutions for cloning plant DNA.

Limited Sau3A digestion
1. *Sau*3A enzyme storage buffer
   10 mM Tris–HCl (pH 7.5)
   50 mM KCl
   0.1 mM Na$_2$EDTA
   1 mM DTT
   500 μg/ml BSA
   50% (v/v) glycerol
2. 5× *Sau*3A reaction buffer
   30 mM Tris–HCl (pH 7.5)
   250 mM NaCl
   30 mM MgCl$_2$
   500 μg/ml BSA
3. 1 U/μl *Sau*3A restriction enzyme (diluted in *Sau*3A storage buffer)[a]
4. 0.5 M Na$_2$EDTA
5. Gel loading dye
   10 mM NaCl
   75% Glycerol
   1 mM Na$_2$EDTA
   0.02% Xylene cyanol
   0.02% Bromophenol blue
6. 25× TEA running gel buffer
   1 M Tris (pH 8.2)
   0.025 M Na$_2$EDTA
   0.5 M Sodium acetate

Size-selection of Sau3A DNA fragments using sucrose gradient sedimentation[b]
1. 2.5× Salts
   2.5 M NaCl
   0.025 M Tris–HCl (pH 7.5)
   0.0125 M Na$_2$EDTA
2. 2 M (68%) Sucrose, filtered and autoclaved
3. 10%, 18%, 25%, 34%, and 40% Sucrose in 1× salts
4. 3 M Potassium acetate (pH 6)
5. 10 mM Tris–HCl (pH 7.5), 5 mM NaCl, 0.1 mM Na$_2$EDTA [TNE]

Preparation of λ Charon 35 vector 'arms'
1. 10× Ligation buffer
   0.66 M Tris–HCl (pH 7.5)
   0.1 M MgCl$_2$
   1 mg/ml BSA
2. 20 mM ATP (pH 7.0)
3. 1 M Dithiothreitol (DTT)
4. 10 mM Tris–HCl (pH 7.5), 0.1 mM Na$_2$EDTA [TE]
5. 5× Cuts all buffer
   500 mM KCl
   100 mM Tris–HCl (pH 7.5)
   35 mM MgCl$_2$
   10 mM β-Mercaptoethanol
   500 μg/ml BSA
6. 0.5 M Na$_2$EDTA

[a] Store at −20°C until ready to use. Diluted enzyme can be stored at −20°C for at least 1 week without losing activity.
[b] DNA fragments can also be purified by preparative gel electrophoresis with DNA extracted from the gel using any one of several published methods. In our experience, however, the yields of large DNA fragments obtained using this method are much lower than those obtained by sucrose gradient sedimentation.

(ii) Set up a test reaction containing 5 μg of plant nuclear DNA and 5 U of *Sau*3A in a total volume of 50 μl of 1× *Sau*3A reaction buffer.

(iii) Incubate reaction at 37°C.

(iv) At 90 sec, 6, 15, 30 and 60 min, aliquot 10 μl from main reaction to a tube containing 1 μl of 0.5 M Na$_2$EDTA to inactivate the enzyme. Add 2 μl of gel loading dye to each aliquot. Store on ice until all time points are completed.

(v) Load the samples on a 0.5% agarose gel in 1× TEA buffer. Run the gel at 25 V (constant voltage) for 16 h to achieve maximum DNA separation. Use well-characterized DNA size markers in the 10–25 kb size range.

(vi) Choose a set of enzyme concentrations that generate the maximum number of DNA fragments in the following size ranges:
(a) 12–22 kb
(b) 20–25 kb
We include 20–25 kb DNA fragments for cloning since large but still clonable fragments are normally not as represented in phage libraries constructed with fragments of size range 12–22 kb only.

(vii) If necessary, repeat steps i–v using another enzyme concentration.

(viii) When you have chosen a set of reaction conditions, scale up the assay reaction 10-fold to generate DNA fragments for cloning. Inactivate the enzyme by adding Na$_2$EDTA to a final concentration of 10 mM. Store at 4°C until ready to fractionate.

3.2.2 *Size selection of Sau3A DNA fragments using sucrose gradient sedimentation*

(i) Prepare gradients by sequentially adding the following solutions of sucrose to a 17 ml polyallomer tube:
(1) 3.5 ml 40%
(2) 6.7 ml 34%
(3) 4.2 ml 25%
(4) 2.0 ml 18%
(5) 0.6 ml 10%

(ii) Let the gradients sit overnight at room temperature to diffuse.

(iii) Carefully pull off about 1.5× the sample volume to be loaded from the top of the gradients.

(iv) Load the samples directly onto the gradients in less than 1 ml of total volume. If necessary, decrease the sample volume by extracting with an equal volume of *sec*-butanol 3–4 times. Load the lower, aqueous layer containing DNAs onto the gradients.

(v) Spin the gradients at 27 000 r.p.m. for 24 h at 20°C. Accelerate to 6000 r.p.m. slowly to minimize disturbance of gradients. When the run is finished, take gradients down slowly without using the automatic brake.

(vi) Drip-fractionate the gradients, taking 10 drop (~200 μl) fractions.

(vii) Analyse the fractions by loading 10 μl of every other fraction of the first 50 fractions on a 0.5% agarose gel (in 1× TEA buffer) using appropriate size markers.

(viii)   Choose and pool the fractions that contain *Sau*3A DNA fragments that are between 15 and 22 kb in length. In our experience 15–22 kb fragments are found in fractions 30–40.

(ix)   Dialyse the pooled fractions against TNE buffer, 5 × 1 litre change, 6 h per change with one 14 h change.

(x)   Determine the volume. Add ⅟₁₀ vol. of 3 M potassium acetate (pH 6) and an equal volume of isopropanol. Precipitate the DNA overnight at −20°C.

(xi)   Spin down the DNA at 20 000 r.p.m. for 45 min at 0°C.

(xii)   Wash the DNA pellet with 70% ethanol.

(xiii)   Repeat step xi.

(xiv)   Decant the supernatant; dry the pellet *in vacuo*.

(xv)   Resuspend the DNA in 50–100 μl of TNE.

(xvi)   Determine the concentration. Store at 4°C until ready to use.

3.2.3 *Preparation of vector DNA*

Our experience is limited to the construction of genomic DNA libraries using the λ Charon vectors. For this reason, our discussion will be restricted to the preparation of λ vector 'arms'.

Lambda Charon replacement vectors [for reviews, see Blattner *et al.* (14) and Loenen and Brammer (15)] contain sequences required for the maintenance of the lytic phage life cycle. In most Charon phages, the sequences for the lytic cycle are contained on the two 'end' or 'arm' fragments and can be separated from the internal 'stuffer' DNA fragments by digestion with an appropriate restriction enzyme followed by fractionation by sucrose gradient centrifugation or agarose gel electrophoresis.

Our procedure for preparing λ Charon vector 'arms' is given below. In our experience, this is the limiting step in the construction of genomic DNA libraries. This method allows for the isolation of large quantities of vector 'arms' virtually free of contaminating phage 'stuffer' DNA fragments. This modified procedure of Maniatis *et al.* (8) involves the annealing of the λ Charon phage 'cos' sites and their ligation prior to restriction enzyme digestion. Vector DNA fragments are then subjected to dephosphorylation with calf intestinal alkaline phosphatase (CIP) followed by fractionation by sucrose density centrifugation. Vector 'arms' are isolated and then subjected to dialysis into a low-salt buffer in preparation for ligation to size-selected plant DNA fragments. We found that ligation facilitates the isolation of vector 'arms' by at least 20-fold and that dephosphorylation of λ 'arms' prior to ligation with size-selected eukaryotic DNA decreases the number of non-recombinant phage generated by greater than 1000-fold (K.D.Jofuku and R.B.Goldberg, unpublished observations).

(i)   Add to 50 μg λ CH35 DNA in TE (contained in two separate test tubes):
      40 μl 10× ligation buffer
      40 μl 20 mM ATP
      6 μl 1 M DTT
Bring to a final reaction volume of 400 μl with H₂O.

(ii)   Add 8 U of T4 DNA ligase to each of the two reactions; mix well.

(iii)   Incubate the reactions at 15°C for 10–12 h.

(iv)   Heat reactions at 70°C for 15 min to inactivate the enzyme; cool on ice; spin reactions briefly in a microfuge to collect the contents.

(v)   Add to the reactions containing ligated phage DNA:
    100 μl 5× cuts all buffer
    100 U *Bam*HI (2 U/μg DNA)

(vi)   Mix well; incubate the reactions at 37°C for 3 h.

(vii) Aliquot 10 μl of pooled reactions into a tube containing 40 μl of H_2O and 10 μl of gel loading dye. Store the main reactions at 4°C until ready to use.

(viii) Heat the aliquots for 10 min at 65°C; quench in liquid nitrogen; thaw on ice-water. Load an aliquot onto a 0.5% agarose gel to test for efficient ligation of 'cos' sites and complete digestion of phage DNA.

(ix)   If digestion is complete, add to main reactions 100 U CIP (Boehringer-Mannheim). Ligation is complete if only the 32 kb 'ligated' right and left arms are visible after heat treatment at 65°C. You should not see the 20.0 or the 10.9 kb right and left 'arm' fragments.

(x)   Incubate at 37°C for 15 min; spin reactions in microfuge to collect contents; add another 100 U of CIP to main reactions; incubate at 37°C for 15 min.

(xi)   Spin to collect the reaction contents. Add 0.5 M Na_2EDTA to a final concentration of 10 mM.

(xii) Load the main reactions directly onto 10–40% sucrose gradients prepared as described in Section 3.2.2.

(xiii) Follow steps v and vi of Section 3.2.2. Analyse 10 μl of every other 200 μl (10 drop) fraction on a 0.5% agarose gel. Choose and pool the fractions that contain only the 32 kb ligated and 20.0 and 10.9 kb 'right' and 'left' arms, respectively (normally these are fractions 1–10).

(xiv) Dialyse and concentrate as in steps ix to xvi of Section 3.2.2.

(xv)   Resuspend the 'arms' in 50 μl of TE. Determine the concentration.

3.3 Ligation of size-selected plant DNA fragments and vector 'arms'

In general, the following considerations should be made regarding the ligation reaction:

(i)   Use a 2:1 molar ratio of vector 'arms' to size-selected plant DNA fragments. This translates into a 3:1 mass ratio of 'arms' to plant DNA if you assume that the average plant DNA fragment length is 15 kb. This step ensures that, on average, each plant DNA fragment will be covalently attached to one vector DNA fragment. This maximizes the efficiency of *in vitro* recombination and decreases the number of non-recombinant phage, as well as the number of phage containing multiple non-contiguous plant DNA inserts.

(ii)   Keep the total DNA concentration high (e.g. >0.1 mg/ml) which is important for successful ligation to occur. In addition, low reaction volume (e.g. <30 μl) is an important consideration if you intend to use commercially prepared packaging extracts where the maximum volume of DNA solution that can be packaged in one reaction is 4 µl.

(iii) *Always* set up a ligation reaction containing only vector 'arms'. Package the resulting ligation products to estimate the number of non-recombinant phage that is generated using vector DNA alone. The reaction conditions are exactly as specified by Maniatis *et al.* (8).

3.4 Packaging of DNA

Several procedures are available for the preparation of *in vitro* phage packaging extracts (16–18). Using one of these procedures, we routinely obtained extracts with packaging efficiencies of at least 10^7 plaque-forming units (p.f.u.) per µg DNA. However, we could rarely obtain extracts of higher efficiencies. For this reason, we now use only commercially prepared *in vitro* extracts from manufacturers that guarantee at least 10^8 p.f.u./µg DNA. In our experience, using these extracts is by far the most efficient and economical way to construct plant genomic DNA libraries.

The procedure we use to package DNA to generate viable recombinant phage is that specified by the manufacturer. The methods we use to grow and maintain recombinant phage in *E.coli* K802 are essentially those of Maniatis *et al.* (8). Once formed, the phage should be stable in the presence of chloroform for at least 1 week.

The number of recombinant phage (N) that is required for a representative genomic library can be calculated using the following equation (19):

$$N = \frac{\ln (1 - P)}{\ln (1 - f)} \tag{1}$$

where P is the probability that a particular DNA sequence is represented in the library and f is the fractional proportion of the genome present in a single recombinant phage.

To determine the total number of phage generated in a single packaging reaction, we infect bacteria with dilutions of the packaged phage. Normally, a 10^{-5}, 10^{-6} and 10^{-7} dilution series is a good range to work in. We then calculate the number of non-recombinant phage by plating dilutions of packaged phage generated from ligated vector 'arms'. We subtract the number obtained from the total to determine the number of recombinant phage present. We can then determine how 'representative' the phage library is with respect to the plant genome of interest by comparing the number of recombinant phage obtained and the number of phage required for a 99.9% 'complete' library.

3.5 Concentration of recombinant phage

Infection of bacteria with packaged phage can present a problem as constituents found in most *in vitro* packaging reactions inhibit bacterial growth. This problem can be circumvented by plating less than ¹⁄₁₀ of the packaging reaction (or $\sim 10^4$ phage) in a single infection onto solid media (e.g. one 150×15 mm agar plate).

Alternatively, the recombinant phage can be concentrated into a smaller volume and/or into a neutral buffer prior to infection and plating.

In our experience, however, phage concentration is generally not necessary as long as the concentration of total DNA packaged is equal to or greater than 2 μg/ml. The concentration of packaged phage should then exceed 2×10^6 p.f.u./ml, sufficient for most purposes. If necessary, however, a procedure for concentrating recombinant phage that is gentle enough to allow for the concentration of phage without affecting phage viability (the ability to infect suitable bacterial hosts) can be found in Maniatis *et al*. (8).

3.6 Amplification of recombinant phage to form a stable library

A procedure for amplifying packaged phage to create stable phage libraries has been published previously (8). It should be noted, however, that amplification may result in the selective loss of phage containing eukaryotic DNA sequences that adversely affect phage replication or favour *in vitro* recombination (8). For this reason, we suggest that only the ligated plant:phage DNA should be stored and that phage should be generated and screened directly for gene sequences of interest using the plaque hybridization procedure of Benton and Davis (20) essentially as described in Maniatis *et al*. (8).

4. ANALYSIS OF PLANT GENE STRUCTURE

4.1 Introduction

Gene structure can be determined using a variety of well-established pro-cedures. It is not our intention to survey and discuss all procedures available but rather to briefly outline those that are rapid, simple, and used routinely in our laboratory. This section is divided into three parts: (i) gene isolation; (ii) gene localization; and (iii) gene characterization. In the first, we discuss procedures for isolating gene sequences from λ phage libraries. We include several modifications for its use with nylon membranes and synthetic oligonucleotide probes as well as a no-fail procedure for the preparation of recombinant phage DNA once phage have been identified and isolated. In the second section, we outline several procedures used in our laboratory to localize gene sequences within cloned DNA inserts. Finally, we discuss several procedures recently established in our laboratory to localize DNA sequences that interact with DNA binding proteins.

4.2 Gene isolation

4.2.1 Isolation of recombinant phage

Plaque hybridization (8) is used for the isolation of recombinant phage from plant genomic DNA libraries. This method, originally described by Benton and Davis (20), has been modified for use with various synthetic membranes as well as with synthetic oligonucleotide probes. We will briefly discuss these modifications.

(i) *Alternate membrane supports.* The use of synthetic membranes has increased dramatically in the past few years for several reasons. Synthetic membranes are less fragile than their nitrocellulose counterparts and are easily stripped of radioactive probe for re-hybridization. In addition, several investigators have reported increased sensitivity using nylon membranes in conjunction with covalent attachment of nucleic acids by UV cross-linking (21).

A number of excellent reviews and procedures are available from commercial manufacturers that recommend the following modifications for use with nylon membranes (22):

(1) Fix the DNA to membranes by UV cross-linking rather than by baking. Church and Gilbert (21) showed that cross-linking for short exposure times reduces the amount of unhybridizable DNA or RNA affixed onto membranes compared to treatment at high temperatures under vacuum.
(2) Change the pre-hybridization/hybridization buffer to:
    6× SSC
    5× Denhardt's solution
    0.5% SDS
    20 µg/ml sheared non-homologous DNA.
(3) Pre-hybridize and hybridize at 65°C.
(4) Perform washes at 65°C using 2× SSC, 0.1% SDS as a buffer.

(ii) *Synthetic oligonucleotide probes.* A major modification to any procedure incorporating synthetic oligonucleotides as hybridization probes is the adjustment of hybridization and wash temperatures to account for the short probe length. The effects of probe length and base pair composition on nucleic acid hybridization kinetics and on the stability of DNA–DNA duplexes were first addressed by Britten *et al.* (23) and have recently been determined for synthetic oligonucleotides (24). The optimum hybridization and wash temperature for use with a specific oligonucleotide can be estimated using the formula $T = 2°C$ (number of A–T bases in the oligonucleotide) + 4°C (number of G–C bases).

The most difficult problems encountered with oligonucleotide probes are high background and low sensitivity. We have alleviated these problems by the following means.

(1) Using the following solutions for pre-hybridization and hybridization:
    Pre-hybridization   6× SSC
                         1× Denhardt's solution
                         0.5% SDS
                         0.05% PP_i
                         100 µg/ml sheared non-homologous DNA
    Hybridization   6× SSC
                         1× Denhardt's solution
                         0.05% PP_i
                         20 µg/ml tRNA
(2) Filtering the hybridization solution containing the end-labelled oligo-

nucleotide through a 0.22 μm Swinex filter which removes any particulate matter that can cause undesirable background.

(3) Incubating filters at 65°C in 3× SSC, 0.1% SDS to remove bacterial debris prior to pre-hybridization. This step appears to be critical for decreasing the amount of non-specific hybridization normally associated with hybridization of oligonucleotide probes to filters containing bacterial colonies and/or phage plaques (25).

4.2.2 *Isolation of recombinant phage DNA*

(i) Small-scale preparations. It is often useful to prepare recombinant phage DNA on a small scale for restriction enzyme analysis or simple cloning experiments. *Table 3* outlines a simple procedure (26) used in our laboratory to isolate phage DNA from a phage stock. Similar to the Birnboim and Doly plasmid DNA isolation procedure (27), this protocol incorporates a potassium acetate–ethanol precipitation step along with a series of phenol and chloroform extractions which together virtually eliminate contaminating nucleases from solution. The DNA obtained is relatively intact and can be used as a substrate in various enzyme reactions without difficulty.

(ii) Large-scale preparations. Several procedures have been published on the preparation of λ phage DNA. *Table 4* outlines the solutions required for the procedure of Blattner *et al.* (14) given below that we have found to be the easiest and most reliable. Using this procedure, we routinely obtain 200–400 μg recombinant phage DNA per litre of culture. We have obtained yields of greater than 1 mg/litre culture for wild-type Charon phages. In brief, this procedure involves infecting bacteria with phage at a low multiplicity of infection (m.o.i.) and allowing multiple rounds of infection to occur. The resultant phage are concentrated, purified by CsCl gradient ultracentrifugation, and subsequently lysed. The phage DNA is then purified by phenol and chloroform extractions.

(i)   Start a culture of a single colony from a fresh (<1 week old) plate in 10 ml NZY broth (*Table 4*). Incubate at 37°C for *exactly* 14 h. Store at 4°C until ready to use.

(ii)   Mix 7 ml of overnight culture with 20 μl of a 10^{10} p.f.u./ml phage stock (2 μl of a 10^{11} p.f.u./ml stock).

(iii)   Pre-absorb the phage to bacteria for 20 min at 37°C.

(iv)   Add the phage:bacteria mixture to a 2 litre Erlenmeyer flask containing 1 litre of sterile pre-warmed (37°C) NZY broth.

(v)   Incubate with shaking at 37°C for 14 h. The media should appear relatively clear with bacterial debris.

(vi)   Add 5 ml of chloroform and 125 μl of a 10 mg/ml DNase I stock to the litre culture. Chloroform promotes lysis; DNase I digests bacterial DNA which interferes with phage resuspension following PEG pre-cipitation.

Table 3. Small-scale preparation of recombinant phage DNA.

Solutions
1. TM [10 mM Tris–HCl (pH 7.5), 10 mM $MgCl_2$]
2. 10% SDS[a]
3. Diethylpyrocarbonate (DEPC; Sigma)
4. 2 M Tris–HCl (pH 8.5), 0.2 M Na_2EDTA
5. 5 M Potassium acetate
    Make a 5 M potassium acetate solution and a 5 M acetic acid solution. Mix the two until the pH of the combined solutions is 6.0. Filter and autoclave.
6. Distilled phenol equilibrated with 10 mM Tris–HCl (pH 7.5), 0.1 mM Na_2EDTA, 0.5 M potassium acetate (pH 6) [TEA]
7. Sevag (24:1 chloroform:isoamyl alcohol) equilibrated with TEA
8. 10 mM Tris–HCl (pH 7.5), 1 mM Na_2EDTA, 1 µg/ml RNase A
    Add: 0.1 ml 1 M Tris–HCl (pH 7.5)
        0.05 ml 0.2 M Na_2EDTA (pH 8)
        1 µl 10 mg/ml heat-treated RNase A[b]
        H_2O to 10 ml.

Procedure
1. Make a fresh phage stock using the plate lysate procedure (ref. 8). Overlay plate containing phage with 5 ml TM, allow phage to diffuse into the TM overnight at 4°C.
2. Take 1 ml of phage solution and transfer to a 1.5 ml microfuge tube.
3. Store the remainder of the phage solution in a sterile screw cap tube as a phage stock. For long term storage, add a few drops of chloroform and $\frac{1}{100}$ vol. of sterile 1% gelatin.
4. Spin the tube containing the 1.0 ml phage solution in a microfuge at maximum speed for 2 min at 4°C. Remove the supernatant carefully, avoiding the white pellet (bacterial debris). Transfer the supernatant to a new 1.5 ml tube.
5. Repeat step 4.
6. Add on ice:
    1 µl DEPC
    10 µl 10% SDS
    50 µl 2 M Tris–HCl (pH 8.5), 0.2 M Na_2EDTA
    (SDS and EDTA serve to lyse the cells and inhibit nuclease activity).
7. Incubate for 5 min at 70°C to lyse the phage.
8. Add 50 µl of 5 M potassium acetate (pH 6); mix; incubate on ice for 30–60 min. Under these conditions, potassium precipitates cell debris, proteins and SDS.
9. Spin for 15 min in the microfuge at 4°C.
10. Transfer the supernatant to a new 1.5 ml tube using a p1000 Pipetman. Do not touch the white pellet. Incubate on ice for 60 min.
11. Repeat step 9.
12. Allow the supernatant to come to room temperature.
13. Add an equal volume of phenol equilibrated with TEA. Extract; transfer upper aqueous phase to new tube containing an equal volume of Sevag equilibrated with TEA.
14. Extract. Transfer the supernatant to a new tube.
15. Precipitate the nucleic acids by adding an equal volume of isopropanol (room temperature). Mix well and immediately spin down the precipitate for 5 min in a microfuge at room temperature. Only nucleic acids precipitate.
16. Pour off the supernatant carefully, as DNA pellet is easily dislodged. Wash the pellet with 70% ethanol. Spin again for 5 min at room temperature.
17. Discard the supernatant and dry the pellet *in vacuo*.
18. Resuspend the pellet overnight in 50 µl of TE + RNase A.
19. The DNA is now ready to use. It is necessary, however, to include KCl in any reaction buffers as the potassium is required to precipitate any residual SDS that may inhibit enzyme activities.

---

[a] Use BDH specially pure SDS that can be purchased from Gallard-Schlesinger Mfg. Corp., Long Island, NY.
[b] To prepare RNase A stock solution, dissolve 20 mg RNase A (Sigma) in 1 ml of 10 mM Tris–HCl (pH 5), 5 mM NaCl; heat for 20 min at 90°C; cool; add 1 ml of 10 mM Tris–HCl (pH 7.6), 5 mM NaCl to neutralize.

Table 4. Solutions for large-scale preparation of recombinant phage DNA.

---

1. NZY broth
   10 g/litre NZ amine
    5 g/l Difco yeast extract
    5 g/litre NaCl
    5 g/litre $MgCl_2 \cdot 7H_2O$
   Autoclave for 20 min to sterilize.
2. 10 mg/ml DNase I (Sigma) in TM
3. 1.5 g/ml CsCl (refractive index between 1.380 and 1.382)
4. Phenol saturated with 10 mM Tris–HCl (pH 7.5), 1 mM Na_2EDTA [TE_{10}]
5. Sevag (24:1 chloroform:isoamyl alcohol) saturated with TE_{10}

(vii)   Add 60 g of NaCl per litre culture, to minimize phage adsorption to debris. Shake for 15 min at 37°C until the salt is dissolved. Place the culture on ice.

(viii)   Spin down the bacterial debris in a polypropylene GSA or GS-3 (Sorvall low-speed, fixed angle rotors) centrifuge bottles at 4°C at 7000 r.p.m. for 10 min. Try to avoid the chloroform when filling the bottles as it will interfere with the subsequent PEG precipitation steps (steps x–xi).

(ix)   After each spin, decant the supernatant containing phage into a sterile 2 litre Erlenmeyer flask containing a Teflon stirbar.

(x)   Add 70 g of polyethylene glycol (PEG 6000) per litre supernatant. Dissolve by stirring at 4°C. PEG in the presence of high salt precipitates phage.

(xi)   Let the flask sit at 4°C for 2–3 h to allow for quantitative precipitation of phage.

(xii)   Gently spin down the phage at 5000 r.p.m. for 10 min at 4°C in one GSA or GS-3 bottle. The phage pellet is bluish-white and will distribute down the bottle side.

(xiii)   Discard the supernatant. Lay the bottle on ice with the pellet side up. Drain for 2 min.

(xiv)   Remove the last of the supernatant with a Pasteur pipette. Add 3.5 ml of *cold* TM buffer (*Table 3*). Let the pellet soak on ice for 5 min.

(xv)   Resuspend the softened pellet by gently squirting the TM through a Pasteur pipette from the top of the pellet (e.g. top of the bottle) down. Transfer the phage solution to a sterile 15 ml Corex tube.

(xvi)   Let the phage resuspend overnight.

(xvii)   Add 0.82 g of CsCl per ml phage solution. Dissolve. Do not vortex. Adjust the density of the solution such that its refractive index is between 1.3800 and 1.3820 ($\rho = 1.5$ g/ml). The phage should band in the middle of the gradient.

(xviii)   Remove the bulk of the contaminating protein and bacterial debris by spinning phage (in CsCl) at 8000 r.p.m. in HB-4 (low speed swinging bucket rotor) at 4°C for 10 min. There should be a fairly large pellet and a protein pellicle.

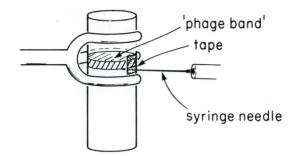

Figure 1. Diagram to show the removal of the 'phage band' from a CsCl gradient.

(xix)   Carefully transfer the supernatant to a cellulose nitrate tube, in which
          bands are easier to see.
(xx)   Spin the gradients at 36 000 r.p.m. for 20 h at 4°C.
(xxi)   Clamp the tube containing the gradient so that it does not move. Affix a
          small piece of tape just under the 'phage band'. Take a 21 gauge needle
          attached to a 3 ml plastic syringe and, with needle bevel side down,
          pierce the tube through the tape just under the 'phage band' (see *Figure
          1*). Re-orient the needle and slowly suck the band into the syringe. The
          phage band should be slightly bluish, avoid harvesting 'strings' running
          through gradient.
(xxii)   Reband the phage as in steps xix–xx or reband in polypropylene tubes
          in a TV865 (Sorvall) vertical rotor for 7 h at 65 000 r.p.m. at 4°C. Do
          not use a vertical rotor for the first gradient, as re-orientation will
          contaminate the phage band with protein pellicle.
(xxiii)   Remove the 'phage band' as in step xxi.
(xxiv)   Dialyse phage against TM buffer at 4°C to remove CsCl. Change the
          buffer once per hour with 1 litre changes for 3 h.
(xxv)   Determine the A_{260} of the phage solution after dialysis. $A_{260} = 1$ is
          equivalent to 10^{12} phage or 50 μg DNA per ml.
(xxvi)   If necessary, adjust the A_{260} to 5 by diluting the phage solution with TM
          buffer. If greater, solution viscosity during phenol extraction will be a
          problem.
(xxvii)   Transfer the phage to a conical glass 15 ml centrifuge tube.
(xxviii) Add an equal volume of phenol saturated with TE. Cover tube with a
          piece of Saran Wrap. Gently rock the tube back and forth to lyse the
          phage and extract the released DNA. Rock tube about 30 times per
          minute for 5–10 min. Excessive force will shear DNA.
(xxix)   Separate the phases by spinning the tube at 500 r.p.m. for 5 min in a
          clinical centrifuge.
(xxx)   Remove the *lower* phenol phase by piercing through the thick pellicle
          with a Pasteur pipette.
(xxxi)   Repeat steps xxviii–xxx at least three more times.

(xxxii) Remove the phenol from the aqueous, DNA-containing phase by dialysing against TNE (*Table 2*) for 72 h at 4°C with several changes. Change the buffer at least twice a day, 1 litre of buffer per change.

(xxxiii) Determine the concentration of the DNA by determining the A_{260} of the solution. The yield of DNA should be between 250 and 400 µg DNA per litre of culture.

(xxxiv) Determine the A_{280} of the solution and calculate the A_{260}/A_{280} ratio. The ratio should be between 1.8 and 1.9. If it is greater than 2.0, there is probably some phenol contamination that may affect enzyme reactions.

4.3 Characterization of cloned DNA inserts

4.3.1 *Restriction enzyme mapping procedures*

In-depth discussions of the properties and uses of restriction enzymes as well as detailed protocols regarding various enzyme reactions are available from the manufacturer or have been published elsewhere. For this reason, we will limit our discussion to experimental strategies for establishing restriction enzyme maps for cloned DNA inserts.

We have used several different restriction enzyme mapping procedures in our laboratory. We have found that the most rapid and unambiguous determination of restriction enzyme site positions without directly determining the DNA sequence is to carry out the following.

(i)   Identify restriction enzymes for which no site occurs within the cloned DNA insert and at least one within the vector DNA.

(ii)   Determine the location of restriction sites that occur once within the insert.

(iii) Utilize these sites to determine the location of restriction sites according to Smith and Birnstiel (28).

In brief, the procedure of Smith and Birnstiel (28) involves digesting an end-labelled DNA fragment under conditions which generate partial digestion products. Since only one end of the fragment is labelled, the lengths of all fragments generated represent the distance between the labelled end and a restriction site cleaved by the enzyme used. In this way, a 'map' of restriction sites can be generated for a given cloned DNA insert using a single substrate. Using this procedure, we have generated a restriction enzyme map for all commercially available enzymes that recognize five and six base pair sequences within two weeks' time. We have included our version of this procedure in *Tables 5* and *6* for convenience. The reaction conditions are designed to maximize 5′ end labelling, using the 76 kd Klenow fragment of DNA polymerase 1, which retains its 5′ to 3′ polymerase activity, but lacks the 5′ to 3′ exonuclease property.

Table 5. End-labelling of DNA fragments.

Solutions
1. 10× Klenow reaction buffer[a]
    500 mM Tris–HCl (pH 7.5)
    60 mM $MgCl_2$
    60 mM β-Mercaptoethanol
    0.05% Gelatin
2. 10 mM dATP, dCTP, dGTP and/or TTP[b]
3. [α-^{32}P]Deoxynucleotide (sp. act. >3000 Ci/mmol)[b]
4. 0.2 M Na_2EDTA (pH 8.0)
5. 10 mM Tris–HCl (pH 7.5), 0.1 mM Na_2EDTA [TE]
6. 3 M Sodium acetate (pH 6)
7. 100% Ethanol
8. Phenol saturated with TE
9. Sevag (24:1 chloroform:isoamyl alcohol) equilibrated with TE

Procedure
1. Digest 1 μg of DNA with the enzyme of choice[c] in a total reaction volume of 25 μl
2. Mix:

		final concentration
10× Klenow buffer	6 μl	1×
[α-^{32}P]dXTP[d]	4 μl	40 μCi
10 mM dXTP (1–3)	1 μl (of those required)	150 μM
digested DNA	20 μl	15 μg/ml
H_2O to total volume, 60 μl.		

3. Add 0.5 U of Klenow fragment; mix well.
4. Incubate for 2–3 min at 10°C[e].
5. Terminate the reaction by adding 10 μl of 0.2 M Na_2EDTA (pH 8.0).
6. Add 180 μl of TE to bring the total volume to 250 μl.
7. Phenol extract.
8. Sevag extract.
9. Add ¹⁄₁₀ vol. 3 M sodium acetate (pH 6) and 2.5 vols 100% ethanol. Mix well.
10. Precipitate the DNA at −70°C for 15 min.
11. Spin down the DNA at 4°C for 15 min in a microfuge.
12. Discard the supernatant (in radioactive waste!)
13. Resuspend the DNA in 250 μl of TE.
14. Re-precipitate to remove further unincorporated label.
15. Spin down the DNA; wash the pellet 4× in 70% ethanol.
16. Dry *in vacuo*.
17. Resuspend in the desired volume of TE. The labelled DNA is now ready for fractionation by gel electrophoresis and isolation of the appropriate DNA fragment.

[a] We omit salt from the 10× reaction buffer as Klenow fragment is relatively insensitive to NaCl concentrations ranging from 0 to 100 mM.
[b] Always try to omit at least one unlabelled dXTP from the reaction to inhibit labelling by nick translation. For example we add only unlabelled TTP and [α-^{32}P]dCTP to fill in a *Xho*1 site:

$$5' \text{———— C} \qquad 5' \text{———— CTC*}$$
$$\xrightarrow{\quad} $$
$$3' \text{———— GAGCT} \qquad 3' \text{———— GAGCT}$$

[c] The only limitation in choosing an enzyme besides the availability of sites is that, upon digestion, it must generate a 3' recessed end. Klenow fragment contains only a 5' to 3' polymerase activity and therefore can only synthesize in a 5' to 3' direction.
[d] We always use [α-^{32}P]dXTP that is at least 3000 Ci/mmol in Tricine buffer.
[e] End-fill is a very fast reaction. Longer incubation times and higher temperatures can lead to DNA degradation and internal labelling at 'nicks'. In addition, labelling of ends that should not label can occur due to the presence of the 3' to 5' exonuclease activity. For example the λ cos sites or 'sticky ends' should not label in the presence of unlabelled TTP, labelled dCTP, and Klenow. Upon prolonged incubation, however, labelling does occur.

Table 6. Smith–Birnstiel restriction enzyme mapping procedure.

Solutions
1. 5× Reaction buffer (according to manufacturer's specifications).
2. Restriction enzyme storage buffer (according to manufacturer's specifications).
3. 5' End-labelled DNA fragment (1–2 μg/ml; 1000–2000 c.p.m./μl; see *Table 5*).
4. 0.2 M Na$_2$EDTA (pH 8.0).

Procedure
1. Add 5 μl of 0.2 M Na$_2$EDTA (pH 8.0) to five 0.5 ml microfuge tubes. Place the tubes on ice.
2. Mix:
    11 μl 5× reaction buffer
    1000–2000 c.p.m. (1–2 ng) DNA fragment
    H$_2$O to a total volume of 55 μl.
3. Dilute restriction enzyme to 0.25 U/μl in *ice cold* enzyme storage buffer.
4. Remove 10 μl from the main reaction and transfer to a tube containing Na$_2$EDTA. This is the '0' time point.
5. To the remainder, add 1 μl of diluted enzyme (0.25 U)[a]; mix well; incubate at 37°C. Begin timing.
6. At 2, 7, 30 and 60 min, transfer 10 μl of the main reaction to a tube containing Na$_2$EDTA. Place on ice.
7. Add 3 μl of loading dye to each tube.
8. Incubate each time point for 10 min at 65°C.
9. Immediately quench on ice.
10. Load samples onto an agarose gel along with ^{32}P-labelled size standards.
11. Dry down the gel onto three pieces of Whatman 3MM paper using a vacuum gel dryer for about 1 h. The gel should be relatively flat but still wet.
12. Wrap the gel in plastic wrap, mount on cardboard and expose the gel to film for at least 12 h[b].

[a] 0.25 U enzyme gives good partial digestion for most enzymes used.
[b] Approximately 40 c.p.m. per band can be visualized by autoradiography at −70°C with intensifying screens within 24 h.

4.4 Gene localization

4.4.1 *Introduction*

It is fairly straightforward to localize a gene within a cloned DNA insert once a restriction enzyme map of the insert has been generated. Several procedures have been used in our laboratory. We will discuss only those procedures that are reasonably easy to use and yield the most information in one experiment. These include procedures used to determine relative gene location (Sections 4.4.2–4.4.3) as well as those that can be used to precisely pinpoint 5' and 3' gene ends (Section 4.4.4).

4.4.2 *Gene localization by DNA or RNA gel blot analysis using radioactively labelled probes*

Relative gene location within a cloned DNA insert can be determined in many ways. One method is to hybridize radioactively labelled DNA fragments to mRNA gel blots (8,29,30). Fragments that hybridize to specific mRNAs contain gene sequences corresponding to those RNAs. In this way, the location of gene sequences can be determined as well as the size of the mRNAs homologous to those gene sequences. In addition, this method can be extremely useful in cases where more than one gene is present within a cloned insert (3).

An alternate method involves hybridizing radioactively labelled cDNA or RNA to DNA gel blots (8). In brief, cloned DNA is digested with a variety of restriction enzymes. The resultant DNA fragments are fractionated by agarose or polyacrylamide gel electrophoresis, blotted onto a nylon membrane, and reacted with radioactively labelled cDNA or mRNA. Only the DNA fragments that contain gene sequences of interest will hybridize to the labelled probes. Obviously, gene location will be more precise if a detailed restriction map is available.

Both of these methods are fairly sensitive and are limited only by the levels at which the genes of interest are expressed. Using these procedures, we can routinely detect genes expressed at fairly low levels [e.g. $10^{-4}\%$ of the poly(A^+) mRNA mass].

All of the procedures described have been published elsewhere. We direct the reader to Maniatis *et al.* (8) for excellent, detailed treatments. Procedures for the isolation of plant mRNA are discussed in Chapter 1.

4.4.3 *R-loop analysis*

An elegant way to visually localize a gene within a DNA insert is by electron microscopy. R-loop analysis was first used in 1979 by Woolford and Rosbash to purify mRNAs and identify structural genes within recombinant DNA molecules (31) and later modified to analyse gene structure (31–35). In brief, double-stranded phage or plasmid DNA is reacted with mRNA under conditions that favour DNA–RNA hybridization. The mRNA displaces the non-complementary DNA strand, forming an 'R-loop', a structure that can be easily recognized in the electron microscope as a 'bubble' in an otherwise double-stranded DNA molecule. In this way, the position of the R-loop, and therefore, the position of the gene, within the DNA molecule can be easily determined. R-looping and DNA–RNA hybridization are considered in detail elsewhere in this series (36).

4.4.4 *Determination of 5' and 3' gene ends*

Several procedures are available for mapping the locations of 5' and 3' gene termini within cloned DNA fragments. These include S1 nuclease mapping (37–39) and various derivatives of the primer extension procedure of Ghosh *et al.* (40,41). All of these methods are relatively easy to perform and require a minimum amount of time. Most are sensitive enough to detect as little as one molecule of RNA per cell (8). For detailed protocols, we refer the reader to Maniatis *et al.* (8), Favaloro *et al.* (39) and Dingerman and Nerke (42).

4.4.5 *Gene sequence determination*

Several methods are available for determining gene sequence. For excellent reviews and detailed protocols, we refer the reader to Maxam and Gilbert (43), Messing *et al.* (44) and Chen and Seeburg (45).

4.5 Gene characterization

4.5.1 *Introduction*

Molecular characterization of a gene once cloned and isolated has become relatively routine. Commercially prepared 'kits' are available for almost all methods used in molecular biology today. Therefore, we focus our discussion on those techniques which are only recently being developed in plant molecular biology laboratories. In particular, we discuss several procedures established in our laboratory to identify DNA sequences that specifically interact with plant DNA binding proteins (46).

4.5.2 *DNA binding protein assays*

(i) *Introduction*. There is an extensive body of literature that deals with the identification of DNA sequences that interact with DNA binding proteins. Techniques used to detect specific protein–DNA interactions include nitro-cellulose filter binding assays (47), gel retardation (48,49), gel blot assays (50–52), and nuclease protection assays (53,54). We have used several of these assays in our laboratory and have incorporated several modifications for work with plant nuclear proteins. This section is divided into three parts. In the first, we will outline procedures for isolating plant nuclear proteins free of contaminating nucleases. In the second part, we will discuss several routine assays used to identify DNA sequences that are involved in specific protein: DNA interactions. Finally, we will briefly outline a modification of published procedures used in our laboratory to identify specific DNA binding proteins by gel blot analysis.

(ii) *Isolation of nuclear proteins.*

Introduction. The isolation of intact nuclear proteins from plant cells is complicated by the presence of endogenous proteases released upon cell disruption. Contaminating nucleases present in crude nuclear protein extracts can present an additional problem when used in reactions with labelled DNA probes. All of these problems are alleviated by first isolating intact nuclei followed by nuclear lysis and nuclear protein isolation. Chapter 1 outlines a detailed protocol used in our laboratory to isolate plant nuclei. In this section we describe how to prepare nuclear protein extracts from intact nuclei (*Table 7*). These procedures allow for the isolation of transcriptionally active nuclei which are then lysed and the proteins, many of which may be involved in active gene transcription, isolated. In brief, nuclei are purified by Percoll density gradient centrifugation (55–57) and then lysed in the presence of high salt (52). The resultant extract is clarified and the supernatant containing total nuclear proteins is then dialysed against a neutral buffer to decrease the salt concentration (52).

Isolation of nuclei. As mentioned previously, a detailed procedure for isolating plant nuclei is outlined in Chapter 1. Yields vary and are dependent on

Table 7. Preparation of nuclear protein extracts.

Solutions
1. 200 mM Spermidine
2. 5 M NaCl
3. Dialysis buffer
   10 mM HEPES (pH 8)
   50 mM NaCl
   1 mM $MgCl_2$
   1 mM Dithiothreitol (DTT)
   5 mM Phenylmethylsulphonylfluoride (PMSF)
   50% Glycerol

Procedure
1. Isolate plant nuclei as described in Chapter 1. Use a nuclear preparation in which the concentration of nuclei exceeds 10^6 nuclei per ml.
2. Thaw the nuclei on ice.
3. Add 200 mM spermidine to a final concentration of 5 mM.
4. Add 5 M NaCl to a final concentration of 0.5 M[a].
5. Incubate for 45 min on ice.
6. Spin down the nuclear debris in microfuge for 10 min at 4°C.
7. Carefully withdraw the supernatant containing soluble nuclear proteins, taking care not to disturb the white pellet.
8. Dialyse the supernatant against at least three changes (500 ml/change) of buffer at 4°C.
9. Divide into 50–100 µl aliquots; store at −20°C[b].
10. Determine the protein concentration using the BioRad Protein Assay Kit[c].

---

[a] High salt lyses the nuclei and presumably dissociates tightly bound protein from DNA.
[b] Frequent freezing and thawing results in loss of biological activity. Long-term storage (>3 months) may result in precipitation of nuclear proteins; if this occurs, gently resuspend proteins before use.
[c] We have found that a typical leaf, stem, or root nuclear preparation (>10^6 nuclei/ml) yields approximately 100–200 µg of total nuclear protein.

the plant tissue used. In general, the highest yields of nuclei per g tissue are obtained from developing seeds, and the lowest yields from roots (L.Walling and R.B.Goldberg, unpublished observations).

It should be mentioned that it is possible to use the nuclear pellet obtained prior to Percoll gradient centrifugation as a source of nuclear proteins (C.Reeves, unpublished results). However, it is also possible that this type of nuclear protein extract will contain undesirably high levels of proteases and nucleases and may be less stable than those obtained from gradient-purified nuclei. We therefore suggest that, if possible, nuclei should be further purified before lysis.

Isolation of nuclear protein. Table 7 outlines a simple procedure for preparing crude nuclear protein extracts (52). The procedure is straightforward and should not pose any major problems. Yields vary and are dependent on the source of nuclei. For example, we consistently obtained yields of 2, 2 and 200 pg protein per soybean leaf, root, and embryo [40 days after flowering (DAF)] cell (46).

(iii) *Characterization of protein–DNA interactions*

Introduction. Our understanding of the molecular mechanisms that regulate plant gene expression and in particular, of the DNA sequences involved, will ultimately require knowledge of protein–nucleic acid interactions. Several quantitative assays have been used to study protein–DNA interactions. These include nitrocellulose filter binding assays (47); gel retardation binding assays (48,49); nuclease (53) and DNA-modifying chemical protection assays (58–61), and protein gel blot analyses (58–60). We cannot possibly review all the techniques that are currently available and will therefore limit our discussion to those procedures used in our laboratory. We direct the reader to Hendrickson (62) for an excellent review.

Gel retardation. Gel retardation assays are based on the observation that protein–DNA complexes can be resolved from uncomplexed DNA as discrete bands by electrophoresis through low ionic strength agarose or acrylamide gels (48,49). This procedure has been used to determine the kinetics of protein–DNA binding in prokaryotes (49,63,64) and more recently, to detect and purify specific DNA binding proteins in eukaryotes (65–68). This type of assay is simple, rapid, and can be modified for use with proteins that are unstable and/or that aggregate (63). In addition it can be used to detect interactions of several proteins binding to a single DNA fragment, a clear advantage over the widely used nitrocellulose filter binding assay. *Table 8* outlines the procedure used in our laboratory to identify DNA sequences that interact with plant DNA binding proteins.

DNase I protection assays. Nuclease and DNA-modifying chemical protection assays provide the most accurate information about the DNA sequences involved in a given protein–DNA interaction. These procedures are based on the

Table 8. Gel retardation assay.

Solutions
1. 5× Binding buffer
    50 mM Tris–HCl (pH 7.5)
    250 mM NaCl
    5 mM Na$_2$EDTA
    5 mM DTT
    25% Glycerol
    Final concentrations: 10 mM Tris–HCl (pH 7.5)
                                    50 mM NaCl
                                    1 mM Na$_2$EDTA
                                    1 mM DTT
                                    5% Glycerol
2. Nuclear protein storage buffer
    10 mM Hepes (pH 8)
    50 mM NaCl
    1 mM MgCl$_2$
    1 mM DTT
    50% Glycerol
3. 2 mg/ml poly(dI–dC)·(dI–dC) (Pharmacia) in 10 mM Tris–HCl (pH 7.5), 0.1 mM Na$_2$EDTA

4. Low ionic strength running gel buffer (25× stock)
    167.5 mM Tris–HCl (pH 7.5)
    82.5 mM Sodium acetate
    25 mM Na$_2$EDTA
    Final concentrations: 6.7 mM Tris–HCl (pH 7.5)
                           3.3 mM Sodium acetate
                           1 mM Na$_2$EDTA
5. 30% Acrylamide (29:1 acrylamide:bis-acrylamide)
6. 0.01% Bromophenol blue: dye to monitor electrophoretic mobility

Procedure
1. Label a DNA fragment of interest at its 5' or 3' end with ^{32}P according to any number of procedures.
2. Label five tubes 0, 1, 2, 3, 4 and 5.
3. Dilute the solution containing ^{32}P end-labelled DNA fragment to 0.5 ng/μl.
4. Mix:

	Tube number	0	1	2	3	4	5
End-labelled DNA fragment solution (0.5 ng)		1 μl	1 μl	1 μl	1 μl	1 μl	1 μl
2 mg/ml Poly(dI–dC)·(dI–dC) solution		0	0.5	1	1.5	2	2.5
5× Binding buffer		5	5	5	5	5	5

   H$_2$O to a final volume of 25 μl (including nuclear protein extract).
5. Add 8 μg of nuclear protein extract. Mix the reaction mixture gently.
6. Incubate at room temperature for 30 min[a].
7. Prepare an agarose or acrylamide gel[b] in 1× low ionic strength running gel buffer.
8. Pre-run the gel for at least 15 min with recirculating buffer[c] at 15 mA (constant current).
9. Add 2 μl of 0.1% bromophenol blue to the samples; load the samples directly onto the gel.
10. Run the gel at 15 mA with recirculating buffer until the dye front has migrated at least 6 cm from the wells.
11. Pre-heat the gel dryer at 60°C for 5 min; turn off heat before applying gel; transfer the gel to two pieces of Whatman 3MM paper and put under vacuum until dry (~2–4 h for a 1.5 cm thick agarose gel; 1 h for a 1.5 mm thick acrylamide gel.
12. Expose to film at −70°C with an intensifying screen.

---

[a] Nuclease activity present in the nuclear protein extract may preclude long incubation times. If this is a problem, decrease the reaction time to 1, 2 or 5 min at room temperature.
[b] The choice of gel medium depends upon the size of the DNA fragment used as a probe. Use a 4% acrylamide gel for DNA fragments <0.5 kb in length and a 1% agarose gel for fragments >0.5 kb.
[c] The running gel buffer is a very poor buffer; consequently, a pH gradient forms during the electrophoretic run that can cause dissociation of protein–DNA complexes. This problem can be circumvented by recirculating the gel buffer during electrophoresis.

observation that proteins bound to DNA 'protect' those sequences from enzymatic or chemical cleavage (53). Our experience is limited to the DNase I protection assay and for this reason, we will only discuss this technique in depth. For detailed protocols and excellent reviews regarding similar assays, we direct the reader to Van Dyke *et al.* (69), Shalloway *et al.* (54) and Elbrecht *et al.* (70).

DNase I protection assays involve reacting an end-labelled DNA fragment with nuclear proteins under previously standardized conditions to allow for maximum binding. The reaction products are then incubated with DNase I under conditions that achieve only partial digestion and are fractionated (68) by gel electrophoresis as described in the next section. Protein–DNA complexes are identified by autoradiography, excised and the DNA purified (43). The

DNA is then fractionated by denaturing gel electrophoresis and compared to DNA digested with DNase I in the absence of protein, using autoradiography. The absence of radioactive bands present in the control reaction without protein are indicative of DNA sequences bound to protein and therefore 'protected' from DNase I digestion.

The procedure outlined in *Table 9* is essentially that of Singh *et al.* (68). Although it appears to be relatively straightforward, it can be one of the most difficult and time-consuming methods used. Difficulties can arise with respect to reproducible enzyme digestion and protein DNA binding activity, as well as yield of DNA following multiple extractions. Perhaps the only way to make this procedure routine is to run test reactions to become familiar with the enzymes and fractionation and purification procedures. We strongly recommend Galas and Schmitz (53) for an excellent introductory article on DNase I protection assays. We have some suggestions based on our own experience for the reader as well.

Table 9. DNase I protection analysis.

Solutions
1. 5× Binding buffer
   50 mM Tris–HCl (pH 7.5)
   250 mM NaCl
   5 mM DTT
   5 mM Na_2EDTA
   25% Glycerol
2. 0.15 M NaCl
3. 0.5 M $MgCl_2$, $CaCl_2$
4. 2.5 mg/ml DNase I solution in cold 0.15 M NaCl: DNase I solution should be stable at 4°C for at least 12 h.
5. Elution buffer
   0.5 M Ammonium acetate (pH 7.5)
   0.1% SDS
   1 mM Na_2EDTA
6. Formamide dye
   80% (v/v) de-ionized or recrystallized formamide
   50 mM Tris–HCl (pH 8.3)
   1 mM Na_2EDTA
   0.1% (w/v) Xylene cyanol
   0.1% (w/v) Bromophenol blue
7. Low ionic strength running buffer (25× stock) [TEA–LIS]
   167.5 mM Tris–HCl (pH 7.5)
   82.5 mM Sodium acetate
   25 mM Na_2EDTA

Procedure
1. Perform a binding reaction—presumably you have already characterized the conditions under which you have specific binding. Scale up the reaction such that you have at least 2000–3000 c.p.m. in a protein–DNA complex. Run a no-protein control reaction.
2. While the binding reaction is incubating, make a 0.025 mg/ml DNase I solution by diluting a 2.5 mg/ml stock solution with 1× binding buffer.
3. After incubation is completed, add ¹⁄₁₀₀ vol. of 0.5 M $MgCl_2$, $CaCl_2$ (final concentration: 5 mM).

4. Add 0.025 mg/ml DNase I to a final concentration of 0.15 μg/μl DNase I[a].
5. Incubate the reaction for 1–15 min[a] at room temperature.
6. Add 0.5 M Na₂EDTA to a final concentration of 5 mM to stop the reaction.
7. Add ¹⁄₁₀ vol. of 0.1% Bromophenol blue.
8. Pre-run a 4% gel in 1× TEA–LIS buffer for 15 min at 15 mA constant current. Load sample. Include a lane containing no-protein control.
9. Run at 15 mA with recirculating buffer until the Bromophenol blue dye front is at the bottom of gel.
10. Cover the gel with plastic wrap and expose to film at 4°C for at least 1 h.
11. Cut out the protein–DNA complex (e.g. bound DNA) and the DNA fragment (e.g. free DNA) from the gel with a scalpel or razor blade.
12. Elute the DNAs into 500 μl of elution buffer overnight at 37°C with shaking.
13. Place the eluted DNAs into a microfuge tube; add 5 μg of carrier tRNA.
14. Extract the DNAs with phenol saturated with 10 mM Tris–HCl (pH 7.5), 0.1 mM Na₂EDTA; repeat until interface is clear.
15. Extract with Sevag (24:1 chloroform:isoamyl alcohol) two times.
16. Precipitate the DNAs twice with either isopropanol or ethanol; wash the pellet with 70% ethanol; dry *in vacuo*; resuspend the DNAs in 10 μl of TE.
17. Count 1 μl in a scintillation counter; add 5.1 μl of formamide due to the remainder; boil for 5 min and cool on ice.
18. Load onto a DNA sequencing gel.

---

[a]The concentration of DNase I and the reaction time at a given DNase I concentration necessary for partial digestion should be determined for each DNA fragment used. We have found that 0.16 μg/ml was an optimum concentration for DNA fragments between 200 and 500 bp in length and at a concentration of 0.02 ng/μl reaction. Reaction times varied between 1 and 15 min.

One of the major problems that we have encountered with this procedure is non-reproducible DNase I digestion. This appears to be correlated with weighing out small quantities of DNase I for a stock solution. We alleviated the problem by making a stock solution of DNase I at high concentration in 50% glycerol and then diluting it to a working concentration just prior to use. The stock solution can then be stored at −20°C for at least 1 week.

Another problem we have had is non-specific protection of DNA fragment 'end' sequences. This is most often seen when a specific DNA binding site lies close to the fragment end and/or when the fragment is extremely small (e.g. <130 bp). This problem can be eliminated by choosing DNA fragments that are larger or that have the putative DNA binding site within the middle of the fragment (J.Gralla, personal communication).

Nuclear protein gel blots. Several procedures have been described for detecting DNA binding proteins by gel blot analysis using labelled DNA or RNA probes (50–52). In general, the methods involve three steps: (i) fractionation of nuclear proteins by gel electrophoresis; (ii) protein transfer to nitrocellulose filters; and, (iii) reaction of the filter-bound proteins with labelled DNA or RNA probes. All of these procedures differ, particularly in the reaction conditions utilized. We use a modified method, incorporating parts of all of these procedures that, in our hands, either increased the sensitivity or decreased the high background associated with the technique. *Table 10* outlines our

Table 10. Nuclear protein gel blots using double-stranded DNA probes.

Solutions
1. Gel transfer buffer
   25 mM Tris
   190 mM Glycine
2. Binding/wash buffer
   10 mM Tris–HCl (pH 8.0)
   50 mM NaCl
   5 mM MgCl$_2$
   1 mM DTT
   1× Denhardt's solution [0.02% Ficoll, BSA, polyvinylpyrrolidone (PVP-40)].
3. Sheared non-homologous DNA: used as a non-specific competitor DNA.

Procedure
1. Size-fractionate 20–50 μg of nuclear proteins/lane on a SDS–polyacrylamide gel. Include pre-stained protein molecular markers[a] to follow the protein transfer to nitrocellulose.
2. Pre-wet the nitrocellulose[b] in binding buffer; let sit overnight at room temperature.
3. Transfer the proteins to the nitrocellulose[c] using a Tris–glycine buffer without methanol[d] at 100 V (0.6 A; constant voltage) for 0.5–1.0 h with cooling. Amperage should increase from 0.6 to 1.5 A during transfer.
4. Remove the nitrocellulose from the gel; place the nitrocellulose blot into 200 ml of binding buffer in a plastic tray; pre-incubate at room temperature with shaking for 1 h.
5. Prepare binding buffer with DNA probe:
   0.3 ml binding buffer/cm^2
   100 ng labelled double-stranded DNA probe (sp. act. $>10^8$ c.p.m./μg DNA; $>10^7$ c.p.m./blot)
   5000 μg sheared salmon sperm DNA (labelled:unlabelled DNA ratio 1:50 000).
6. React the nuclear protein gel blot with labelled DNA on plastic wrap[e]:
   a. Lay a piece of plastic wrap on a bench; secure the corners of the plastic with tape, creating a flat surface for reacting the gel blot with the DNA probe.
   b. Place 1–2 ml of binding buffer (without labelled DNA) onto the plastic in two vertical rows.
   c. Place the blot onto buffer, protein side up, so that there is buffer between the plastic and the blot. There should be only enough buffer to cover the bottom surface of the blot; buffer should not flow onto the top surface of the blot.
   d. Place the binding buffer with probe onto the top surface of the blot, taking care to cover entire surface with buffer. There should be only enough buffer to create a 'mound' of buffer due to surface tension.
   e. React the DNA probe with blot for 30 min at room temperature.
7. Place the blot in a tray containing 150 ml of binding buffer (without probe). Wash with shaking for 30 min at room temperature.
8. Repeat step 7 three to four times.
9. Expose the blot to film at −70°C with an intensifying screen for at least 2 h.

[a] Pre-stained protein molecular weight markers can be purchased from Bethesda Research Laboratories (BRL).
[b] We use 0.2 μm nitrocellulose from Schleicher and Schuell.
[c] A technical note: make sure that there is a minimum of buffer between the nitrocellulose and gel; otherwise, bands can be fuzzy and sensitivity can drop dramatically. This is in contrast to the manufacturer's instructions; however, we have found that this step is critical for good results. Normally, we place the gel on a piece of pre-wetted Whatman 3MM paper, drain the nitrocellulose of buffer, then lay the nitrocellulose onto the gel, taking care that no air bubbles are trapped.
[d] We omitted methanol from the transfer buffer originally because we thought that it would facilitate re-folding of the nuclear proteins necessary for DNA binding activity. However, this does not appear necessary in all cases (R.Deans, personal communication) and methanol facilitates transfer of proteins to nitrocellulose.
[e] The use of plastic wrap as a support medium for hybridization was first suggested by Jack *et al.* (51). In our experience, background problems are minimized by using plastic wrap instead of plastic bags for the hybridization.

modified procedure. Our method is simple and rapid, the entire procedure completed within one working day. We would like to stress, however, that any deviations from the procedure shown usually give less than adequate results.

5. ACKNOWLEDGEMENTS

We gratefully acknowledge the tireless efforts of our colleagues, both past and present, without which this manuscript could never have been written. We give special thanks to Drs Bob Fischer, John Harada, Linda Walling and Bill Timberlake for their critical scientific insight and contributions to this chapter. This work was supported by a US Department of Agriculture Grant to R.B.G.

6. REFERENCES

1. Fischer,R.L. and Goldberg,R.B. (1982) *Cell*, **29**, 651.
2. Goldberg,R.B., Hoschek,G. and Vodkin,L.O. (1983) *Cell*, **33**, 465.
3. Okamuro,J.K., Jofuku,K.D. and Goldberg,R.B. (1986) *Proc. Natl. Acad. Sci. USA*, **83**, 8240.
4. Barker,S.J., Harada,J.J. and Goldberg,R.B. (1988) *Proc. Natl. Acad. Sci. USA*, **85**, 458.
5. Harada,J.J. and Goldberg,R.B. Manuscript in preparation.
6. Jofuku,K.D. and Goldberg,R.B. Manuscript in preparation.
7. Dellaporta,S.L., Wood,J. and Hicks,J.B. (1983) *Plant Mol. Biol. Rep.*, **1**, 19.
8. Maniatis,T., Fritsch,E.F. and Sambrook,J. (1982) *Molecular Cloning: A Laboratory Manual*. Cold Spring Harbor Laboratory: Cold Spring Harbor, NY.
9. Kaiser,K. and Murray,N.E. (1985) *DNA Cloning—A Practical Approach*. Glover,D.M. (ed.), IRL Press, Oxford, Vol. 1, pp. 1.
10. Meyerowitz,E.M., Guild,G.M., Prestridge,L.S. and Hogness,D.S. (1980) *Gene*, **11**, 271.
11. Ish-Horowicz,D. and Burke,J.F. (1981) *Nucleic Acids Res.*, **9**, 2989.
12. Grosveld,F.G., Dahl,H.-H.M., deBoer,E. and Flavell,R.A. (1981) *Gene*, **13**, 227.
13. Loenen,W.A.M. and Blattner,F.R. (1983) *Gene*, **26**, 171.
14. Blattner,F.R., Williams,B.G., Blechl,A.E., Deniston-Thompson,K., Farber,H.E., Furlong, L.-A., Grunwald,D.H., Kiefer,D.O., Moore,D.D., Sheldon,E.L. and Smithies,O. (1977) *Science*, **196**, 161.
15. Loenen,W.A. and Brammar,W.J. (1980) *Gene*, **10**, 249.
16. Sternberg,N., Tiemeier,D. and Enquist,L. (1977) *Gene*, **1**, 255.
17. Farber,H.E., Kiefer,D. and Blattner,F.R. (1978) Application to the National Institutes of Health for EK2 certification of a host-vector system for DNA cloning. Supplement X. Data on *in vitro* packaging method.
18. Hohn,B. (1979) In *Methods in Enzymology*. Wu,R. (ed.), Academic Press, New York, Vol. 68, p. 299.
19. Clarke,L. and Carbon,J. (1976) *Cell*, **9**, 91.
20. Benton,W.D. and Davis,R.W. (1977) *Science*, **196**, 180.
21. Church,G.M. and Gilbert,W. (1984) *Proc. Natl. Acad. Sci. USA*, **81**, 1991.
22. Protocols available from Amersham, Schleicher and Schuell, ICN, New England Nuclear.
23. Britten,R.J., Graham,D.E. and Neufeld,B.R. (1974) In *Methods in Enzymology*. Grossman,L. and Moldave,K. (eds), Academic Press, New York, Vol. 29, p. 363.
24. Wallace,R.B., Shaffer,J., Murphy,R.F., Bonner,J., Hirose,T. and Itakura,K. (1979) *Nucleic Acids Res.*, **6**, 3543.
25. Woods,D. (1984) *BRL Focus*, **6**, 1.
26. Davis,R.W., Thomas,M., Cameron,J., St.John,T.P., Scherer,S. and Padgett,R.A. (1980) In *Methods in Enzymology*. Grossman,L. and Moldave,K. (eds), Academic Press, New York, Vol. 65, p. 404.
27. Birnboim,H.C. and Doly,J. (1979) *Nucleic Acids Res.*, **7**, 1513.
28. Smith,H.O. and Birnstiel,M.L. (1976) *Nucleic Acids Res.*, **3**, 2387.
29. Botchan,M., Topp,W. and Sambrook,J. (1976) *Cell*, **9**, 269.
30. Jeffreys,A.J. and Flavell,R.A. (1977) *Cell*, **12**, 429.
31. Woolford,J.L.,Jr and Rosbach,M. (1979) *Nucleic Acids Res.*, **6**, 2483.
32. Lomedico,P., Rosenthal,N., Efstratiadis,A., Gilbert,W., Kolodner,R. and Tizard,R. (1979) *Cell*, **18**, 545.

33. Hereford,L., Fahrner,K., Woolford,J.,Jr, Rosbash,M. and Kaback,D.B. (1979) *Cell*, **18**, 1261.
34. Woolford,J.,Jr, Hereford,L. and Rosbash,M. (1979) *Cell*, **18**, 1247.
35. Early,P.W., Davis,M.M., Kaback,D.B., Davidson,N. and Hood,L. (1979) *Proc. Natl. Acad. Sci. USA*, **76**, 857.
36. Higgins,S.J. and Hames,B.D. (1986) *Nucleic Acids Hybridization—A Practical Approach*, IRL Press, Oxford.
37. Casey,J. and Davidson,N. (1977) *Nucleic Acids Res.*, **4**, 1539.
38. Berk,A.J. and Sharp,P.A. (1977) *Cell*, **12**, 721.
39. Favaloro,J., Freisman,R. and Kamen,R. (1980) In *Methods in Enzymology*. Grossman,L. and Moldave,K. (eds), Academic Press, New York, Vol. 65, p. 718.
40. Ghosh,P.K., Reddy,V.B., Swinscoe,J., Lebowitz,P. and Weissman,S.M. (1978) *J. Mol. Biol.*, **126**, 813.
41. Murphy,W.J., Watkins,K.P. and Agabian,N. (1986) *Cell*, **47**, 517.
42. Dingerman,T. and Nerke,K. (1987) *Anal. Biochem.*, **162**, 466.
43. Maxam,A.M. and Gilbert,W. (1980) In *Methods in Enzymology*. Grossman,L. and Moldave, K. (eds), Academic Press, New York, Vol. 65, p. 499.
44. Messing,J., Crea,R. and Seeburg,P.H. (1981) *Nucleic Acids Res.*, **9**, 309.
45. Chen,E.Y. and Seeburg,P.H. (1985) *DNA*, **4**, 165.
46. Jofuku,K.D., Okamuro,J.K. and Goldberg,R.B. (1987) *Nature*, **328**, 734.
47. Riggs,A., Suzuki,H. and Bourgeois,S. (1970) *J. Mol. Biol.*, **48**, 67.
48. Garner,M. and Revzin,A. (1981) *Nucleic Acids Res.*, **9**, 3047.
49. Fried,M.G. and Crothers,D.M. (1981) *Nucleic Acids Res.*, **9**, 6505.
50. Bowen,B., Steinberg,J., Laemmli,U.K. and Weintraub,H. (1980) *Nucleic Acids Res.*, **8**, 1.
51. Jack,R.S., Gehring,W.J. and Brack,C. (1981) *Cell*, **24**, 321.
52. Miskimins,W.K., Roberts,M.P., McClelland,A. and Ruddle,R.H. (1985) *Proc. Natl. Acad. Sci. USA*, **82**, 6741.
53. Galas,D.J. and Schmitz,A. (1978) *Nucleic Acids Res.*, **5**, 3157.
54. Shalloway,D., Kleinberger,T. and Livingston,D.M. (1980) *Cell*, **20**, 411.
55. Luthe,D.S. and Quatrano,R.S. (1980) *Plant Physiol.*, **65**, 305.
56. Luthe,D.S. and Quatrano,R.S. (1980) *Plant Physiol.*, **65**, 309.
57. Walling,L., Drews,G.N. and Goldberg,R.B. (1986) *Proc. Natl. Acad. Sci. USA*, **83**, 2123.
58. Ross,W., Landy,A., Kikuchi,Y. and Nash,H. (1978) *Cell*, **18**, 297.
59. Sibenlist,U., Simpson,R. and Gilbert,W. (1980) *Cell*, **20**, 269.
60. Becker,M. and Wang,J.C. (1984) *Nature*, **309**, 682.
61. Ptashne,M., Jeffrey,A., Johnson,A., Maurer,R., Meyer,B., Pabo,C., Roberts,T. and Sauer,R. (1980) *Cell*, **19**, 1.
62. Hendrickson,W. (1985) *Biotechniques*, **3**, 198.
63. Hendrickson,W. and Schleif,R. (1984) *J. Mol. Biol.*, **178**, 611.
64. Bushman,F.D., Anderson,J.E., Narrison,S.C. and Ptashne,M. (1985) *Nature*, **316**, 651.
65. Strauss,F. and Varshavsky,A. (1984) *Cell*, **37**, 889.
66. Piette,J., Kryszke,M.-H. and Yaniv,M. (1985) *EMBO J.*, **4**, 2675.
67. Carthew,R.W., Chodosh,L.A. and Sharp,P.A. (1985) *Cell*, **43**, 439.
68. Singh,H., Sen,R., Baltimore,D. and Sharp,P.A. (1986) *Nature*, **319**, 154.
69. Van Dyke,M.W., Hertzberg,R.P. and Dervan,P.B. (1982) *Proc. Natl. Acad. Sci. USA*, **79**, 5470.
70. Elbrecht,A., Tsai,S.Y., Tsai,M.-J. and O'Malley,B.W. (1985) *DNA*, **4**, 233.

CHAPTER 3

Isolation and analysis of chloroplasts

COLIN ROBINSON and LINDA K.BARNETT

1. INTRODUCTION

The main purpose of this chapter is to describe protocols for analysing *in vitro* the biogenesis of chloroplast proteins. It has been known for many years that chloroplast proteins are synthesized in two distinct compartments in the cells of higher plants and green algae (1,2). Approximately 20% of chloroplast proteins are encoded by chloroplast DNA and synthesized within the organelle; the remainder are imported, after synthesis in the cytoplasm. In this article we describe methods for studying the synthesis and assembly of both classes of protein. We also discuss some of the approaches used to fractionate chloroplasts into component membrane and soluble phases. These methods can be used either to determine the precise location of a chloroplast protein, or, in scaled-up form, in the purification of proteins from chloroplasts.

The procedures described below have been successfully used in our laboratory to study aspects of chloroplast biogenesis. In each case, chloroplasts have been isolated from pea seedlings (*Pisum sativum*, var Feltham first), and it should be emphasized that problems may be encountered if other species are used as the starting material.

2. ISOLATION OF CHLOROPLASTS

We describe below two procedures ('rapid' and 'Percoll gradient') for the isolation of higher plant chloroplasts. Both protocols have been adapted from previously published methods (3,4). The nature of the experiment to be carried out dictates the choice of method. The rapid method, as its name implies, is used to prepare chloroplasts in a short space of time. This is, for example, an important factor when isolating chloroplasts for *in vitro* protein synthesis studies. The rapid isolation method is also commonly used when purifying stromal or thylakoidal extracts from large quantities of chloroplasts. The second method involves further purification of the chloroplasts by Percoll gradient centrifugation. This step efficiently separates broken and intact chloroplasts, and is used for protein transport studies and envelope membrane isolation.

2.1 Growth conditions and handling of chloroplasts

In order to isolate intact chloroplasts reliably, it is essential that the plant growth conditions are carefully regulated.

(i) The illumination regime is a critical factor; pea seedlings should be grown under a 12 h photoperiod at a low light intensity (45 μmol photons/m^2/sec). Higher light intensities lead to the production of starch granules which rupture the envelope membrane during chloroplast isolation.

(ii) The growth temperature is not so critical but should be between 15 and 25°C.

(iii) The growth time depends on the purpose of the experiment, and will be discussed in later sections.

(iv) When handling chloroplasts it is important to avoid subjecting the organelles to undue stress. Chloroplasts are easily lysed by shearing stresses and should be handled gently. It is advisable to avoid pipetting chloroplast suspensions whenever possible; when pipetting is essential, it is important to use wide-bore pipettes and to proceed slowly. When using small volumes, we favour widening the apertures of blue or yellow pipette tips by cutting off the ends.

(v) We find it convenient to resuspend chloroplast pellets by gentle use of a sterile surgical swab (cotton-wool rolled on the end of a small piece of wood).

2.2 Rapid isolation of chloroplasts

Plant material: for bulk preparation of stromal or thylakoidal fractions it is often convenient to use mature pea seedlings with fully expanded leaves. Under typical growth conditions such seedlings may be 10–14 days old, depending on growth temperature. For the analysis of protein synthesis by isolated chloroplasts, younger seedlings with immature leaves should be harvested.

The method can also be used for the isolation of chloroplasts from spinach and tobacco leaves. However, we have found that very low yields of intact organelles are obtained from leaves of cereals such as wheat or barley.

The basic isolation protocol is detailed below. The grinding medium is placed at −20°C prior to homogenization until an ice slurry forms. Homogenization is carried out in a Perspex container which is also kept at −20°C prior to use. In our laboratory, a Polytron homogenizer is routinely used, but many domestic homogenizers can also give satisfactory results. After the initial homogenization step, all steps should be carried out at 0°C.

(i) Harvest leaves from pea seedlings and mix with semi-frozen grinding medium (0.35 M sucrose, 25 mM Hepes, 2 mM EDTA, pH 7.6), in a pre-cooled Perspex container. Use 100 ml of grinding medium per 20 g (fresh weight) of leaves.

(ii) Homogenize the leaves with 2 × 3 sec bursts at 75% full speed using a Polytron homogenizer (Northern Media Supplies Ltd). The leaves should be ground into very small pieces by this procedure.

(iii) Strain the homogenate through eight layers of muslin to remove debris.

(iv) Pellet the chloroplasts by centrifugation at 4000 g for 1 min at 0°C.

(v) Discard the supernatant and wipe the inside of the tubes before resuspending the pellet(s) in 10–20 ml of grinding medium.

(vi)   Centrifuge at 4000 *g* for 1 min. The resulting pellet will be referred to as 'washed chloroplasts'.

In our hands, the rapid isolation procedure yields washed chloroplast suspensions which are 40–50% intact. The degree of intactness can be estimated by phase-contrast microscopy; intact organelles appear light green with a form of 'halo' surrounding them. Broken chloroplasts appear a darker shade of green, more opaque and lack the halo.

Chloroplasts isolated in this way can be used in the preparation of stromal and thylakoidal extracts; methods are given in Section 4 for the fractionation of chloroplasts into component membrane and soluble phases.

2.3 Isolation of chloroplasts by Percoll gradient centrifugation

This method offers a quick and effective means of separating broken and intact chloroplasts, and can yield chloroplast suspensions which are 90–95% intact. This is therefore the method of choice when isolating chloroplasts for protein transport studies or envelope membrane preparations; both types of experiment require predominantly intact chloroplasts as starting material. The drawbacks of this procedure are that the gradients are time-consuming to prepare, and are expensive when used on a large scale. Percoll is supplied by Pharmacia Fine Chemicals.

(i)   Prepare the Percoll gradients by layering successively 80%, 65%, 45%, 25% and 10% (v/v) Percoll in 50 mM Hepes–KOH, 0.33 M sorbitol; the pH of the buffer should be 7.5 for envelope membrane preparations and 8.4 for protein import studies. Two 10 ml gradients are generally sufficient for most protein import studies (Section 3.2) whereas several 30 ml gradients are required for envelope preparations (Section 4).

(ii)   Resuspend washed chloroplast pellets (Section 2.2) in the above Hepes–sorbitol buffer of the appropriate pH; 7.5 for envelope preparations and 8.4 for import experiments. For import experiments, washed chloroplast pellets from a 200 ml homogenization mixture should be resuspended in a total of 4 ml Hepes–sorbitol, and layered on to two 10 ml Percoll gradients. For envelope preparations, washed chloroplasts from large homogenization mixtures are layered onto several 30 ml gradients (see Section 4.3).

(iii) Centrifuge at 1500 *g* for 15 min at 0°C (brake off). Intact chloroplasts migrate further down the gradient than lysed organelles (*Figure 1*).

(iv)   Remove the intact chloroplast band using a wide-bore pipette and mix with five volumes of Hepes–sorbitol buffer. Centrifuge at 4000 *g* for 2 min at 0°C, discard the supernatant, and wash the pellet with a further 10 ml Hepes–sorbitol to remove Percoll from the chloroplast preparation. The resulting pellet contains 'Percoll-purified chloroplasts'.

3. THE SYNTHESIS IN VITRO OF CHLOROPLAST PROTEINS

In the field of plant molecular biology, many types of experiment require the synthesis *in vitro* of chloroplast proteins.

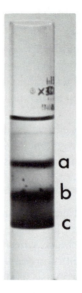

Figure 1. Separation of broken and intact chloroplasts by Percoll gradient centrifugation. **Bands (a)** and (**b**) contain broken and intact organelles, respectively. **Band (c)** contains organelles which are aggregated.

These include studies involving the analysis of:

(i)   the sites of synthesis of chloroplast proteins,
(ii)   levels of translatable mRNA,
(iii) transport of proteins into chloroplasts,
(iv)   assembly of oligomeric proteins,
(v)   hybrid-selected translation of mRNA using cloned DNA probes.

Most chloroplast proteins are synthesized initially in the cytoplasm, and this class of proteins will be considered first. The synthesis of proteins by isolated chloroplasts is covered in Section 3.3.

3.1 The synthesis *in vitro* of nuclear-encoded chloroplast proteins

A number of methods have been employed to study the synthesis *in vitro* of nuclear-encoded chloroplast proteins. In each case, mRNA, prepared either from plant extracts or by transcription of cloned DNA, is used to programme a cell-free translation system. Two translation systems have been extensively used: the rabbit reticulocyte lysate (5) and wheat-germ lysate (6). Since both systems synthesize very low quantities of protein, translation is carried out in the presence of a radioactively labelled amino acid. The translation products are visualized by SDS–polyacrylamide gel electrophoresis followed by fluorography.

We favour the wheat-germ lysate for several reasons.

(i)   Preparation of the lysate is rapid, cheap and relatively straightforward. The main problem often lies in obtaining a good source of wheat-germ (discussed in ref. 6).

(ii)   In our experience the wheat-germ system gives a higher rate of incorporation of label into protein than does the reticulocyte lysate.

(iii) The wheat-germ system often produces very low background translation of endogenous mRNAs. A problem with the reticulocyte lysate is that most preparations synthesize appreciable amounts of globin.

(iv)   The wheat-germ system is recommended for use in protein transport experiments. The reticulocyte lysate has been found to cause lysis of isolated chloroplasts in several laboratories.

The major drawback of the wheat-germ system is that it is relatively poor at synthesizing high molecular weight translation products; the reticulocyte lysate is superior for translation of proteins of molecular weight above 60 000.

A commonly used method of synthesizing primary translation products is to programme a cell-free translation system with mRNA extracted from appropriate plant material. In the case of chloroplast proteins, this often involves the extraction of total leaf RNA, followed by oligo-dT cellulose chromatography to prepare total polyadenylated mRNA. Many extraction procedures have been published, depending on the species and tissue under study. The method used in the authors' laboratory for the preparation of pea leaf mRNA (7) has been found to be suitable for wheat, spinach and tobacco leaf mRNA. Efficient translation in cell-free systems is achieved only after the oligo-dT cellulose purification step; total RNA translates poorly (or not at all). Since plastid mRNAs are not polyadenylated, this means that translation of polyadenylated plant mRNA leads to the synthesis of nuclear-encoded, but not chloroplast-encoded, chloroplast proteins. The products obtained by translation of pea leaf polyadenylated mRNA in wheat-germ or reticulocyte lysate systems are shown in *Figure 2*. Many translation products are apparent, including major polypeptides of 32 kd and 20 kd. These are the primary translation products of the most abundant chloroplast proteins, the light-harvesting chlorophyll binding protein (a thylakoidal protein) and the small subunit of ribulose bisphosphate carboxylase (a stromal protein) respectively. Like all nuclear-encoded chloroplast proteins studied to date, these proteins are initially synthesized as larger precursors; in both cases, the pre-sequences have molecular masses of 6 kd. The presequences of other imported chloroplast proteins range in size from 2 to 11 kd.

In order to analyse a single translation product, it is necessary to subject the translation mixture to immunoprecipitation using a monospecific antiserum raised against the appropriate chloroplast protein. Analysis of the immunoprecipitate by SDS–polyacrylamide gel electrophoresis and fluorography should then reveal a single polypeptide representing the precursor of the imported protein (8).

In our laboratory we have found that this procedure is easily applied to the study of abundant and middle-abundant mRNAs. For example, precursors of

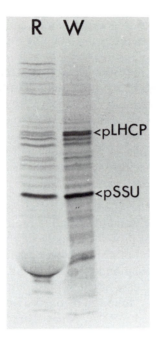

Figure 2. *In vitro* translation of pea leaf mRNA. Isolated pea leaf poly(A)-containing mRNA was translated in the presence of [^{35}S]methionine in reticulocyte lysate (R) and wheat-germ (W) cell-free systems. The translation products were subjected to SDS–polyacrylamide gel electrophoresis and fluorography. Bands representing the precursors of the light-harvesting chlorophyll binding protein (pLHCP) and the small subunit of ribulose bisphosphate carboxylase (pSSU) are indicated.

ribulose bisphosphate carboxylase small subunit (pre-SSU) and the light-harvesting chlorophyll binding protein (pre-LHCP) can be readily immuno-precipitated from 10 µl of wheat-germ mix containing pea leaf mRNA translation products. In both cases, the appropriate band can be seen after an overnight exposure of the fluorogram to X-ray film. For other translation products, larger volumes of translation mix are subjected to immunoprecipit-ation. Aliquots of 50 µl have been found to be sufficient for the identification of precursors to plastocyanin, ferredoxin, ferredoxin–NADP reductase, δ-subunit of the CF$_1$, ATPase, and the 34 kd and 23 kd subunits of the photosystem II oxygen-evolving complex. In all of the above examples, translation was carried out in the presence of [^{35}S]methionine (sp. act. >800 Ci/mmol).

If a full-length cDNA clone encoding the chloroplast protein of interest is available, it is possible to synthesize the precursor by a simpler method. The cDNA sequence can be inserted into a suitable vector and transcribed *in vitro* using SP6 or T7 RNA polymerase. The synthetic transcripts will automatically direct translation of a single polypeptide in a wheat-germ or reticulocyte lysate system (9).

3.2 Transport of proteins into isolated chloroplasts

The import of cytoplasmically-synthesized chloroplast proteins can be reconstituted *in vitro* by incubating isolated chloroplasts with chloroplast protein precursors synthesized *in vitro*. Under the appropriate conditions, precursors are rapidly taken up, converted to the mature size, and localized in the correct sub-organellar compartment. The transport process takes place post-translationally (1,2) and therefore the precursors are mixed with chloroplasts after translation is complete. The mechanism also requires ATP which can either be supplied exogenously or synthesized endogenously via photophosphorylation if the chloroplasts are illuminated (10).

The protein transport and processing mechanisms appear to be highly conserved within higher plants. Consequently, pea chloroplasts have been found to be capable of correctly importing precursors of other dicotyledonous and also monocotyledonous chloroplast proteins. Indeed, pea chloroplasts are capable even of importing precursors of *Chlamydomonas* chloroplast proteins; however, the imported proteins appear to be incorrectly processed (11).

The protocol is detailed below. Percoll-gradient purified chloroplasts are required because broken chloroplasts are still capable of binding precursors with high affinity, effectively sequestering a proportion of precursor molecules. After the incubation period, the chloroplasts are subjected to limited trypsin digestion to remove bound translation products. Other proteases, such as chymotrypsin or thermolysin, can be substituted if a particular protein under study is resistant to trypsin.

(i)   Set up a wheat-germ translation of the required volume to prepare radioactively labelled translation products. After translation is complete, mix the following components:
      3 vols of translation products
      2 vols of 200 mM methionine
      1 vol. of 5 × Hepes–sorbitol buffer, pH 8.4 (Section 2.3)
      The volume of translation products used depends of course on the abundance of the labelled polypeptide of interest. Cold methionine is included in the mixture to 'dilute' the free [^{35}S]methionine in the translation mixture. The labelled methionine is otherwise imported into the chloroplasts and incorporated into protein by the chloroplast protein synthetic machinery.

(ii) Prepare Percoll gradient-purified pea chloroplasts from young seedlings (as described in Section 2.3) from 2 × 10 ml gradients. Resuspend the final pellet in 1 ml Hepes–sorbitol buffer, pH 8.4, and measure the chlorophyll concentration as follows: remove a small volume (e.g. 20 μl) of chloroplast suspension into 80% aqueous acetone and measure the absorbance at 645 and 663 nm.

$$[chlorophyll] = (20.2 \times A_{645}) + (8.02 \times A_{663})$$

(iii) Adjust the concentration of the chloroplast suspension to 1 mg/ml chlorophyll with Hepes–sorbitol. To each 50 μl of import mixture (translation mix + methionine + 5 × Hepes–sorbitol buffer) add 100 μl of chloroplast suspension. Mix gently and incubate at 25°C in an illuminated water bath (100 µmol photons/m^2/sec). Occasionally shake the mixtures to prevent the chloroplasts from settling out. Incubation periods of 30–60 min are generally sufficient for efficient import of precursor molecules.

(iv) At the end of the incubation period, add 25 µl of trypsin (1 mg/ml in Hepes–sorbitol buffer) per 150 µl of import incubation mixture.

(v)   Incubate on ice for 30 min and then add 25 µl of soybean trypsin inhibitor (1 mg/ml in Hepes–sorbitol buffer). Add 5 ml of Hepes–sorbitol buffer and pellet the chloroplasts by centrifugation at 4000 *g* for 2 min. If problems arise in the use of trypsin, thermolysin is an effective alternative. The same amount of protease is added, but in Hepes–sorbitol containing 10 mM CaCl$_2$. After incubation on ice as above, the protease is inhibited by the addition of EDTA (in Hepes–sorbitol) to a final concentration of 50 mM.

(vi) The location of the imported polypeptide(s) can be analysed by fractionation of the pelleted chloroplasts. Lyse the pellet in a small volume (e.g. 60–100 µl) of 10 mM Tris–HCl, pH 7–8, 1 mM phenyl methyl sulphonyl fluoride (PMSF), and centrifuge the lysate in a microfuge for 10 min to generate stromal and thylakoidal fractions. Analyse aliquots of these fractions by SDS–polyacrylamide gel electrophoresis followed by fluorography.

3.3 Synthesis of proteins by isolated intact chloroplasts

The synthesis of proteins which are encoded by chloroplast DNA can be analysed by incubating isolated chloroplasts in the presence of a labelled amino acid (usually [^{35}S]methionine). Under these conditions, label is incorporated into a number of stromal and thylakoidal proteins (12). It should be emphasized that chloroplasts do not carry out translation for a long period of time after isolation. It is therefore essential that the isolation is completed as quickly as possible.

(i)   Isolate chloroplasts from 15 g of young (8–9 days old) apical pea leaves using the rapid method described in Section 2.2.

(ii)   Resuspend the pellet of washed chloroplasts in a small volume of ice-cold 0.33 M sorbitol, 50 mM Tricine–KOH, pH 8.4, to a chlorophyll concentration of 50–200 µg/ml. Place the mixture on ice in an acid-washed glass tube.

(iii) Add [^{35}S]methionine to a final concentration of 50–100 µCi/ml. Use label of high specific activity, for example 1000 Ci/mmol.

(iv)   Place the tube immediately in a glass-bottomed water bath at 20°C, illuminated from below. The light intensity at the bottom of the tube should not be below 600 µmol photons/m^2/sec. Swirl the contents of the tube at 10 min intervals to resuspend settled chloroplasts. Incubation should be for 30–60 min.

(v)   After incubation is complete, the chloroplasts can be collected by centrifugation (4000 *g* for 2 min). Preparation of stromal and thylakoidal extracts is as described in Section 3.2. Analysis of these fractions by SDS–polyacrylamide gel electrophoresis and fluorography reveals numerous labelled polypeptides in each case. By far the most heavily labelled stromal protein is the large subunit of ribulose bisphosphate carboxylase (55 kd). In the thylakoids, the most prominent labelled band corresponds to the 32 kd 'herbicide-binding protein' of photosystem II.

Another method used to analyse the synthesis of plastid DNA-encoded polypeptides involves programming an *in vitro* transcription–translation system with fragments of chloroplast DNA. The method is more complex but can be used to study chloroplast DNA-encoded proteins of plant species other than pea, for example wheat (13) and spinach (14). The transcription–translation system is derived from *Escherichia coli* extracts and can be obtained commercially.

4. THE FRACTIONATION OF CHLOROPLASTS

The chloroplast is a structurally complex organelle containing three membranes (two envelope membranes plus the thylakoidal membrane) which enclose three discrete soluble phases (inter-envelope membrane space, stroma and thylakoid lumen). For technical reasons, very little is known about the contents (if any) of the inter-envelope membrane space, but methods have been developed to fractionate chloroplasts into the remaining component compartments.

4.1 Preparation of stromal extracts

Preparation of crude stromal extracts is very straightforward. A chloroplast pellet is resuspended in hypotonic buffer at 4°C and left for 5 min to ensure complete lysis. Many commonly used buffers (e.g. Tris, Hepes, Tricine) at concentrations of 10–50 mM have been used. Total membranes are subsequently pelleted by centrifugation at 40 000 *g* for 20 min to yield a stromal supernatant which is usually slightly brown or yellow in colour. Analysis of this fraction by SDS–polyacrylamide gel electrophoresis reveals two major bands of apparent M_r 55 000 and 14 000; these are the large and small subunits of ribulose bisphosphate carboxylase, which is by far the most abundant stromal protein (*Figure 3*).

4.2 Preparation of thylakoid membranes

The thylakoid membranes of higher plants are complex structures which show a high degree of internal organization. For example, many proteins are restricted to either the granal (stacked) or stromal (single) vesicles in the thylakoidal network. In this section we have detailed a simple procedure for the preparation of thylakoidal membranes essentially free of stromal contamination. The reader is referred to recent reviews (15,16) which cover in greater detail the techniques which can be employed in the preparation and fractionation of thylakoidal membranes.

(i)   Resuspend a pellet of washed chloroplasts (Section 2.2) or Percoll-purified chloroplasts (Section 2.3) in 10 mM Tris–HCl, pH 7.0. Incubate on ice for 5 min.

(ii)   Centrifuge at 5000 g for 5 min at 4°C. This step pellets a high proportion (but not all) of the thylakoid membranes but leaves the majority of envelope membranes in the supernatant.

(iii) Wash the thylakoidal pellet twice with 10 mM Tricine–NaOH, pH 7.0, 300 mM sucrose, 5 mM $MgCl_2$. This washing procedure removes the vast majority of loosely bound stromal protein, as judged by the disappearance of ribulose bisphosphate carboxylase. A typical SDS–polyacrylamide gel of total pea thylakoidal protein is shown in *Figure 3*.

4.3 Isolation of pea chloroplast envelope membranes

In recent years, methods have been developed for the preparation of spinach (17) and pea (18) chloroplast envelope membranes which are essentially free of thylakoids. This is a difficult task because the envelope membranes account for only about 1% of total chloroplast membranes. In this Section we describe a

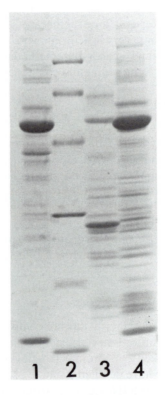

Figure 3. Stained SDS–polyacrylamide gel of pea chloroplast stromal (**track 1**), thylakoidal (**3**) and envelope protein (**4**). **Track (2)** contains molecular weight markers of M_r 96, 67, 43, 30, 20 and 12 kd. The major bands of apparent M_r 55 and 14 kd in **tracks (1)** and (**4**) represent the large and small subunits of ribulose bisphosphate carboxylase.

modification of these methods which has proved effective for the isolation of pea chloroplast envelopes.

(i)   Prepare pellets of washed chloroplasts as described in Section 2.2 from a total of approximately 450 g (fresh weight) of mature pea leaves (11 day old seedlings).

(ii)   Resuspend the pellets in a total of 40 ml of 0.33 M sorbitol, 50 mM Hepes–KOH, pH 7.5, and layer 5 ml over each of eight 30 ml Percoll gradients. Prepare pellets of Percoll-purified chloroplasts as described in Section 2.3.

(iii)   Lyse the chloroplasts in a total of 11 ml of 10 mM Tricine–KOH, pH 7.8, 4 mM $MgCl_2$, 1 mM PMSF. Add the PMSF from a freshly-made 100 mM stock in ethanol. Gently resuspend the chloroplast pellets with a cotton-wool swab, followed by a brief vortex.

(iv)   Pool the lysed chloroplasts and make up to a final volume of 12 ml with the lysis buffer given above. Add 1.8 ml of 80% (w/v) sucrose and mix briefly.

(v)   Layer the mixture on to a discontinuous sucrose gradient of 0.98 M and 0.6 M sucrose, buffered in each case with 10 mM Tricine–KOH, pH 7.8, 4 mM $MgCl_2$. Centrifuge at 90 000 g for 2 h, after which the gradient appears as shown in *Figure 4*.

(vi)   Remove the envelope band (after removing the stroma) and dilute with 5–10 vols of lysis buffer. Centrifuge at 120 000 g for 90 min, and resuspend the envelopes as required.

It should be emphasized that very small quantities of envelope protein are obtained in these preparations. For example, a typical preparation as described above may yield only 200 µg of total envelope protein. The pattern of envelope proteins obtained by SDS–polyacrylamide gel electrophoresis is shown in *Figure 3*. The membranes are invariably contaminated with ribulose bisphosphate

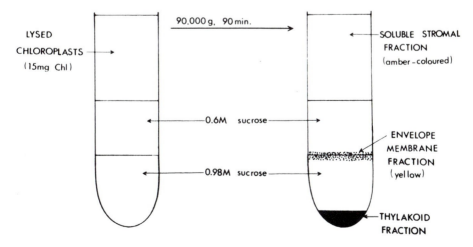

Figure 4. Isolation of pea chloroplast envelopes. See text for details of the methodology.

carboxylase (subunit mol. wts 55 kd and 14 kd in pea); no procedures have been published which avoid contamination by this stromal protein. The polypeptide pattern shown represents both inner and outer envelope proteins; the two membranes can be separated to an extent by preparing them from chloroplasts that have been ruptured by freezing and thawing under hypertonic conditions (18).

5. REFERENCES

1. Schmidt,G.W. and Miskind,M.L. (1986) *Annu. Rev. Biochem.*, **55**, 879.
2. Ellis,R.J. and Robinson,C. (1987) *Adv. Botan. Res.*, **14**, 1.
3. Blair,G.E. and Ellis,R.J. (1973) *Biochim. Biophys. Acta*, **319**, 223.
4. Morgenthaler,J.J., Marden,P.F. and Price,C.A. (1975) *Arch. Biochem. Biophys.*, **168**, 289.
5. Pelham,H.R.B. and Jackson,R.J. (1976) *Eur. J. Biochem.*, **67**, 247.
6. Anderson,C.W., Straus,J.W. and Dudock,B.S. (1983) In *Methods in Enzymology.* Wu.R., Grossman,L. and Moldave, K. (eds), Academic Press Inc., London, Vol. 101, p. 635.
7. Highfield,P.E. and Ellis,R,J. (1978) *Nature*, **271**, 420.
8. Grossman,A.R., Bartlett,S.G., Schmidt,G.W., Mullet,J.E. and Chua,N.-H. (1982) *J. Biol. Chem.*, **257**, 1558.
9. Smeekens,S., Bauerle,C., Hageman,J., Keegstra,K. and Weisbeck,P. (1986) *Cell*, **46**, 365.
10. Grossman,A.R., Bartlett,S.G. and Chua,N.-H. (1980) *Nature*, **285**, 625.
11. Mishkind,M.L., Wessler,S.R. and Schmidt,G.W. (1985) *J. Cell. Biol.*, **100**, 226.
12. Ellis,R.J. (1983) In *Subcellular Biochemistry.* Roodyn,D.B. (ed.), Plenum Press, Vol. 9, p. 237.
13. Howe,C.J., Bowman,C.M., Dyer,T.A. and Gray,J.C. (1982) *Mol. Gen. Genet.*, **186**, 525.
14. Westhoff,P., Alt,J. and Herrmann,R.G. (1983) *EMBO J.*, **2**, 2229.
15. Andersson,B. and Anderson,J.M. (1985) In *Modern Methods of Plant Analysis, New Series.* Linskens,H.F. and Jackson,J.F. (eds), Springer-Verlag, Berlin, Vol. 1, p. 231.
16. Barber,J. (1983) *Plant Cell Environ.*, **6**, 311.
17. Douce,R. and Joyard,J. (1982) In *Methods in Chloroplast Molecular Biology.* Edelmann,M., Hallick,R.B. and Chua,N.-H. (eds), Elsevier Biomedical Press, Amsterdam, p. 239.
18. Cline,K., Andrews,J., Meusey,B., Newcombe,E.H. and Keegstra,K. (1981) *Proc. Natl. Acad. Sci. USA*, **78**, 3595.

CHAPTER 4

Isolation and analysis of plant mitochondria and their genomes

WOLFGANG SCHUSTER, RUDOLF HIESEL, BERND WISSINGER, WERNER SCHOBEL and AXEL BRENNICKE

1. INTRODUCTION

Molecular analysis of the plant mitochondrial genomes has been comparatively slow in comparison with the information obtained on animal and fungal mitochondrial DNAs. This lag-phase has been partly due to the technical difficulties encountered in isolating mitochondria and their nucleic acids from plant cells. The other inherent problem is posed by the size and configuration of plant mitochondrial genomes that usually consist of several different molecules with a unique sequence complexity of 200–2500 kb. Substantial progress has been achieved in recent years with the availability of standard isolation procedures for plant mitochondrial nucleic acids, proteins and *in organello* labelling techniques.

Many of these techniques have been described in detail in the recent edition 'Plant Molecular Biology', volume 118 of *Methods in Enzymology* (1). Here we want to give a general idea of the techniques available for the analysis of plant mitochondrial genomes with emphasis on the methods employed in the analysis of dicotyledons in tissue culture. We will use the term 'mitochondria' here without distinction between the organelles *in vivo* and the artificial vesicles purified by the disruption of cells and the different manipulation steps exhibiting the typical enzymatic activities associated with mitochondria. A comprehensive review of the current status of research on plant mitochondrial genomes can be found in ref. 2.

2. ISOLATION OF MITOCHONDRIA

Isolation and purification of mitochondria is the first prerequisite to obtain reasonably enriched mitochondrial material. Efficient isolation of plant mito-chondria is much more difficult than their isolation from animal tissues. Plant cells are difficult to break, and the mechanical forces necessary to disrupt a plant cell will normally also lead to breakage of many organelles, nuclei, plastids and mitochondria.

Differential centrifugation of disrupted plant cells, with subsequent DNase digestion of the crude mitochondrial preparation and a density gradient purification, yields mitochondrial DNA (mtDNA) sufficiently pure for many purposes. Gradient purification of mitochondria, however, is necessary for the isolation of mitochondrial RNA (mtRNA) and mitochondria for protein analyses.

2.1 Plant material

Virtually all the different living tissues from plants have been used successfully as a source of mitochondria for the isolation of nucleic acids. These include green leaves, storage tissues such as potatoes and beets, root cells, activated embryos, dark grown shoots and of course *in vitro* grown cells either from suspension or callus cultures as a source of continuously available homogenous material. In the following we will give some specific advice and caveats for each of the different cell types, for which the general protocol listed in *Table 1* must be individually modified both with regard to the cell source and the species investigated.

2.1.1 *Callus tissue cultures*

We have found that callus cultured cells offer many advantages, especially in long-term investigations for which a continuous supply of constant material is essential. However, in many plant species the mitochondrial genome is unstable during the transition into and out of tissue cultures, and the mtDNA organization must be compared carefully between native tissues such as leaves and the cultured cells.

Callus cultures generally contain large, elongated cells with huge vacuoles that are easily disrupted by mechanical forces. Suspension culture cells are more compact and have developed considerably stronger cell walls, often requiring treatment in a French press or glass bead homogenizer to be broken.

The protocol given in *Figure 1* and *Table 1* applies directly for callus culture cells, but should be tested and adapted for each species. The buffer recipe given works equally well for *Oenothera* and carrot culture cells and will yield mitochondria from most species and types of tissue. Optimal yields, however, will require some individual buffer compounds that need to be tested individually for each species and tissue.

2.1.2 *Green leaves*

Preparations of mitochondrial nucleic acids from green tissues will generally be more heavily contaminated by chloroplast DNA (cpDNA) than preparations prepared from pale tissues. Contamination can often be observed directly as faint bands in a restriction digest of mtDNA, co-migrating with cpDNA fragments. Low degrees of contamination may only be detectable by hybridization. The amount of cpDNA varies between different preparations and tissue sources. mtDNA from leaves has been reported to be purer than from green fruits, although the cpDNA content of leaves is much greater (1). This may be due to tissue specific parameters playing a role in the isolation efficiencies.

Isolation procedures are the same as described for tissue cultures, but the

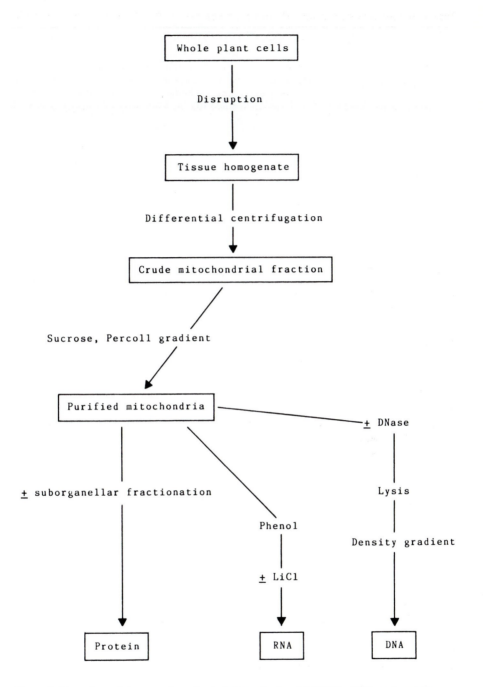

Figure 1. Flow chart of the procedures for isolating mitochondrial DNA, RNA and proteins.

Plant mitochondria and their genomes

Table 1. Isolation of mitochondria from tissue cultures.

---

1. Harvest between 500 and 700 g of freshweight of tissue culture cells from healthy log phase cells by collecting the cells in a glass beaker on ice. Avoid excessive contamination of agar; small amounts, however, do not seem to interfere with the purification procedure.

2. Prepare buffers the day before the extraction, keep them in the cold room overnight and put them on ice before starting the extraction. Extraction buffer contains 300 mM mannitol, 5 mM $MgCl_2$, 10 mM KH_2PO_4, 50 mM Tris–HCl, 5 mM EDTA, 4 mM mercaptoethanol (added just before use) pH 7.4. DNase buffer is the same without EDTA and the wash buffer after DNase treatment is the extraction buffer without $MgCl_2$.

3. Add approximately five times the cell weight of ice-cold extraction buffer to the cells. The exact ratio of cells to buffer depends on the source (i.e. species) and the type of cells. Light, almost white cells seem to contain less phenolic and complex sugar compounds and thus require less buffer to give a liquid suspension of the disrupted cells. *Oenothera* cells, for example, contain excessive amounts of such high molecular weight secondary substances, and sparingly used buffer will result in a thick paste that can neither be poured nor successfully centrifuged. There is no upper limit for the amount of buffer to be used; some tissues will require up to ten times the volume of cells.

    The deleterious effects of phenolic compounds are counteracted by agents such as PVP and bovine serum albumin, mercaptoethanol and octanol and you should adjust a mixture of these additives in your isolation buffer individually for the cells you use. Omission will still give you mtDNA, albeit very low yields.

4. Disrupt the cells in a pre-cooled Waring blender 3–5 times for 5 sec each at high speed with 10 sec breaks in between. This time schedule works well for *Oenothera* cells, but should be tested in different cell types for disruption efficiency under a microscope. Not all cells will be broken, thus timing is a balance between increasing disruption of cells and mitochondria with increasing mechanical forces.

All the following steps must be done in the cold.

5. Centrifuge in 250 ml tubes in a Sorvall GSA-type rotor at 4000 g for 15 min. Take care to avoid spill over of the loose pellet when transferring the supernatant to new centrifuge tubes. All centrifuge tubes should be used for the preparation of mtDNAs only, tubes (or any other containers) previously used for the preparation of bacterial plasmids will lead to considerable contamination of the mtDNA.

    You can increase the mtDNA yield considerably (almost double) by re-homogenizing the pelleted cell debris and intact cells after the addition of fresh extraction buffer. This is, however, only recommended for the preparation of mtDNA.

6. Re-centrifuge supernatant as in step (5) after adding extraction buffer to the centrifuge tube volumes or pooling.

7. Sediment mitochondria from the supernatant at 12 000 g and pour off the supernatant by inverting the centrifuge tubes.

8. Take up the mitochondrial pellet in isolation buffer with a Pasteur pipette and homogenize the mitochondrial aggregates carefully by repeated slow passage through the pipette. Fine homogenization can be done in a glass homogenizer with several slow strokes. Alternatively you can use soft painters' brushes or cotton wool wrapped around a glass rod to suspend the mitochondrial pellet. This suspension is now a crude mitochondrial preparation, contaminated with attached ribosomes and of course substantial amounts of nuclear DNA.

To purify further the mitochondrial vesicles, density gradient purification of mitochondria is necessary, especially for the preparation of clean mtDNA from species where the buoyant densities of nuclear and mtDNAs are very similar. The preparation of reasonably pure mtRNA always requires introduction of a density gradient step. See *Table 2* for gradient procedures.

---

relative amounts of cells and buffer should be varied with regard to the cell compounds (see *Table 1*).

After disruption of the cells the resulting homogenate must be filtered through four layers of muslin to remove large leaf parts. The crude mitochondrial supernatant from the first low speed centrifugation step should be filtered through one layer of Miracloth or similar nylon gauze.

2.1.3 *Shoots*

Plant species with large seeds such as wheat, maize and sunflower offer an easy way of isolating comparatively pure mitochondria from etiolated shoots. This source has been extensively applied in the study of maize and wheat mtDNA and detailed protocols have been described (1,2). Seeds must be surface sterilized, germinated with sterile water and allowed to grow for several days in the dark. Hypocotyls, coleoptiles and leaves are harvested and cut into slices before homogenization in a blender or mortar. Bacterial contamination must be avoided at all steps (3).

2.1.4 *Other tissues (storage tissues, fruits)*

Potatoes and beets have been used as a source of mitochondrial nucleic acids in the study of potato and sugar beet mtDNA and mtRNA. This material offers the advantage of being storable and thus available at any time—if enough has been grown during the growth season. Since the tissue is very tough it needs to be sliced and/or minced well before the cell homogenization step.

2.2 Cell disruption

Disruption of cells must be done in the cold to minimize degradation of organelles and macromolecules released from the different compartments.

The most commonly used procedure to break plant cells is by mechanical forces in Waring blendor type machines. This procedure is used for larger amounts of cells for tissue cultures, shoots and leaf cells. Small samples can be ground in a mortar with quartz sand to break the cells.

An alternative enzymatic cell lysis makes use of a fungal enzyme preparation called Novozym (Novo Chemicals, Copenhagen), often used to prepare protoplasts. The recommendations of the manufacturer should be followed strictly, especially concerning precautionary measures since this concoction has very unpleasant side effects on humans.

2.3 Differential centrifugation

Enrichment of mitochondria in the cell lysate is generally done by differential centrifugation steps making use of the different sedimentation rates of unbroken cells, nuclei, plastids and mitochondria.

Several rounds of low speed centrifugation remove most of the cell wall debris, nuclei and plastids. Pelletting mitochondria removes much of the soluble unwanted compounds and reduces the cytosolic ribosomal contamination. This

Table 2. Gradient purification.

We routinely run sucrose gradients that are either continuous or have several concentration steps.

1. Prepare the continuous gradient either in a gradient former or obtain it in a semi-continuous form from a multiple step gradient that has been left to diffuse overnight at 4°C. The steps follow the protocol of Leaver *et al.* (3) and are generally made with 3 ml of 2.0 M, 7.5 ml of 1.45 M, 7.5 ml of 1.2 M, 7.5 ml of 0.9 M and 4 ml of 0.6 M sucrose in sucrose gradient buffer [0.05 M Tris–HCl, pH 8; 0.1% bovine serum albumin (BSA); 0.04 M β-mercaptoethanol] in 50 ml centrifuge tubes. A continuous gradient can be prepared in standard gradient formers by mixing 70% and 20% sucrose solutions in 0.05 M Tris–HCl, pH 8, 0.1% BSA and 0.04 M mercaptoethanol. Two step gradients of 1.6 M and 0.6 M sucrose have also been used and will not result in multiple bands of mitochondria as do the other gradients.
2. Carefully layer the mitochondrial suspension obtained after differential centrifugation (*Table 1*) on top of the gradient and centrifuge for 90 min in a Sorvall G-50 rotor at 20 000 r.p.m. or an SW 27 for 30 min at 18 000 r.p.m. or their equivalents.
3. Multiple bands of yellowish mitochondria will usually be observed in the gradients, which all contain mtDNA. Collect the bands at 1.25–1.35 M sucrose in these preparations.
4. Dilute the isolated mitochondrial fraction (~5 ml taken from the gradient) to 50–200 ml total volume with extraction buffer. Slow addition of the buffer during 15–20 min is advisable to minimize the osmotic shock, but has not been found to be totally disastrous if done quickly.
5. Pellet the mitochondria by centrifugation at 10 000 *g* and use the pellet for DNase treatment or direct isolation of mitochondrial nucleic acids or *in organello* labelling of proteins.

Tissue culture cells will usually yield 1–5 g of purified mitochondria per 500 g of cells depending on the status of the cells and the species.

crude mitochondrial fraction will be green, if isolated from green tissues and DNA or RNA preparations isolated directly from this fraction will always be heavily contaminated with nuclear and plastid nucleic acids due to many broken plastids, nuclei and co-sedimenting ribosomes. DNase digestion of this fraction will result in mtDNA preparations sufficiently pure for many purposes.

2.4 **Gradient centrifugation**

Banding the mitochondria in a density gradient will be necessary in most procedures to obtain sufficiently clean mitochondria to ensure that most of the macromolecules isolated are indeed associated with mitochondria.

Sucrose and Percoll are the two most commonly used gradient materials. The osmotically neutral Percoll is only necessary for special purposes, where functional mitochondria are required. Sucrose gradients are widely used and yield mitochondria sufficiently intact even for *in organello* synthesis of proteins. *Table 2* details protocols for continuous and discontinuous sucrose gradients. Recipes for Percoll gradients are found in ref. 3.

Figure 2 shows a schematic drawing of a continuous sucrose gradient after banding of the crude mitochondrial fraction. Four or five diffuse bands can be discerned of which the first, white band below the top of the gradient and the next one down, of yellowish colour, contain most of the mitochondria. For mtDNA purification both fractions can be used, for cloning mtDNA and for RNA preparations only the top band should be recovered.

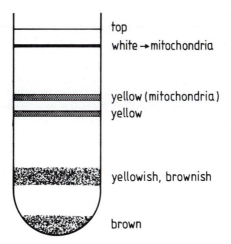

Figure 2. Schematic drawing of a sucrose gradient with the relative position of mitochondria indicated as described in *Table 2*.

3. ISOLATION OF MITOCHONDRIAL DNA, RNA AND PROTEINS

3.1 **Isolation of mtDNA**

To purify mostly linear, that is broken, mtDNA molecules for restriction enzyme analysis or other investigations, for which intact native molecules are not essential, mitochondria should be DNase I-treated prior to CsCl density gradient centrifugation (*Table 3*).

3.1.1 *Linear or main band mtDNA*

The relative position of the linear nuclear and mtDNAs varies between the different species due to the different G+C contents of the nuclear DNA. Higher plant mtDNA has a remarkably constant G+C percentage of about 1.706 g/cm^3. Sometimes, several bands of mtDNA are detected in CsCl gradients (*Figure 3*). No explanation for this difference in the apparent buoyant densities has been found; these DNAs appear to be identical in restriction digests.

3.1.2 *Closed circular mtDNA*

Occasionally a second band consisting of closed circular DNA might be detectable below the linear DNA, if the DNase incubation has been gentle and the mitochondrial vesicles have been closed. This, however, is not reproducible in our hands and is probably dependent on as yet unclear conditions. Always collect the region where closed circular molecules should be, even if no band is visible, as very often enough DNA can be isolated to visualize in a gel.

Without DNase incubation of the mitochondrial homogenate at least three bands in the gradient should be detectable. The uppermost consists of the contaminating nuclear DNA, immediately below will be the linear mtDNA, and

Table 3. Isolation of mitochondrial DNA.

---

1. Suspend the pellet obtained from the crude mitochondrially enriched fraction at the end of *Table 1*, in extraction buffer without EDTA and treat with DNase to remove attached nuclear DNA contaminants. Perform this digestion step in 10–15 ml of buffer with 50–75 µg DNase I at 4°C for 1 h.

2. Add 200 ml of extraction buffer with EDTA and pellet the mitochondria again. Repeat this washing procedure once or twice to remove the DNase as completely as possible.

3. Resuspend the final mitochondrial pellet in 5–8 ml of STE[a] and transfer the suspension to ultracentrifuge tubes. Use 13 ml tubes fitting the Beckman Ti 50 or Ti 70.1 fixed angle rotors. Since the DNA bands must be visible in UV and the tubes must be punctured, neither polyethylene nor ultra clear tubes are generally suitable, polyallomer tubes are recommended.

4. Add 150 µl of a 25% SDS solution to 8 ml of mitochondrial suspension per tube and mix gently. Vigorous shaking will lead to physical disruption of the large molecules and should thus be avoided.

5. Add 200 µl of ethidium bromide (EtBr) per tube from a stock solution of 10 mg/ml in TE[b].

6. Add solid CsCl in a ratio of 0.9 g/ml final suspension and slowly dissolve by gently rolling the tubes. This will require some time. Allow the tube to equilibrate to room temperature to dissolve the salt. You should also avoid excess exposure to light from now on to minimize nicking of the DNA molecules by the combination of EtBr and UV light. The final volume will now be approximately 12 ml. Close the tube lids and fill up the void volume with liquid paraffin.

7. The density gradient forms by centrifugation at 110 000 g for 48 h.

8. At least one faint band containing the mtDNA should be visible under UV illumination. In the case of incomplete DNase digestion nDNA will be observed as a second, broad band. Some preparations will give different bands of mtDNA, which yield the same restriction fragment pattern and presumably contain molecules in different associations. Their physical structure is as yet unclear, however, and they are not obtained in consistent relative stoichiometric amounts (*Figure 3*).

9. Recover the mtDNA band from the gradient by either piercing the tube from the side or the bottom with a syringe or by taking off the gradient from the top with a pipetman. For some purposes (e.g. cloning mtDNA) reband the mtDNA in a second CsCl gradient by adding a CsCl solution of 0.9 g/ml in TE and some EtBr and re-centrifuging.

10. Remove EtBr by three or more extractions with isopropanol saturated over CsCl-saturated TE buffer.

11. Add 2 vols of water and precipitate the DNA after the addition of 2 vols of absolute ethanol (of the final aqueous volume) and overnight incubation at −20°C.

12. Pellet the DNA by centrifugation at 10 000 g for 20 min and wash the mtDNA once with 70% ethanol.

13. Take up the pellet in 20 µl of TE. This DNA can be used for restriction analysis with most enzymes.

This DNA will consist of mostly mtDNA, but will always be contaminated by plastid and nuclear DNA to a varying degree. The average yield will be about 0.1 µg of linear mtDNA per g tissue.

---

[a] STE is 50 mM Tris–HCl, pH 8, 10 mM EDTA, pH 8, 100 mM NaCl.
[b] TE is 10 mM Tris–HCl, pH 8, 1 mM EDTA.

below that the supercoiled molecules. Sometimes additional bands will be visible, in an as yet unknown configuration of mitochondrial molecules, that yield the same restriction fragment pattern as do the linear or closed circular fractions (*Figures 3, 4* and *5* and Section 3.2.1).

Substantial amounts of closed circular DNA cannot be isolated from all plant species, this depends on the *in vivo* configuration of the mitochondrial genome,

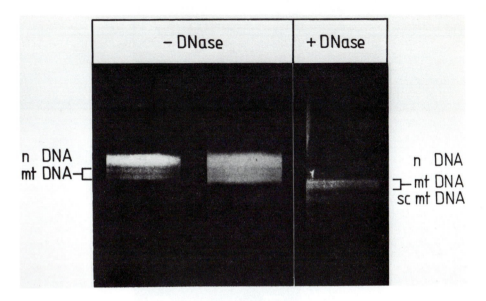

n DNA
mt DNA—

n DNA
—mt DNA
sc mt DNA

Figure 3. CsCl gradients of mtDNA preparations from *Oenothera*. Two bands of mtDNA can be detected just below the nuclear DNA in untreated mitochondria. DNase digestion of isolated mitochondria removes the nuclear DNA contaminant. Closed circular mtDNA molecules can be discerned as a faint band below the linear DNA.

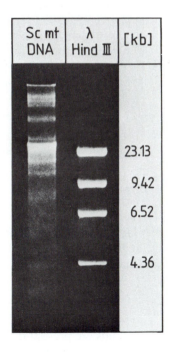

Sc mt DNA	λ Hind III	[kb]
		23.13
		9.42
		6.52
		4.36

Figure 4. Closed circular mtDNA from *Oenothera* can be separated in agarose gels into distinct size classes.

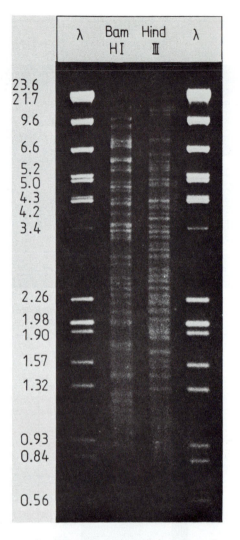

Figure 5. Restriction digests of mtDNA from carrot tissue cultures show a typical profile of the complexity of plant mitochondrial DNA.

which can consist of only very large molecules, that are too fragile to survive the extraction procedure intact, as in wheat for example. The relative amount of closed circular DNA isolated thus varies according to the tissue used and the species investigated.

3.1.3 *Mitochondrial plasmids*

In some mitochondria DNA plasmids are present that usually co-purify with the main mtDNA. Detailed protocols for the isolation and characterization of these molecules can be found through ref. 4.

3.1.4 *Contamination with plastid and nuclear DNAs*

Probing for plastid contamination with cloned plastid DNA is straightforward, if restricted mtDNA and plastid DNA are run side by side on a gel. Fragments of sizes identical between the two DNAs in the mtDNA preparation will indicate the degree of contamination. Lesser degrees of contaminating plastid DNA can be detected by hybridization with total plastid DNA and/or cloned plastid DNA fragments. However, due to plastid sequences integrated in the mtDNA, there will also be genuine mtDNA fragments hybridizing with sizes generally differing from the plastid DNA fragments.

The degree of nuclear DNA contamination will be firstly visible as a background smear on the gel and could secondly of course be analysed by similar hybridization experiments. Usually this nuclear DNA contamination does not interfere with the mtDNA identification; clone libraries, nevertheless, will of course contain fragments of nuclear origin.

3.1.5 *Mini-preparation method*

For many practical purposes (e.g. checking the derivation of mitochondrial genomes in genetic crosses) diagnostic restriction digests are required from single plants or even less material. *Table 4* describes one such method convenient for extracting mtDNA from small amounts of tissues and yielding mtDNA sufficient for a few digests.

3.2 **Isolation of mtRNA**

3.2.1 *RNA from total mitochondrial nucleic acids*

Crude preparations of mitochondrial nucleic acids usually contain enough specific RNAs for blot hybridization analyses of transcript sizes with the mtDNA giving a weak high molecular weight signal. LiCl precipitation is necessary to separate RNA and DNA when quantification of the RNA concentration is required in the experiment. Such an RNA preparation is usually sufficiently enriched in mitochondrial nucleic acids to yield clear signals for the mitochondrial gene transcripts and to determine their sizes in Northern-type hybridization and nuclease protection experiments.

Such extracts are obtained from the mitochondrial fraction of the sucrose gradient by repeated phenol/chloroform extractions. The detailed protocol is given in *Table 5a*.

Pure RNA for detailed studies can be obtained by pelleting the RNA in a CsCl gradient (*Figure 6* and *Table 5b*).

3.2.2 *Isolation of specific RNA sequences*

(i) *rRNAs*. 18S and 5S rRNAs will be the most prominent bands in gel analyses in most preparations. The rRNAs usually constitute about 80% of the total RNA preparation. To obtain a good yield of the 26S rRNA too vigorous shaking during the extraction procedures must be avoided, since the large rRNA is easily

Table 4. Minipreparation method of mtDNA and plastid DNA.

For rapid screening of mitochondrial genome types minipreparation methods have been developed to characterize the mtDNA from parts of single plants without killing the donor. These methods usually require a hybridization step, as total cellular nucleic acids are isolated, restricted, run out on a gel and probed with labelled mtDNA to visualize only the mtDNA fragments.

A recently described method for the simultaneous preparation of mt- and cpDNA makes use of the fragility of protoplasts to prepare organelle DNAs in amounts sufficient to be detectable by simple EtBr staining in gels. We will outline the protocol here as described by Charbonnier *et al.* (5) for tobacco, sunflower and other dicotyledons.

1. Wash protoplast preparations from 5 g of cells of plantlets, grown under controlled room conditions by two centrifugations in 330 mM KCl, 14 mM CaCl$_2$ and resuspend the final pellet in ice-cold extraction buffer[a]. Incubate on ice for 20 min.
2. Disrupt the protoplasts by two filtrations through 25 μm nylon gauze and remove the nuclei by three filtrations through a 10 μm nylon gauze.
3. Dilute the organelle suspension with 4 vols of extraction buffer and pellet chloroplasts by centrifugation for 10 min at 1500 *g* in 2°C.
4. Pellet the mitochondria by centrifugation at 12 000 *g* for 20 min at 2°C.
5. Lyse the organelles separately by resuspension in 4 ml of lysis buffer[b] and incubation at 37°C for 2 h.
6. Extract the proteins by addition of 1 vol. of phenol saturated over lysis buffer and 1 vol. of chloroform at room temperature and centrifuge at 12 000 *g* for 20 min.
7. Precipitate the nucleic acids with ethanol (1/10 vol. of 3 M sodium acetate, 2 vols of ethanol) for 20 min at −70°C or overnight at −20°C. Wash the nucleic acid pellet with 70% ethanol before drying.
8. Dissolve the mitochondrial pellet in 20 μl of TE (*Table 3*), the chloroplast pellet in 50–100 μl TE. Restrict 1/5 to 1/10 of the chloroplast nucleic acid preparation and the total mtDNA for one gel slot each.

[a]Extraction buffer is 330 mM KCl, 50 mM Tris–HCl, pH 8, 20 mM EDTA, 0.1% BSA, 5 mM β-mercaptoethanol.
[b]Lysis buffer is 50 mM Tris–HCl, pH 8.3, 100 mM EDTA, 50 mM NaCl, 1% SDS, 1% *N*-lauroylsarcosine, 200 μg/ml proteinase K.

fragmented. Any excess heating of the RNA should be avoided wherever possible, as plant mtRNA will be rapidly degraded by boiling. The isolation of rRNAs for direct sequence analysis has been described for wheat mitochondria.

(ii) *mRNAs*. Specific mRNAs can be enriched and selected from the mtRNA preparations by standard methods such as hybrid release and nuclease protection.

(iii) *tRNAs*. The preparation of large quantities of mitochondrial tRNAs is required for identification, direct sequence analysis and probing for specific genes. Defining the genuine mitochondrial tRNAs has to take into account the plastid sequences present in plant mitochondrial genomes, many of which contain tRNA genes. Some of these imported tRNA genes have been shown to be transcribed, the tRNA will then be identical to the plastid tRNA species. These cross-homologies complicate differentiation between chloroplast con-taminants and tRNAs used in the mitochondrion.

Table 5. Isolation of RNA from total mitochondrial nucleic acids.

Two methods of mitochondrial RNA preparation are described; the first (Part A) involves purification of total mitochondrial nucleic acids and selective precipitation of mtRNA by LiCl. This RNA will be sufficiently clean for Northern hybridization analyses, but can be too crude for clean nuclease protection, primer extension, and cDNA synthesis. The second method (Part B) involves pelleting of the RNA through a CsCl gradient and will yield RNA suitable for most modifications.

Part A.
1. Isolate mitochondria by differential centrifugation and sucrose gradient purification (*Tables 1* and *2*).
2. Extract the mitochondrial suspension firstly with 1 vol. of phenol containing 1% SDS equilibrated over 1 M Tris–HCl, pH 8.0, by vigorous shaking, then add 1 vol. of chloroform and mix before centrifugation at 12 000 *g* for 20 min.
3. Re-extract the aqueous phase three to four times with the same procedure, only omitting the SDS. Subsequent extraction with chloroform will remove phenol contaminants.
4. Precipitate the nucleic acids by addition of $^1/_{10}$ vol. of 3 M sodium acetate and 2 vols of ethanol. Leave at −20°C overnight.
5. Dissolve the RNA pellet in 1–5 ml of TE (*Table 3*) and deproteinize by several rounds of phenol/chloroform extractions until the interface material has been entirely removed.
6. To obtain RNA free from DNA, LiCl precipitation is necessary. Take up the dried pellet in TE and precipitate the large RNA molecules selectively by addition of an equal amount of 4 M LiCl, incubation at 4°C for 12 h and centrifugation. Rinse the pellet with cold 2 M LiCl and 80% ethanol. Re-dissolve the dried pellet and repeat the LiCl precipitation. Extract this RNA twice with phenol/chloroform (50:50, v/v), twice with chloroform, re-precipitate with ethanol and wash with 70% ethanol before drying.
7. Check the purity and intactness of the prepared RNA on a short (10–20 cm long) 3.2% polyacrylamide denaturing gel. Many distinct bands should be visible without background smear of degraded molecules in a clean preparation.
8. To prepare RNA blots glyoxal and formaldehyde gels (*Figure 6*) are routinely used. Shortening of the modification reaction performed at 50 or 65°C and/or slightly lowering the temperature substantially improves the signal to noise ratio in the hybridization. Avoid excessive heating or even boiling the RNA at any stage since this will lead to rapid degradation of the larger RNAs.

Part B.
1. Isolate mitochondria by differential centrifugation and sucrose gradient purification (*Tables 1* and *2*). Mitochondria must be sucrose gradient-purified to be able to dissolve the RNA pellet after CsCl gradient purification.
2. Treat as steps 3–7 of *Table 3*.
3. Take up the coloured pellet in 1 ml of sterile TE (*Table 3*). Remove EtBr by two extractions with 1 vol. each of water-saturated butanol, the butanol phase is on top.
4. Extract the RNA twice with phenol/chloroform (50:50, v/v), twice with chloroform, precipitate with $^1/_{10}$ vol. of 3 M sodium acetate and 2 vols of absolute ethanol and wash with 70% ethanol before drying.
5. Since single-stranded DNA co-precipitates with the RNA in the CsCl gradient, LiCl precipitations as described in (A) step 6 are necessary to obtain clean RNA preparations.
6. If the RNA is pelleted through a CsCl gradient without EtBr the butanol extractions can be omitted. DNA can then be recovered by re-centrifugation of the clear supernatant with EtBr.

This RNA preparation procedure will yield approximately 0.5–1 µg of mtRNA per g of cells. This RNA is, in our hands, the best starting material for enzymatic and chemical modifications for detailed analyses.

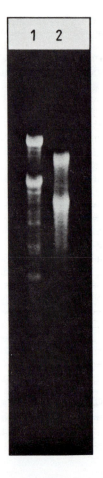

Figure 6. mtRNA (left lane) has been size fractionated in a formaldehyde gel with *Escherichia coli* 16S and 23S rRNAs as size markers in the right lane.

The procedure of isolating large quantities of mitochondrial tRNAs has been established in the laboratory of Dr J.-H. Weil and is described in detail in ref. 6.

(iv) *RNA plasmids*. Single- and double-stranded RNAs have been found in higher plant mitochondria. Their phylogenetic derivation and their contribution to mitochondrial function is as yet unresolved. Isolation, purification and analytical procedures are found in refs 1 and 7.

Table 6. *In organello* protein labelling.

1. Resuspend mitochondria prepared by differential centrifugation followed by purification on sucrose gradients (important for removing bacterial contamination) in resuspending buffer[a] and add the suspension to the incorporation mixture described below. Incorporation with each of the three energy supplies given below is normally carried out for each mitochondrial preparation.

2. Energy mixes for incorporation and controls.
   Acetate. This non-oxidizable substrate can be utilized by bacteria but not by mitochondria. Hence amino acid incorporation in its presence gives an indication of bacterial contamination. Incorporation in the presence of acetate should be less than 5% of the incorporation in the presence of other energy supplies.
   Creatine phosphate/creatine phosphokinase. This energy source supplies externally generated ATP. Incorporation is linear for up to 90 min.
   Succinate/ADP. Mitochondria utilize succinate as a substrate from which to generate their own ATP by oxidative phosphorylation. Incorporation slowly reaches a plateau after 30–45 min.

3. Solutions.
   Store the solutions used in incorporation separately and mix them in their final proportions just before use. Autoclave the stock solutions for the salts mixture separately and store at 4°C. Make up the salts mixture from the stock solutions and store in aliquots of about 1 ml at −20°C. Autoclave the 1 M sodium acetate solution and store at 4°C. Prepare the 1 M sodium succinate solution freshly on the day of the experiment using sterile double distilled water. All other solutions should also be made up in sterile double distilled water and/or filter sterilized and stored at −20°C.

   (i) Prepare the salts solution according to the following recipe by mixing the stock solutions: 4.7 ml of 0.8 M mannitol; 1.875 ml of 1 M KCl; 1.05 ml of 0.2 M Tricine, pH 7.2; 0.5 ml of 0.2 M KPO_4, pH 7.2; 0.41 ml of 0.5 M $MgCl_2$; 0.165 ml of 0.1 M EGTA, pH 7.2; add sterile double distilled water to 10 ml final volume. Adjust the pH of the final salts mixture to pH 7.2 with sterile 0.5 N KOH.

   (ii) The amount of L-[^{35}S]methionine used in an incorporation can be varied within the range of 0.185–0.74 MBq. The figures below assume the use of 0.74 MBq per incorporation. If you require a different amount you should note that the volume of radioactive methionine and water should always add up to 14.5 μl per incorporation. The concentration of L-[^{35}S]methionine is assumed to be 0.185 MBq/μl.

   (iii) You should prepare a slight excess of the incubation mixture to provide samples for background and total c.p.m. counting, half of one incubation mix is adequate for this purpose. Per incubation mix: 120 μl salts solution; 2.5 μl of 50 mM GTP; 0.5 μl of 1.0 M DTT; 12.5 μl of a 0.5 mM amino acid mix of 19 essential amino acids without methionine; 10.5 μl sterile double distilled water; 4.0 μl L-[^{35}S]methionine (0.148 MBq) for a total incubation mix volume of 150 μl.

   (iv) Prepare the energy cocktails as follows: dilute 5 μl of 1 M sodium acetate with 45 μl of sterile double distilled water; mix 4.0 μl of 0.5 M creatine phosphate, 5 μl of 5 mg/ml creatine phosphokinase, 6.25 μl of 0.24 M ATP and 34.75 μl sterile double distilled water; mix 2.5 μl of 1 M succinate freshly made by dissolving 0.675 g of sodium succinate in 125 μl of 0.2 M KPO_4, pH 7.2 and filled to 2.5 ml with sterile double distilled water; 5.0 μl of 54 mM ADP and 42.5 μl of water.

4. Perform the incorporation experiment using the following protocol:
   (i) Pipette into each siliconized test tube on ice 50 μl of any *one* of the three above energy mixes and 150 μl of incubation mix.
   (ii) Pipette 50 μl of mitochondrial suspension (5–15 mg/ml in resuspending buffer) into successive tubes and incubate at 25°C with vigorous shaking for aeration.
   (iii) At the end of the incorporation (60–90 min) remove 2 × 5 μl samples from each incorporation to estimate TCA precipitable counts. Also take 4 × 5 μl samples from the leftover incubation mix to provide samples for total and background counting.
   (iv) Squirt the incorporated mitochondria into 1.0 ml of ice-cold resuspending buffer containing ¹⁄₂₄ (v/v) of 0.12 M L-methionine in an Eppendorf tube.
   (v) Pellet mitochondria for 5 min in an Eppendorf centrifuge at 4°C, remove the radioactive supernatant and freeze the pellet on dry ice. Store the pellet at −80°C.
   (vi) These samples are now ready for analysis by denaturing polyacrylamide gels (11; *Figure 7*).

---

[a] Resuspending buffer is 0.4 M mannitol, 10 mM Tris–HCl, pH 7.2, 1 mM EGTA, pH 7.2.

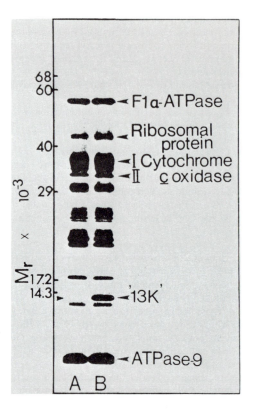

Figure 7. *In organello*-labelled mitochondrial proteins of maize have been separated in a SDS–polyacrylamide gel. The positions of the cms type-T specific 13 kd and 25 kd polypeptides are indicated. (Photograph by kind permission of Dr C.J.Leaver.)

3.3 Isolation of mitochondrial proteins

Preparation of proteins and gel analysis techniques from isolated mitochondria will essentially be standard procedures of protein and enzyme chemistry. The procedure for solubilizing total plant mitochondrial proteins involves a detergent solution step prior to gel analysis as described in *Table 6* for *in organello* synthesized polypeptides.

Subfractionation of mitochondrial inner and outer membranes and the intramitochondrial cisternae will be required for the localization of mitochondrial polypeptides and for import and transport studies of proteins synthesized in the cytosol. To clarify the mechanism of cytoplasmic male sterility (cms) induced by aberrant mitochondrial polypeptides their exact location and *Wirkungsort* (site of function) will be required. The association of the cms T specific 13-kd protein with the mitochondrial membrane for example has been shown by subfractionation of mitochondria (8).

Isolation of DNA and RNA polymerases and modifying enzymes has been done from total mitochondrial lysates and does not require prior subfractionation (9,10). One of the major problems of mitochondrial protein studies will be the small amounts of material obtainable from plant mitochondria. Another difficulty is the varying degree of membrane adherence of different proteins, that require individual adjustments of detergent conditions to purify.

4. ANALYSIS OF ISOLATED MITOCHONDRIA

4.1 *In organello* synthesis of proteins

The synthesis of proteins in isolated mitochondria with specific labelling of the mitochondrially encoded polypeptides has proven to be a powerful tool in the identification of plant mitochondrial gene products.

This procedure has been a key factor in the identification of aberrant protein products in mutant mitochondrial genomes responsible for cytoplasmic male sterility. The method has been developed by Dr C.J.Leaver's laboratory (3), who succeeded in identifying a number of the mitochondrially encoded proteins. We here give their recipe with kind permission of Dr C.J.Leaver, Dr P.G.Isaac and co-workers (*Table 6*).

4.2 *In organello* synthesis of RNA

The usefulness of transcription in isolated mitochondria is currently being examined by specifically labelling the RNAs made in the suspension. Some of the pitfalls have already been discovered; for example contamination with plastid pieces active in transcription and co-purifying with the mitochondria has to be taken into consideration when interpreting the data (12).

4.3 *In organello* synthesis of DNA

Study of mtDNA replication with emphasis on some of the small DNA species present in many plant mitochondria might be approached by *in organello* labelling techniques. Most of the *in organello*-labelled nucleic acids seem to be derived from repair synthesis since all restriction fragments of the mtDNA appear to be labelled more or less uniformly (13).

4.4 Cloning of mtDNA

4.4.1 *Cosmid clones*

The large size and the heterogenous population of mtDNA molecules has required cloning of large continuous fragments in cosmid libraries; these techniques have been described by Drs D.A.Lonsdale's laboratory and Drs F.Quetier, B.Lejeune and their co-workers (1). To establish a cosmid bank, at least 50 μg of pure, high molecular weight mtDNA are required.

4.4.2 *Plasmid clones*

Cloning of plant mtDNA into plasmid vectors is straightforward and usually does not present any specific problem. It is always advisable to re-band mtDNA in a second CsCl gradient to obtain better yields in fragment cloning. The smaller mtDNA genomes from *Brassica* species have been mapped without cosmid clones by aligning classic restriction fragment libraries. For the average sized plant mtDNA two different libraries of 600–900 clones will usually contain all the sequences present in mtDNA. Rare sequence rearrangements present in subliminal amounts in the mtDNA will, however, require special efforts.

4.5 **Mapping of mitochondrial genomes**

Small and simple genomes consisting of only one or a few genomic molecules like several *Brassica* species can be mapped by multiple restriction digests and corresponding hybridizations.

More complex genomes will require cloning and walking experiments with continuous control of the walk in total mtDNA to establish the various linkage groups in the different molecules. The highly asymmetric stoichiometries of the different molecules observed in several plant mitochondrial genomes probably preclude the establishment of the ultimately 'complete' map for some species (2,14).

5. IDENTIFICATION OF MITOCHONDRIAL GENES

A number of the classic mitochondrial genes encoded in the mitochondrial genomes of most species have already been identified in plant mitochondrial genomes (*Table 7*). Plant mtDNA, however, encodes some genes not found in other mitochondria, which are currently being identified. The α subunit of the ATPase is one example of those specific genes. Identification of new genes in plant mitochondria therefore requires some special efforts, which are outlined in this section.

5.1 **Heterologous probes**

Most of the standard mitochondrial genes like cytochrome *b* and cytochrome oxidase subunits are generally well conserved in their sequence and could thus be identified with heterologous probes from other species, mostly yeast. This approach has been highly successful in the past, but appears now to be of less importance with only the less well conserved genes and/or plant mitochondrial specific genes left to be identified.

All of the genes analysed so far are highly conserved in their primary sequence between the different plant species and can easily be identified in all higher plants by heterologous hybridization.

Since probes are now freely available from several higher plant species for the genes identified so far, these will be easily identified by hybridizations under standard conditions with slightly lowered stringency in any new species to be investigated.

Table 7. Genes identified in plant mitochondria.

		Reference[a]
Ribosomal RNAs	5S	15,16
	18S	17,18
	26S	19
tRNAs	several listed in	2
Cytochrome oxidase subunits	I	20
	II	21
	III	22
ATPase subunits	6	23
	9	24
	α	25
Cytochrome *b*		26
NADH subunits	1	27
	3	37
	5	28
Ribosomal proteins	S12	37
	S13	29
	S14	38
ORF	25 kd protein	30
ORF	13 kd protein (cms)	30
ORF	Reverse transcriptase	31

[a] Only the reference for the original identification of each gene in plant mitochondria is given.

Heterologous hybridization experiments to identify rRNAs can only be done with genuine plant mitochondrial rRNA genes under stringent conditions and must even then be interpreted with care due to the presence of plastid and nuclear rRNA sequences integrated in the plant mitochondrial genomes and also with respect to plastid DNA contaminations (31).

5.2 End-labelled mtRNA

Several plant mitochondrial genes have been identified using the presence of high levels of steady state transcripts. mtRNA has been 5′ end-labelled with polynucleotide kinase and hybridized to clone libraries of mtDNA. Abundantly transcribed regions have been thus identified and sequenced, having led to the identification of several tRNAs and ATPase subunits.

5.3 Reverse-transcribed mtRNA

Using the same rationale, while being more sensitive, is the reverse transcription of total mtRNA primed with a mixture of random oligonucleotides, with the incorporation of one or more radioactive nucleotides. This approach also allows the detection of the less abundantly transcribed loci in mtDNA clones. The risk is the same nevertheless of having to sequence the entire hybridizing region and relying on computer searches in the sequences available in data banks for the identification of the encoded information.

5.4 cDNA clones

Selecting random cDNA clones from a library has been used to identify several plant mitochondrial genes. *Table 8* gives a recipe for the construction of cDNA clones from plant mitochondrial transcripts. Care must be taken that the clone to be investigated is indeed of mitochondrial origin before a detailed analysis is undertaken, since many cDNA clones will be derived from nuclear and/or plastid RNA contaminations. This approach is theoretically more sensitive again to detect low abundance transcripts, but is also very labour intensive since a lot of sequencing is involved. The cDNA clone and the genomic sequences need to be determined, as most of the cDNA clones contain the 3' non-coding regions and varying portions of the actual gene sequence.

5.5 Sequence analysis of entire mitochondrial genomes

With the improvement of sequencing techniques it will not be long before one or more of the smaller mitochondrial genomes will be sequenced in their entirety.

Table 8. cDNA cloning.

Preparation of cDNA clones from mtRNAs requires CsCl-purified RNAs that have been purified from DNA by one or more LiCl precipitations. Prior to priming first strand synthesis oligo(A) tails need to be added since plant mitochondrial transcripts are not polyadenylated. The following procedure is taken from ref. 33.

1. Purify mtRNA following the protocol of *Table 5B*.
2. Add poly(A) tails to the mtRNAs (5 μg) with the ATP:RNA adenyltransferase as recommended by the manufacturer (BRL) in a volume of 50 μl. Monitor the reaction with trichloroacetic acid (TCA) precipitated aliquots and terminate after the addition of approximately 50 A-residues (reaction time 1 h). This RNA can now be used for first strand synthesis after one phenol/chloroform (50:50 v/v) and two chloroform extractions and an ethanol precipitation.
3. Synthesize the first strand with actinomycin D and RNasin in the reaction mixture[a].
4. After 60 min add an additional 50 μl of AMV reverse transcriptase and incubate for another 30 min at 43°C.
5. Quantify first strand synthesis by comparing TCA aliquots before and after the reaction.
6. One phenol/chloroform extraction series purifies the nucleic acids sufficiently for second strand synthesis. AMV reverse transcriptase will yield approximately 150 ng and M-MuLV reverse transcriptase (Pharmacia) about 600 ng of total cDNA synthesized. The latter enzyme also gives better incorporation during first strand synthesis when omitting actinomycin from the reaction buffer.
7. Synthesize the second strand according to the procedure described by Gubler and Hoffmann (34).
8. Adjust tails to 15 (insert) and 20 nucleotides (vector: *Pst*I site in pUC 8) according to Deng and Wu (35). Annealing and transformation of JM 83 is done following standard procedures (36).

---

[a] Reaction mixture is 50 mM Tris–HCl, pH 8.3 at 43°C; 8 mM $MgCl_2$; 10 mM DTT; 128 mM KCl; 50 μg/ml actinomycin D; 20 μCi [^{32}P]dCTP (3000 Ci/mmol); 1 mM dNTPs; 500 U/ml RNasin; 1000 U/ml AMV reverse transcriptase.

This effort will however be left to the larger laboratories, since it is still quite a task to identify 200 000 odd nucleotides.

5.6 Sequences of plastid or nuclear origin

The origin of cloned fragments must always be thoroughly checked and connections with known mtDNA marker genes need to be established before any new clone or sequence is labelled as mitochondrial. Many sequences also present in plastid DNA have been identified in plant mtDNA by hybridization and sequence comparisons. Some of these regions are also identified by probing mtDNA with labelled mtRNA due to the unavoidable contamination of these preparations with plastid and nuclear RNAs. These must be taken into account when trying to estimate coding capacities and the like of plant mitochondrial genomes.

5.7 Assignment of unknown reading frames

Many of the as yet unknown plant mitochondrial genes will undoubtedly be identified by computer homology searches of sequence libraries, but some also have no corresponding similarities to any of the known genes. Then only the tedious route of localizing the protein product with antibodies and the assignment of enzymatic and biological function will lead to identification. This has not yet been done and is beyond the scope of this text.

5.8 tRNA analysis

Several tRNAs have been identified in plant mitochondria more or less accidently during sequence analyses of genomic regions. Specific investigations of tRNAs have been carried out and described in detail by Dr J.H.Weil's laboratory and have led to the identification of a number of tRNA species. The problems associated with plastid tRNA sequences transcribed in plant mito-chondria have been mentioned in Section 3.3.3.

6. IDENTIFICATION OF LESIONS IN ABERRANT ORGANELLES

The cytoplasmic male sterility (cms) trait is caused by sequence rearrangements and mutations in the mtDNA. Several examples have so far been analysed, but we will need more detailed descriptions of the relationships between specific lesions in the mtDNA and the cms phenotype.

The economic importance of this trait in plant breeding and hybrid seed production has created considerable interest in this field. Many investigations however have been restricted to restriction fragment comparisons of normal and sterile mtDNAs. The varying success of such polymorphisms in the identification of mtDNA lineages is due to the high variability of the mtDNA arrangement in different nuclear backgrounds and the transitions into or out of tissue culture. We therefore emphasize the need to look in detail at all levels of mitochondrial genome expression (transcription and translation) to establish the cause–effect relationship and to allow the actually responsible mutant allele to be screened for.

6.1 **Restriction fragment comparison**

Only the rarest events of the many sequence rearrangements in plant mtDNA will actually be connected with the failure to develop mature pollen and it is a lottery against high odds to identify the specific restriction fragment connected with cms from the numerous polymorphisms observed in cms versus fertile mtDNAs. The cms trait may be lost without any concomitant change in the restriction pattern, which is therefore not enough to predict cms in new breeds of cms mitochondria and different nuclear genomes. Nevertheless, restriction fragment comparison of mtDNA from normal, cms and cms-restored plants is an essential part of the investigation.

6.2 **Transcript analysis**

Genomic sequence rearrangements leading to mutant polypeptides that are involved in the cms trait will generally also produce different transcripts that need to be analysed to establish the link between genome and phenotype. The second step in the analysis is therefore the comparison of specific transcripts again between normal, cms and cms-restored lines.

6.3 *In organello* **protein synthesis**

The *in organello* protein synthesis method has been found to be of extreme usefulness in the identification of mutant polypeptides in cms mitochondria and must be included in all such analyses of the molecular bases of the cms phenotype. Quantitative and/or qualitative changes of the protein pattern synthesized in mitochondria isolated from cms, normal and cms-restored lines will give an indication for the polypeptide involved in the specific type of cms. *Figure 7* shows the aberrant 13 kd protein synthesized in maize cms-T mitochondria.

7. ANALYSIS OF TRANSCRIPTION SIGNALS

7.1 **Transcript termini analysis**

mtRNA isolated as described above is sufficiently pure for nuclease protection and primer extension experiments by standard methodology. The hybridization temperatures must of course be individually adjusted, necessitating sequence analysis beforehand. The conserved G+C content of higher plant mitochondria does not allow universal hybridization conditions due to the differential distribution of nucleotides.

The identification of rRNA precursors depends on the rate of processing which might vary between species and may thus be feasible in one but not in another plant. 3' transcript termini can be defined by cDNA analysis and nuclease protection experiments.

7.2 *In vitro* capping

Plant mitochondrial transcripts are not capped *in vivo* and the termini of primary transcripts might be identified by adding the cap structure *in vitro* and analysing the termini. This method has recently been established and shown to work reliably in several species (39).

7.3 *In vitro* transcription

No *in vitro* transcription or translation system like those prepared from chloroplasts has yet been established for plant mitochondria. All mitochondrial extracts contain high activities of nucleases capable of rapidly degrading the added templates. The purification of nucleic acid synthesizing enzymes away from the nucleases will be the major problem in these investigations. Studies of specific nucleic acid modifying enzymes have begun and promising approaches have been described.

8. ACKNOWLEDGEMENTS

We are grateful to Drs P.G.Issac and C.J.Leaver FRS for unpublished material, the photograph of the protein gel and many discussions throughout. This work was supported by generous grants from the Deutsche Forschungsgemeinschaft and a Heisenberg Fellowship to A.B.

9. REFERENCES

1. The authors of Section V. Mitochondria (1986) in *Methods in Enzymology*, Colowick,S.P. and Kaplan,N.O. (eds), Academic Press, Orlando, FL, Vol. 118, p. 437.
2. Lonsdale,D.M. (1987) In *The Biochemistry of Plants*. Springer-Verlag, Berlin, in press.
3. Leaver,C.J., Hack,E. and Forde,B.G. (1983) In *Methods in Enzymology*, Fleischer,S. and Fleischer,B. (eds), Academic Press, Orlando, FL, Vol. 97, p. 476.
4. Pring,D.R. and Lonsdale,D.M. (1985) *Int. Rev. Cytol.*, **97**, 1.
5. Charbonnier,L., Primard,C., Leroy,P. and Chupeau,Y. (1987) *Plant Mol. Biol. Rep.*, **4**, 213.
6. Guillemaut,P., Burkard,G. and Weil,J.H. (1972) *Phytochemistry*, **11**, 2217.
7. Finnegan,P.M. and Brown,G.G. (1986) *Proc. Natl. Acad. Sci. USA*, **83**, 5175.
8. Dewey,R.E., Timothy,D.H. and Levings,C.S.III (1987) *Proc. Natl. Acad. Sci. USA*, **84**, 5374.
9. Christophe,L., Tarrago-Litvak,L., Castroviejo,M. and Litvak,S. (1981) *Plant Sci. Lett.*, **21**, 181.
10. Ricard,B., Eccheverria,M., Christophe,L. and Litvak,S. (1983) *Plant Mol. Biol.*, **2**, 167.
11. Hames,B.D. and Rickwood,D. (eds) (1987) *Gel Electrophoresis of Proteins—A Practical Approach*. IRL Press Ltd, Oxford.
12. Makaroff,C.A. and Palmer,J.D. (1987) *Nucleic Acids Res.*, **15**, 5141.
13. Bedinger,P. and Walbot,V. (1986) *Curr. Genet.*, **10**, 631.
14. Small,I.D., Isaac,P.G. and Leaver,C.J. (1987) *EMBO J.*, **6**, 865.
15. Spencer,D.F., Bonen,L. and Gray,M.W. (1981) *Biochemistry*, **20**, 4022.
16. Chao,S., Sederoff,R.R. and Levings,C.S.III (1983) *Plant Physiol.*, **71**, 190.
17. Spencer,D.F., Schnare,M.N. and Gray,M.W. (1984) *Proc. Natl. Acad. Sci. USA*, **81**, 493.
18. Chao,S., Sederoff,R.R. and Levings,C.S.III (1984) *Nucleic Acids Res.*, **12**, 6629.
19. Dale,R.M.K., Mendu,N., Ginsburg,H. and Kridl,J.C. (1984) *Plasmid*, **11**, 141.
20. Isaac,P., Jones,V.P. and Leaver,C.J. (1985) *EMBO J.*, **4**, 1617.
21. Fox,T.D. and Leaver,C.J. (1981) *Cell*, **26**, 315.
22. Hiesel,R., Schobel,W., Schuster,W. and Brennicke,A. (1987) *EMBO J.*, **6**, 29.
23. Dewey,R.E., Levings,C.S.III and Timothy,D.H. (1985) *Plant Physiol.*, **79**, 914.
24. Dewey,R.E., Schuster,A.M., Levings,C.S.III and Timothy,D.H. (1985) *Proc. Natl. Acad. Sci. USA*, **82**, 1015.

25. Isaac,P.G., Brennicke,A., Dunbar,S. and Leaver,C.J. (1985) *Curr. Genet.*, **10**, 321.
26. Dawson,A.J., Jones,V.P. and Leaver,C.J. (1984) *EMBO J.*, **3**, 2107.
27. Stern,D.B., Bang,A.G. and Thompson,W.F. (1986) *Curr. Genet.*, **10**, 857.
28. Wissinger,B., Hiesel,R., Schuster,W. and Brennicke,A. (1988) *Mol. Gen. Genet.*, **212**, 56.
29. Bland,M.M., Levings,C.S.III and Matzinger,D.F. (1986) *Mol. Gen. Genet.*, **204**, 8.
30. Dewey,R.E., Levings,C.S.III and Timothy,D.H. (1986) *Cell*, **44**, 439.
31. Schuster,W. and Brennicke,A. (1987) *EMBO J.*, **6**, 2857.
32. Stern,D.B., Hodge,T.P. and Lonsdale,D.M. (1984) *Plant Mol. Biol.*, **3**, 355.
33. Hiesel,R. and Brennicke,A. (1987) *Plant Sci.*, **51**, 225.
34. Gubler,U. and Hoffmann,B.I. (1983) *Gene*, **25**, 263.
35. Deng,G. and Wu,R. (1981) *Nucleic Acids Res.*, **9**, 4173.
36. Maniatis,T., Fritsch,E.F. and Sambrook,J. (1982) *Molecular Cloning: A Laboratory Manual.* Cold Spring Harbor Press, Cold Spring Harbor, New York.
37. Gualberto,J.M., Wintz,H., Weil,J.-H. and Grienenberger,J.-M. (1988) *Mol. Gen. Genet.*, in press.
38. Wahleithner,J.A. and Wolstenholme,D.R. (1988) *Nucleic Acids Res.*, **16**, 6897.
39. Mulligan,R.M., Maloney,A.P. and Walbot,V. (1988) *Mol. Gen. Genet.*, **211**, 373.

CHAPTER 5

Subcellular localization of macromolecules by microscopy

CHRIS HAWES

1. INTRODUCTION

Plant scientists are confronted with a wealth of techniques which they can apply at the subcellular level, to obtain data on the distribution of the macromolecular components of cells. Apart from the well established cytochemical and autoradiographic techniques, the last decade has seen the emergence of the powerful new technologies of immunocytochemistry and *in situ* hybridization, the full potential of which is only now being realized and exploited in the study of plant cells. To document in one chapter all of the procedures applicable to plant cells is a daunting task and as such I have concentrated on the more recently developed techniques, with only a brief mention of enzyme cyto- chemistry and autoradiography. For further details on these latter methods the reader is referred to the texts cited.

2. ENZYME CYTOCHEMISTRY

Enzyme cytochemistry is used to locate specific enzymes at their site of action within cells. This is normally carried out by the formation of an insoluble reaction product, which can be located in sections by light microscopy or electron microscopy (EM). Two basic procedures are employed depending on the substrates used:

(i)   the enzyme/substrate interaction results in an insoluble reaction product as in the localization of peroxidases (see Section 2.1);
(ii)   the enzyme/substrate interaction produces a primary soluble reaction product, which must be simultaneously captured by a trapping reagent, to give a final insoluble product, as in the lead salt procedure for phosphatases (1,2).

Whilst the use of immunolabelling techniques has in many respects super- seded the use of cytochemical techniques for the localization of enzymes in tissues, it should be remembered that the former only shows the presence or absence of the particular enzyme, whilst the latter gives information on the activity of the enzyme at a particular location in the cell or tissue.

2.1 **Localization of peroxidase**

The methodology for the various enzyme cytochemical techniques is well documented elsewhere (1–3). However, the localization of peroxidase will be described, as it is often the enzyme of choice for visualizing enzyme-conjugated antibodies in immunocytochemical procedures (see Section 4.2.5) and is one of the few enzyme cytochemical procedures originally developed specifically for plant tissues. Peroxidases catalyse the reduction of hydrogen peroxide to water, the most popular electron donor being 3,3′-diaminobenzidine (DAB) which gives a brown reaction product that can be seen directly with the light microscope. Alternatively, the reaction product can be used to reduce osmium during a post fixation step for EM localization of the enzyme. This latter reaction gives an intense deposit of electron opaque osmium over sites of enzyme activity. The protocol for peroxidase localization is given in *Table 1* and an example of the localization of peroxidase on the surface of maize root protoplasts is shown in *Figure 1*.

Table 1. Localization of peroxidase activity.

1. Fix blocks or thick hand cut sections of tissue for 30 min, in a mixture of 1% glutaraldehyde and 1% paraformaldehyde in 0.1 M sodium cacodylate buffer, pH 6.9[a].
2. Wash twice for 10 min in cacodylate buffer.
3. Wash twice for 10 min in 50 mM Tris–HCl, pH 7.6.
4. Incubate in DAB medium[b]. Try a range of times from 10 to 60 min.
5. Wash twice for 10 min in 50 mM Tris–HCl, pH 7.6.
6. Post fix in 1% OsO_4 in sodium cacodylate buffer for 45 min.
7. Wash for 10 min in cacodylate buffer.
8. Wash twice for 10 min in distilled water.
9. Dehydrate in a water/ethanol series, 10%, 20%, 30%, 50%, 70%, 90%, absolute ethanol, absolute ethanol (stored over drying agent such as a molecular sieve) for 10 min each change.
10. Infiltrate with an alcohol/Spurr resin mixture (3:1) for 30 min[c].
11. Infiltrate with alcohol:Spurr resin (1:1) for 30 min.
12. Infiltrate with alcohol:Spurr resin (1:3) for 30 min.
13. Infiltrate with Spurr resin for 1 h.
14. Infiltrate with Spurr resin for 12 h.
15. Infiltrate with fresh resin and polymerize for 9 h at 70°C in suitable moulds.

Controls: (1) Omission of H_2O_2 (beware of reaction from endogenous tissue peroxide). (2) Heat deactivate at 95°C for 5–15 min. (3) Inhibit enzyme activity with 10–100 mM KCN in the incubation medium.

[a] Peroxidase activity is in general reasonably resistant to fixation.
[b] Incubation medium is 10 ml of 50 mM Tris–HCl buffer, pH 7.6, 5 mg of DAB. Warm the DAB solution to aid solubility, cool, filter and add 0.1 ml of H_2O_2 [dilute 30% (100 vols) stock solution] just prior to use.
[c] Spurr resin is routinely used in the author's laboratory as it is of low viscosity and infiltrates hard tissue readily. However, it is toxic and should only be handled with extreme care using gloves and in a fume cupboard. Other resins such as Epon and the Epon substitutes can be used with modifications to the infiltration times. The infiltration times given will change depending on the nature of the tissue being embedded.

3. AUTORADIOGRAPHY

Autoradiography localizes radioactive molecules within a section of tissue by the exposure of a layer of photographic emulsion placed adjacent to it. During the exposure, radiation emitted from these radioactive molecules will form a latent image of silver halide crystals in the emulsion. When the emulsion is developed, a pattern of silver grains will be left on the section, which reflect the original distribution of the radioactive probe. Success with autoradiography depends on careful choice of the precursor to be radiolabelled, so that it is only incorporated into the macromolecule being studied and that it is not broken down or re-distributed during the course of the experiment. As amino acids, sugars and nucleotides can be radiolabelled, a wide range of macromolecules can be studied. However, for investigations into the distribution of specific macro-molecules, this technique is rapidly being overtaken by the use of antibody probes (see Section 4) which give faster results at a higher resolution. At the EM level, autoradiographic emulsions can take many weeks to expose and, as the relationship of the developed silver grain to the source depends on the energy of the particular isotope and thickness of the emulsion layer, great care has to be taken in the analysis and interpretation of results. However, the technique is still invaluable for studying metabolic events by microscopy, especially when pulse–chase experiments are involved as, for example, in the investigation of the incorporation of Golgi-packaged polysaccharides into the cell wall.

For full details of the various autoradiographic procedures see refs 4 and 5.

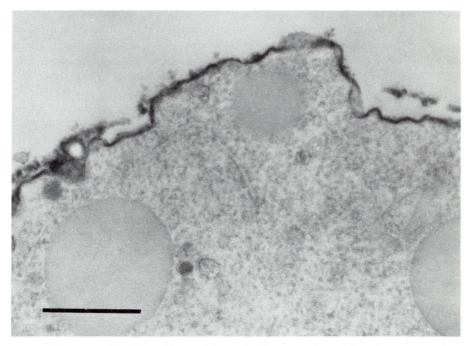

Figure 1. Demonstration of peroxidase activity on the plasma membrane of a maize root protoplast by the DAB reaction. Thin section electron micrograph. Bar = 1 μm.

4. IMMUNOCYTOCHEMISTRY

4.1 Basic principles

The use of antibodies as direct probes for specific macromolecules in cells and tissues is now a well established and extremely powerful technique. There are numerous excellent texts on the uses of antibodies in cell biology and on immunocytochemistry (6–10) and the reader is recommended to consult these before undertaking immunocytochemical projects. However, the majority of texts relate to the use of the antibodies in animal cell biology and biomedicine. In this section I have attempted to select procedures and techniques that will enable the plant scientist to apply this technology to their particular problem.

4.1.1 *Raising antibodies: polyclonals versus monoclonals*

It is not the intention in this chapter to discuss in any detail the raising of antibodies, as this is well documented elsewhere (10–13). In plant immunocyto-chemical studies, both polyclonal and monoclonal antibodies have been extensively used and both forms have their merits. When producing polyclonal antibodies, it is important that the antigen is as pure as possible. In many cases this will entail considerable effort, especially if it is not exceptionally abundant in the cell or tissue. However, as a polyclonal serum is in effect a mixed family of antibodies to a range of epitopes on the antigen, the likelihood of retaining cross-reactivity after the various preparative procedures for microscopy is high. Monoclonals have the advantage that the titre of immunizing antigen can be very low and the preparation need not be pure. A family of antibodies is then raised which can be screened for cross-reactivity against the antigens of interest, both biochemically and cytochemically. This has the added advantage that clones can be selected that are not affected by the preparative techniques for microscopy. However, it is also possible that monoclonal antibodies may be less specific, if the epitope on the antigen which they recognize also occurs on other macromolecules in the cell. This has been shown to be a problem with plant cells where many hybridomas, obtained from mice immunized with plant extracts, secreted antibodies to arabinose and galactose (14). As these two sugars are found in most plant extracts and as residues on glycoproteins, there can be problems with the specificity of monoclonal antibodies to plant macromolecules. It is therefore advisable to deglycosylate protein extracts before immunization.

It is also worth noting that many antibodies raised against animal cell proteins show good cross-reactivity with plant cells and much important information has been obtained this way. For example, *Figure 2* shows fluorescence of coated vesicles in a carrot suspension culture cell after staining with an anti-clathrin antibody raised against bovine brain clathrin.

4.1.2 *Basic procedure for immunolabelling plant tissue*

The basic protocol for immunolabelling of antigens in plant tissue is given below. This scheme should be modified with regard to the tissue type, the antigen/antibodies used and the method of visualizing the antibody/antigen complex by either light or electron microscopy.

(i)   If sufficient quantities of the antigen can be obtained, check the cross-reactivity of the antibody with the antigen. This can be done simply by dot blotting, micro-ELISA or Western blotting. The latter would also permit specificity to be confirmed. If possible, also assess the effect of fixation on the antigen.

(ii)   Fix the tissue. The choice of fixatives and buffers will vary depending on whether light microscopy or EM observations are employed and will also vary with the sensitivity of the antigen to fixation.

(iii) Prepare the specimens for immunolabelling by exposing antigenic sites for labelling: i.e. permeabilize whole cells or cryosection for light microscopy, embed in resin for EM.

(iv)   Wash the specimens in buffer with blocking agents to occupy any non-specific protein binding sites.

(v)   Stain with the primary antibody raised against the antigen to be labelled. (For direct staining procedures this antibody is conjugated to the probe of choice, prior to the staining step).

(vi)   Wash the specimens in blocking buffer.

(vii) Stain the specimens with the secondary antibody raised to the primary immunoglobulin. This secondary antibody should be tagged with a suitable probe to allow visualization with the light or electron microscope.

(viii) Counterstain if necessary and mount the specimen for microscopy.

4.2 Light microscopy

It is stressed that the preparative schedules given here should be used only as guides and that the potential immunochemist will have to work out the most suitable preparation and staining procedure for the combination of tissue, antigen/antibody and probe to be used.

4.2.1 *Preparation of tissue*

There are three major factors to be taken into consideration when preparing tissue for the immunocytochemical localization of macromolecules:

(i)   The susceptibility of the antigen to fixation and the other preparative procedures;

(ii)   Permeabilization of the cell to the antibody and the attached probe (i.e. use of permeabilized whole cells or sectioning techniques);

(iii) Choice of probe for visualizing the antigenic sites.

(i) *Effect of fixative on antigenicity*. It is possible that many of the stages in the preparation of the tissue for microscopy will adversely affect the antigenicity of the macromolecule to be probed. Antigens are particularly susceptible to fixation and heat during polymerization of resins. It may therefore be the case that immunocytochemistry is restricted to unfixed razor-cut sections or cryostat sections of the particular tissue.

It is possible to assess the effect of the fixative used on the antigen, using an *in vitro* enzyme-linked immunosorbent assay (ELISA) (11). The wells of ELISA plates are coated with the antigen and then reacted with the fixatives of choice for set times. After thorough washing, the plates are cross-reacted with the primary antibody and the degree of cross-reactivity assayed, using peroxidase-conjugated secondary antibodies (see Sections 2.1 and 4.2.5).

Paraformaldehyde is the most popular fixative for light microscope immuno-cytochemistry and concentrations of 2–5% in a suitable buffer should be used. Add paraformaldehyde powder to distilled water to make up about half the final required volume, heat to 70°C and add drops of 1 M NaOH until all the powder has dissolved. Allow to cool and add concentrated buffer to give the final correct concentration of paraformaldehyde fixative and buffer. Glutaraldehyde should be used only when fluorescent probes are not employed (see Section 4.2.4). If possible use the same buffer as that in which the antibodies are made up. The commonest buffers used are Tris–HCl, phosphate-buffered saline (PBS) and Pipes (pH 6.9). Additions may be made to the buffers. For example in the localization of tubulin, a microtubule stabilizing buffer is used composed of 0.1 M Pipes, 1 mM $MgSO_4$, 2 mM EGTA.

(ii) *Permeabilization of cells.* Plant cells offer two main barriers to the entry of antibodies, the cell wall and the plasma membranes. The immunocytochemist is left with two options, either to section the specimen, thus exposing antigenic sites to the probes, or to enzymatically release whole cells from the tissue and permeabilize them prior to antibody staining. This latter method, known as the 'square protoplast technique', was originally developed for the study of the microtubule cytoskeleton (15,16), but has now been used for the subcellular localization of many different antigens.

(i)   Fix the tissue in 3% paraformaldehyde in a suitable buffer for the staining procedure.

(ii)   Thoroughly wash the tissue to remove all traces of fixative.

(iii)   Incubate the tissue in wall-digesting enzymes. The enzyme mixture will depend on the tissue being used. In the author's laboratory, the release of cells from root tips is achieved using a mixture of 2% 'Onozuka' R10 Cellulase (Yakult Biochemicals Co., Nishinomiya, Japan) and 1% pectinase in buffer for 20–30 min at 30°C.

(iv)   Gently transfer the tissue to slides [8-well multitest slides (Flow Laboratories) are convenient] or coverslips (acid or acetone cleaned) which have been previously coated with poly-L-lysine. [Use 1–3 mg poly-L-lysine (mol. wt ~90 000) dissolved in distilled water. Treat slide for 1 h, pipette off the poly-L-lysine (this can be re-used), wash briefly with distilled water and air dry.]

(v)   Tap the tissue with the rounded end of a glass rod into a drop of buffer, to release the cells. Allow the cells to settle. Remove the largest fragments of tissue.

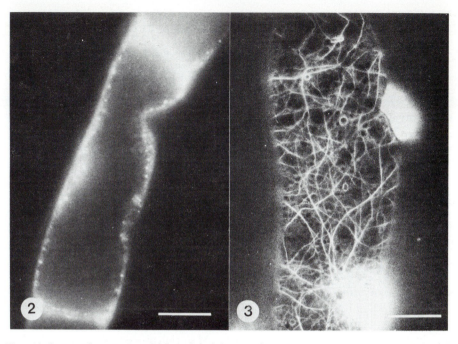

Figure 2. Immunofluorescence staining of clathrin-coated vesicles in a carrot suspension culture cell using a polyclonal antibody to bovine brain clathrin. Bar = 15 μm.

Figure 3. Fluorescence of the actin cytoskeleton in a carrot cell permeabilized using Triton X-100 and stained with rhodamine-labelled phalloidin. Bar = 15 μm.

(vi)   Permeabilize the plasma membranes. Three methods have been em-
       ployed with plant cells.
       (a) Treat the slides with pre-chilled methanol at −10°C for 6–8 min.
       (b) Air dry the slides.
       (c) Treat with 0.5–0.1% Triton X-100 in buffer for 1–5 min.
(vii) Wash the slides/coverslips with buffer and perform the immunostaining.

The above protocol can also be modified for use on naturally occurring single cells or cells from suspension culture (*Figure 2*).

4.2.2 *Sectioning techniques*

(i) *Cryosectioning*. For immunofluorescence microscopy, probably the best method for cutting sections and retaining good antigenicity is to use a cryostat or a cryoultramicrotome. In this way, any problems associated with dehydration and resin embedding are not encountered. Cryostats can produce sections up to 12 μm for immunostaining. Sections should be picked up onto glass slides that have been coated with gelatin or subbed as described in *Table 2*.

Sections up to 2 μm thick can easily be cut with a cryoultramicrotome. To prepare specimens after aldehyde fixation, the tissue is perfused with sucrose, increasing the concentration until it has reached 2 M. This acts both as a

Table 2. Subbing solution for coating glass slides.

1. Dissolve 2 g of gelatin in 400 ml of warm distilled water.
2. Add 0.5 g of chrome alum [CrK(SO$_4$)12H$_2$O] and then sodium azide to make a 0.02% solution.
3. Add four drops of Teepol or similar detergent.
4. Warm and filter the solution before using to coat the slides.

cryoprotectant and a support for the tissue during sectioning. Small blocks of specimen are then mounted on the cryo-stubs, plunged into liquid nitrogen and inserted into the chuck of the microtome. Ribbons of sections are picked up on a drop of 2 M sucrose held in a wire loop. This is then touched onto the surface of a poly-L-lysine-coated slide [see Section 4.2.1 (ii)] and immunolabelled. With highly vacuolate tissue, fragmentation of the cryosections is often a major problem. This can often be obviated by infusing the tissue with 10% (w/v) gelatin in the 2 M sucrose solution for 30 min at 37°C. *Figure 4A* shows the staining of microtubules in a cryosectioned maize root tip meristem with an antibody to yeast α-tubulin.

(ii) *PEG embedding*. If access is not available to cryostats or a cryoultramicrotome, the polyethylene glycol (PEG)-embedding technique often proves a satisfactory substitute (17,18). Material is fixed and embedded in high molecular weight PEG (*Table 3*) and sectioned dry. Material processed by this schedule can usually be cut at 1 μm. However, final embedding in a mixture of 80% PEG 4000 and 20% PEG 1500 can make thicker sectioning easier. Sections are then picked up on a sucrose drop as described in Section 4.2.2 (i) and immunostained. This technique is suitable for the detection of antigens that are not denatured by heat (PEG solutions have to be kept liquid at 60°C). Problems may be encountered when sectioning highly vacuolate tissues.

(iii) *Resin embedding*. It is possible to carry out immunofluorescence procedures on semi-thick (1–2 μm) resin sections. Prior to staining, in order to expose sufficient antigenic sites, either the surface of the section must be etched by removal of the resin, or the resin totally removed from the section (see *Table 4*). Resin can be removed from sections by dipping the slide in sodium methoxide or ethoxide made by adding 2.5 g of dry sodium piece by piece to 100 ml of methyl alcohol or ethanol. Other workers use 25 ml of alcohol and dilute the resultant solution with an equal volume of benzene.

4.2.3 *Antibody labelling*

It is now common practice to use the indirect antibody labelling technique (*Table 5*) as this has several advantages over the direct technique. Stocks of labelled secondary antibodies are readily obtainable commercially and can be stored for long periods in the freezer, and the level of staining is enhanced by multiple binding of the secondary antibody to the primary. It is possible to use Protein A (a cell wall protein from *Staphylococcus aureus* which binds to immunoglobulins) instead of the secondary antibody, but bear in mind that

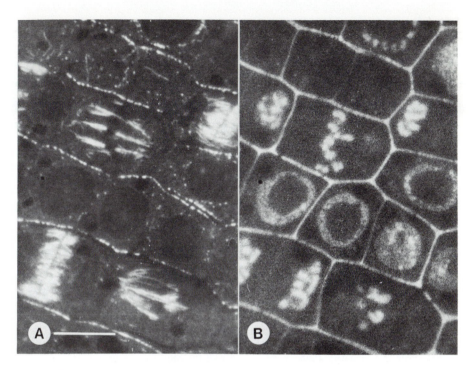

Figure 4. Cryo-section of a maize root tip meristem double-stained with (**A**) an α-tubulin antibody which localizes the cortical and spindle microtubules and (**B**) with Hoechst 33342 for DNA to locate the chromosomes. Bar = 10 μm.

Table 3. Schedule for PEG embedding.

1. Melt PEG (mol. wt 4000) in an oven or water bath at 60°C and prepare the following solutions with 100% ethanol: 20%, 50% and 70% (all v/v).
2. Fix the material and wash it in buffer, followed by distilled water.
3. Dehydrate the material in a water/ethanol series (see *Table 1*).
4. Bring all the PEG solutions to 60°C in a water bath.
5. Slowly add 20% PEG 4000 dropwise to the samples in absolute ethanol to bring to 10% PEG.
6. Infiltrate with 20, 50 and 70% PEG 4000 at 60°C for 30 min each solution.
7. Infiltrate with neat PEG 4000, two times for 1 h at 60°C.
8. Embed in neat PEG 4000. It is best to use silicon rubber moulds pre-heated to 60°C which will allow time for orientation of the specimen before the PEG polymerizes.

Protein A has a very weak cross-reaction with some mouse and rat monoclonal antibodies. This latter problem may be overcome by using an intermediate step with rabbit anti-mouse or anti-rat IgGs.

The concentration of the primary antibody to be used is best determined by trial and error as this will depend on the antigenicity of the tissue and the procedures used in preparation of the material. Prior to this step blocking reagents may have to be used to prevent non-specific binding of the antibodies.

111

Table 4. Immunostaining of resin sections.

1. Fix the tissue for immunostaining.
2. Dehydrate in water/ethanol series (see *Table 1*).
3. Embed in the resin of choice (see *Table 1*).
4. Cut thick sections (1–2 μm) and float these on a drop of distilled water on a glass slide (or in the wells of a multiwell test slide).
5. Dry the slides at 60°C.
6. Dip in sodium methoxide or ethoxide (the time will depend on the degree of resin etching required).
7. Wash in 90% ethanol followed by 75% ethanol, 5 min each.
8. Rinse with distilled water.
9. Wash with the buffer to be used in the immunostaining schedule.
10. Antibody label (see Section 4.2.3).

Table 5. Immunolabelling for light microscopy.

1. Prepare the material for immunolabelling by one of the techniques described in Sections 4.2.1 and 4.2.2.
2. Apply any necessary blocking reagents to prevent non-specific binding, e.g. 0.5 mg/ml sodium borohydride if glutaraldehyde fixation is used.
3. Incubate with the primary antibody (e.g. antigen raised in rabbit). Try a range of concentrations and times, e.g. 0.1–1.0 μg/ml in buffer for 10 min to 2 h. Into this mixture can be added more blocking agents such as BSA (10 mg/ml) and non-immune serum (6%).
4. Wash in buffer containing 10 mg/ml BSA, three times for 10 min.
5. Stain with the secondary antibody conjugated with the chosen probe (e.g. FITC sheep anti-rabbit antibody). For most commercial antibodies a dilution of 1:40 in buffer for 30–60 min will suffice.
6. Wash in buffer, three times for 10 min.
7. Counterstain if necessary or stain with another antibody to a different antigen.
8. Mount for microscopy in a mountant which retards quenching of the fluorescence (see Section 4.2.4).

Controls: It is important in all immunocytochemical experiments that adequate controls are performed. These should be devised for each individual experiment and will depend on the nature of the antigens or antibodies used. A few examples are given.

1. Stain an equivalent tissue which is known not to contain the antigen.
2. Remove the antigen from the tissue prior to staining, e.g. for tubulin treat with anti-tubulin drugs.
3. Omit primary antibody from the staining procedure.
4. Use non-immune serum from the same species in which the primary antibody was raised, instead of the primary antibody.
5. For monoclonal antibodies replace the primary antibody with:
   (a) a monoclonal antibody raised against an antigen not found in the tissue to be stained (e.g. an antibody raised against an animal protein) and
   (b) unused tissue culture medium.

If glutaraldehyde has been used as a fixative, reactive aldehyde groups can be blocked with sodium borohydride (0.5 mg/ml in the buffer) for about 10 min. Non-specific protein binding can be blocked with bovine serum albumin (BSA) (10 mg/ml) and by treatment with non-immune serum.

4.2.4 *Fluorescence microscopy*

(i) *Choice of fluorochrome.* By far the most popular method of visualizing antibodies at the light microscope level is to conjugate the primary (direct staining) or secondary (indirect staining) antibody to a fluorochrome which will fluoresce when excited with light of a specific wavelength. The most popular fluorochromes used in immunocytochemistry are:

(i)   fluorescein isothiocyanate (FITC) with an optimal excitation wavelength of 480 nm (blue) and an emission peak at 525 nm (green) and

(ii) tetramethylrhodamine isothiocyanate (TRITC) with optimal excitation at around 550 nm (green) and emission at 580 nm (red).

Other probes are available, such as Texas Red (Molecular Probes) with similar characteristics to TRITC. This latter compound shows brighter fluorescence than rhodamine derivatives and fades less rapidly on excitation. As FITC and TRITC have non-overlapping excitation and emission spectra, they can be used in the same preparation, tagged to different antibodies, to stain different antigens and can be used alongside other markers that are excited by wavelengths in the UV end of the spectrum such as Calcofluor white and DAPI (see Sections 6.2 and 6.3).

With plant cells, the choice of fluorochrome may be dictated by autofluorescence of the tissue. This often occurs in cell walls, lignified tissue and in tissue rich in polyphenols. The red fluorescence of chlorophyll, will also preclude the use of rhodamine probes for labelling chloroplast antigens. It has also been noted that the use of glutaraldehyde greatly increases the level of autofluorescence in plant cells to an extent which may mask the immunofluorescence. If at all possible, this fixative should not be used for immunofluorescence procedures.

Unless very intense staining occurs, it is not advisable to use transmitted-light fluorescence in immunocytochemistry. If at all possible incident-light fluorescence should be used with an epifluorescence condenser, in conjunction with dichroic beam splitting mirrors and narrow waveband filters reflecting the spectral characteristics of the fluorescent probes used. This system has the great advantage that with increased magnification of the objective lens, combined with increased numerical aperture, the intensity of the incident excitation beam will greatly increase, as will the intensity of the fluorescence.

(ii) *Choice of photographic film.* The major problem experienced during the photographic recording of fluorescing specimens is the quenching of the fluorescence over the time required for satisfactory exposure. This quenching can be retarded by mounting the specimens in a glycerol–PBS mixture (8:2 v/v) containing *para*-phenylene-diamine (1–10 mg/ml) or 5% *n*-propyl gallate. Proprietary mountants such as Mowiol with added *n*-propyl gallate or Citifluor (Citifluor Ltd, City University, London, UK) can also be used. However, it is still often necessary to expose for considerable periods of time (30–120 sec) and a reasonably fast film is recommended. For black and white photography, try

Ilford HP5, Kodak Tmax or Kodak Technical Pan film, the latter rated at 200 ASA. For colour microscopy a high speed slide film such as Ektachrome 400 or Fuji 400 is recommended.

4.2.5 *Peroxidase labelling*

It is possible to use an enzyme such as peroxidase as a marker for antibody staining. If peroxidase is complexed to an anti-peroxidase antibody, this can be bound to the tissue antigen/antibody complex via a bridging antibody (e.g. use a rabbit anti-antigen, bridge with a porcine anti-rabbit and stain with peroxidase rabbit anti-peroxidase). This whole complex can then be rendered visible by standard enzyme cytochemical procedures for peroxidase localization, at either the light microscope level with DAB and hydrogen peroxide, or at the EM level by the reduction of osmium by the oxidation product of the enzyme substrate (see Section 2.1). The peroxidase anti-peroxidase complex (PAP) can be also obtained commercially, thus reducing the number of steps in the staining procedure (6).

This above procedure is considered to be more sensitive than using a covalently-linked horseradish peroxidase/antibody complex as an indirect label. It must also be remembered that many plant tissues have a naturally high level of peroxidase activity, which should be taken into account (or blocked) and careful control experiments must be performed.

4.2.6 *Gold labelling for light microscopy*

The immunogold technology developed for the localization of antigens by electron microscopy (see Section 4.3.4) can be used at the light microscope level. Procedures are as for immunofluorescence labelling, but the secondary antibody or protein A is conjugated to colloidal gold particles which should be 20 nm or larger. As colloidal gold in solution is deep red in colour, localization of the antigen will be seen by red staining (19). This technique is restricted in use to cryo-sections, PEG sections or wall-free cells, such as protoplasts or some endosperm cells, where heavy labelling of the antigen and entry of the gold sol into the cell is possible (20). To visualize gold probes at the light microscope level on resin sections, it is necessary to silver-enhance the gold particles.

4.2.7 *Silver enhancement*

Although it is possible to visualize gold probes directly at the light microscope level as discussed in Section 4.2.6, silver enhancement procedures can be used to significantly increase the signal from the gold probes (21,22). Up to a thousand-fold increase in signal can be obtained in 10 min. In this procedure the metallic gold on the specimen is used to catalyse the reduction of silver lactate in the presence of the reducing agent hydroquinone. This results in a shell of metallic silver building up around each individual gold particle. It is beneficial in this procedure to use small (5 nm) gold probes as this gives a more intense silver staining due to the higher number of gold particles labelling the antigen.

Janssen Life Science Products also market a silver enhancement kit (IntenSE11) that allows the silver development to be carried out in daylight. This has the advantage that the degree of silver enhancement can be monitored with the light microscope during the development process and interrupted by fixation when necessary.

Figure 5A shows an example of the localization at the light microscope level of nodule-specific uricase in soybean root nodules prepared as described in *Table 6* (22). The uricase is located in peroxisomes which can easily be detected by silver enhancement of sections which were stained with an anti-uricase serum and gold labelled.

4.3 **EM immunocytochemistry**

Two different approaches have been developed for immunocytochemistry at the electron microscope level. With pre-embedding labelling, the antigen is labelled with the antibody and probe before the tissue is embedded in a resin and sectioned, whilst with post-embedding labelling, the tissue is processed and thin sectioned prior to antibody labelling. The former technique is useful for the study of cell surface antigens and antigens in cells which can be permeabilized to both the antibody and probe, as for example in the localization of cytoskeletal proteins in detergent-extracted animal cells. Unfortunately, due to the presence of the cell wall and the subsequent difficulty in exposing the cell surface or permeabilizing the cell, the great majority of immunocytochemical investigations with plant cells have used post-embedding labelling techniques. The protocols discussed here will relate to the latter procedure. Pre-embedding techniques are restricted to the study of the surface of the plasma membrane in protoplasts, or to the surface of the cell wall itself.

As with light microscope immunocytochemistry, the ultimate choice of preparative protocol for the tissue being studied will depend on a variety of factors, including the effect of fixation and embedding procedures on tissue antigenicity and the quality of ultrastructural preservation that is acceptable. For instance, paraformaldehyde fixation may give a higher retention of antigenicity than glutaraldehyde fixation, but the quality of preservation at the ultrastructural level will be compromised. It may also be acceptable to lose most of the antigenicity of the protein being studied and still achieve a high enough level of labelling to give satisfactory results as, for example, in the localization of highly concentrated storage proteins in cotyledons.

4.3.1 *Fixation*

The ideal fixative should preserve the antigenicity of the tissue, immobilize the antigen and prevent its extraction during subsequent processing, as well as giving a satisfactory preservation of cytoplasmic ultrastructure. This is an almost impossible goal to achieve and compromises have to be accepted. The range of fixatives and buffers used in the various immunocytochemical procedures developed for plant cells is far too large to be discussed in detail here. *Figure 5B* shows the localization of nodule specific uricase in soybean root nodules and

115

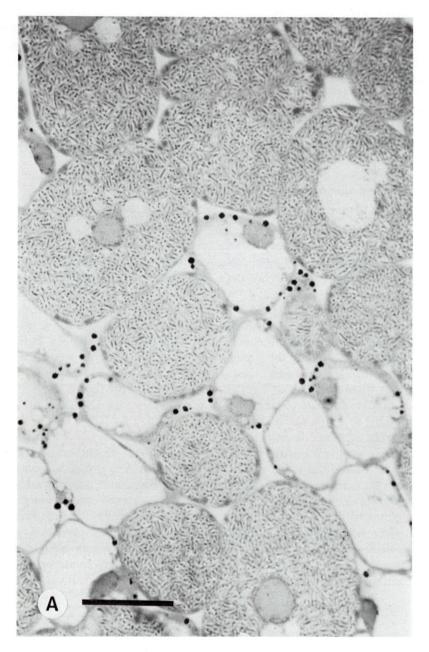

Figure 5. Immunocytochemical localization of nodule-specific uricase in Spurr resin sections of soybean root nodules. (**A**) Light micrograph of a silver-enhanced section labelled with 5 nm protein A-gold. Peroxisomes containing uricase can be seen as black spheres in the uninfected cells. Bar = 20 μm.

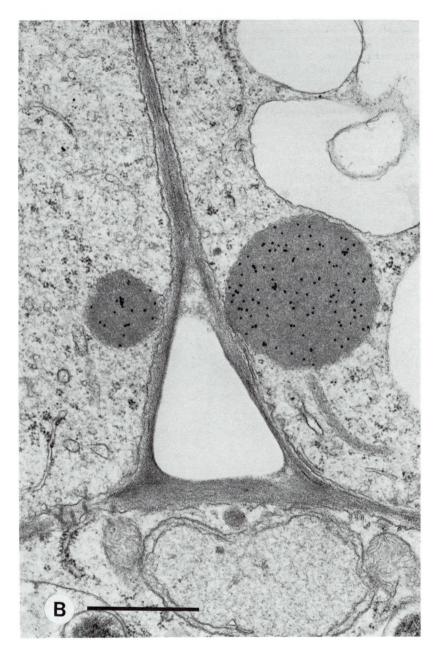

(**B**) Electron micrograph showing labelling over peroxisomes with 20 nm protein A-gold. Material was fixed in glutaraldehyde and osmicated. Sections were periodate etched prior to staining with primary anti-uricase at 60 μg/ml in Tris–HCl, pH 7.2. Bar = 1 μm.

Table 6. Silver enhancement of gold-labelled sections.

---

1. Aldehyde-fix the specimens by whichever technique you have devised for optimal retention of antigenicity for immunogold labelling.
2. Dehydrate the specimens in water/ethanol series without osmium post-fixation (see *Table 1*).
3. Embed in a suitable resin (e.g. LR White at 50–55°C for 48 h).
4. Cut semi-thin (0.5–1.0 μm) sections with an ultramicrotome.
5. Mount the sections on chrome alum/gelatin-coated slides (see *Table 2*).
6. Incubate the sections in primary antibody, including any necessary blocking procedures.
7. Label the primary antibody with 5 nm gold-conjugated secondary antibody[a].
8. Immerse the slides in silver developer[b] for 5–10 min under photographic safe-light conditions.
9. Stop the development by fixing for 3 min in fixer (e.g. Janssen-fixing solution).
10. Wash the slides for 10 min in running tap water.
11. Counterstain the sections with 1% Azure 11, 1% methylene blue in 1% sodium borate.
12. Mount for microscopy.

---

[a] Use 5 nm light microscopy grade gold-conjugated antibodies available from Janssen Life Science Products.
[b] Silver developer contains 77 mM hydroquinone and 5.5 mM silver lactate in 200 mM citrate buffer, pH 3.85.

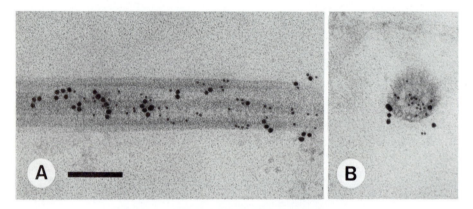

Figure 6. Immunogold staining of Lowicryl K4M sections of *Chlamydomonas* flagella with 5 nm gold detecting β-tubulin and 20 nm gold detecting α-tubulin. (**A**) Longitudinal and (**B**) transverse sections. Bar = 300 nm.

Figure 6 the staining of tubulin in flagella of *Chlamydomonas*. *Figure 7* shows the immunogold localization of serine glyoxylate amino transferase in pumpkin cotyledons. All these specimens were prepared with different combinations of fixatives and buffers (see figure legends). In general 2–3% glutaraldehyde (with or without paraformaldehyde) in a suitable buffer is used as the primary fixative. If osmium post-fixation is employed, it is necessary to unmask antigenic sites by treatment of the sections with periodate (0.56 mM aqueous sodium meta-periodate for 1 h) (22,23) before satisfactory immunolabelling can be achieved.

4.3.2 Resin embedding

A wide variety of resins have been used in immunocytochemical studies of plant antigens, ranging from the commonly used Epon and Spurr epoxy resins (*Figure 5*)

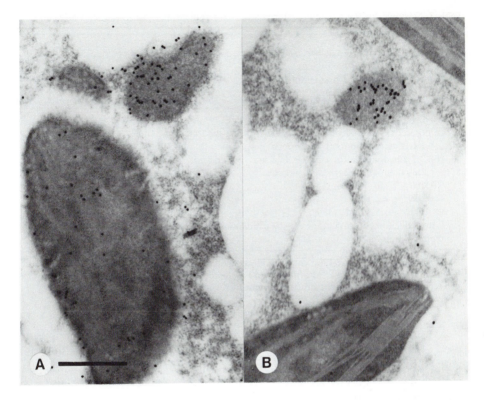

Figure 7. Immunogold labelling of SGAT in LR White section on pumpkin cotyledons. Sections were reacted with a rabbit anti-SGAT IgG followed by 15 nm gold-labelled anti-rabbit IgG. Buffer containing just 1% (w/v) BSA gives heavy background labelling (**A**) whilst addition of 0.5 M NaCl and 0.1% (w/v) Tween-20 eliminates most of the non-specific labelling (**B**). For complete protocol see *Table 8*. Bar = 500 nm.

to the specialist low temperature methacrylates developed specifically for immunocytochemistry such as Lowicryl K4M (see *Table 7* and *Figure 6*). In general, antigenicity is best preserved at lower temperatures, but use of low temperature resins precludes the use of osmium as a fixative. Thus the quality of cell ultrastructure is often poor, with membranes badly preserved or seen in negative contrast. The acrylic resin LR White (London Resin Company) may also be used at low temperatures, in a fashion similar to Lowicryl and polymerized with the use of a catalyst instead of external heat (24). The heating of the specimen by the exothermic reaction of the catalyst with the resin is minimized by cooling the unpolymerized resin to −80°C immediately after adding the catalyst and raising the blocks to room temperature over a period of 6 h in an atmosphere of nitrogen. In one study of tubulin labelling in sectioned flagella (*Figure 6*) it was found that low temperature UV curing of LR White resulted in a 3-fold increase in gold labelling over heat-cured blocks and Lowicryl K4M embedding resulted in a 3-fold increase in labelling over any other resin (Wells and Roberts, personal communication).

119

Table 7. Schedule for the low-temperature embedding of plant material in Lowicryl K4M[a].

---

1. Fix the material.
2. Dehydrate the material at low temperature[a] using the following series of steps:
    (a) 30% ethanol for 1 h on ice;
    (b) 50% ethanol for 1 h at −20°C;
    (c) 70% ethanol for 1 h at −35°C;
    (d) 95% ethanol for 1 h at −35°C;
    (e) two changes 100% ethanol for 1 h at −35°C.
3. Infiltrate the material with Lowicryl K4M using the following series:
    (a) 1:1 ethanol/resin for 1 h at −35°C.
    (b) 1:2 ethanol/resin for 1 h at −35°C.
    (c) 1:3 ethanol/resin for 1 h at −35°C.
    (d) 100% resin 1 h at −35°C.
    (e) 100% resin overnight at −35°C.
    (f) 100% resin 8 h at −35°C.
4. Polymerize for 24 h under indirect UV irradiation at −40°C followed by 48 h under UV at room temperature.

---

[a]Dehydration can be carried out in freezers and polymerization can be carried out in a freezer modified to carry a UV light. Alternatively a low temperature tissue handling device (25) for the whole procedure can be purchased from Thames Cryogenic Ltd, Oxford, UK.

As a general rule, if low temperature embedding techniques are not being used, the best tissue ultrastructure is obtained with the epoxy resins, but antigenicity is best preserved with acrylic resins such as LR White (22,23). The embedding procedures for epoxy resins are the same as would be used for conventional electron microscopy of the particular tissue (see also *Table 1*). The protocol for low temperature embedding in Lowicryl K4M is given in *Table 7* and embedding in LR White is described in *Table 8*.

4.3.3 *Antibody labelling*

The basic procedures and controls for antibody labelling for EM are the same as given for light microscopy immunocytochemistry (see Section 4.2). The indirect procedure is most often used, the secondary antibody (or protein A) being conjugated to an electron opaque marker or an enzyme.

4.3.4 *Choice of label*

Colloidal gold is the label of choice for the majority of immunocytochemists, although ferritin- and peroxidase-labelled secondary antibodies (including the PAP technique) can be used. The latter are visualized at the EM level by the DAB reaction and subsequent reduction of osmium (see Section 2.1). The use of colloidal gold particles conjugated to antibodies or protein A has several marked advantages over the other probes. Due to their high atomic number, gold particles are extremely electron opaque, they can easily be conjugated to proteins and can be manufactured in different sizes. This latter property permits double or even triple labelling of different antigens on a single section (*Figure 6*). The most commonly used sizes of colloidal gold in EM immunocytochemistry

Table 8. Complete schedule for the immunolocalization of SGAT in pumpkin cotyledons.

1. Cut the cotyledons into 1 mm^3 blocks under fixative [1% (w/v) paraformaldehyde, 1% (v/v) glutaraldehyde, 2% (w/v) sucrose, 0.05% (w/v) $CaCl_2$ in 0.1 M Pipes buffer at pH 6.9].
2. Fix for 10–15 min under vacuum until the blocks cease floating.
3. Fix for a further 45 min at atmospheric pressure.
4. Wash twice for 10 min with Pipes buffer.
5. Wash twice for 10 min with distilled water.
6. Dehydrate in a water/ethanol series 10%, 20%, 30%, 50%, 70%, 90%, absolute ethanol for 10 min in each concentration.
7. Transfer to a 2:1 (v/v) mixture of LR White:absolute ethanol in an unstoppered vial for 1 h.
8. Infiltrate with pure LR White for 1 h followed by 12 h in fresh resin. Rotate or agitate gently during this period.
9. Embed in gelatin capsules tightly capped and polymerize at 50°C for 24 h, or in open moulds under a nitrogen atmosphere.
10. Cut ultrathin sections and pick up onto 200 mesh nickel or gold grids.
11. Float the grids on droplets of 20 mM Tris–HCl buffer, pH 8.2 with 0.5 M NaCl, 0.1% (w/v) Tween-20, 1% (w/v) globulin-free BSA three times for 5 min to block non-specific staining[a,b].
12. Incubate for 1 h in primary antibody (rabbit anti-SGAT dilute IgG preparation 1:1000) in blocking buffer[c].
13. Wash six times for 30 sec in blocking buffer[d].
14. Incubate for 30 min in goat anti-rabbit IgG conjugated with 15 nm colloidal gold (GAR 15) diluted 1 in 40.
15. Wash with buffer six times for 30 sec.
16. Wash with distilled water six times for 30 sec.
17. Stain the sections with 0.5% aqueous uranyl acetate followed by lead citrate.

[a] All incubations must be carried out at room temperature in a humid chamber. Grids are floated section side down onto 10 µl droplets of reagents on wax sheets.
[b] Pre-immune serum may also be used for blocking protein binding sites (see Section 4.2).
[c] Controls: (1) Omission of primary rabbit IgG; (2) Replacement of primary IgG with pre-immune rabbit serum; (3) Omission of blocking step 12.
[d] Grids should be gently agitated during washing steps.

are between 5 and 20 nm in diameter. Recent unpublished results (Roberts and Wells, personal communication) showed that when 10 nm probes were compared with 5 nm probes, there was four times more labelling with the smaller probe, although detection of the larger probe was easier.

When using gold probes below 5 nm in diameter, or ferritin as the label, care must be taken not to mask the probe by overstaining the labelled sections with heavy metal salts.

It is relatively easy to make colloidal gold sols in the laboratory and conjugate them to the antibody of choice (26). Alternatively they may be purchased already conjugated from a variety of suppliers. Janssen Life Science Products produce a range of gold sols together with a free booklet on the methodology of gold conjugation to antibodies and protein A. One other advantage in the use of colloidal gold is that quantification of the labelling is easily carried out by counting the number of gold particles in a given area of the section. It is therefore easy to assess the efficiency of the various preparative procedures being used when working out a protocol for the immunocytochemical localization of a particular antigen.

4.3.5 *Non-specific binding*

A major problem with the immunogold staining of resin sections is non-specific binding of both the primary and secondary antibodies. Two approaches should be made to eliminate or reduce this as far as possible. Firstly, the optimum concentration of primary antibody should be assessed. Use of too high a concentration (>1 mg/ml) often results in excessive background staining with no increase in the amount of antigen labelling. Some workers have actually reported a reduction in labelling density with high antibody concentrations. Therefore to ascertain the correct concentration to use, a series of dilutions of the primary antibody should be tested over different staining times. It is often better to use a very dilute primary serum and incubate the section in it for long periods, rather than short staining in more concentrated antibody. Secondly, sections should be pre-conditioned with a blocking buffer to prevent non-specific binding of the antibodies, and subsequent washing steps should be carried out in this buffer. A common pre-conditioning buffer is Tris–HCl (0.1–0.2 M, pH 7.2–8.2) with 0.5 M NaCl, 1% (w/v) globulin-free BSA and Tween-20 (polyoxethylene sorbitan monlaurate) (0.1–0.3%) but 1% ovalbumin or 5% normal serum may be used instead of BSA. In addition 10 mM glycine and 0.1% gelatin have been used in buffers.

The effectiveness of a complete pre-conditioning buffer is demonstrated in *Figure 7a* and *b* which shows the labelling of pumpkin cotyledon glyoxysomes with an antiserum to serine glyoxylate amino transferase (SGAT). When treated with a Tris buffer containing only BSA, there is considerable non-specific background labelling on the section (*Figure 7A*), but with the addition of salt and Tween-20 to the buffer, background staining is almost eliminated, yet the level of label over the glyoxysome remains unaltered (*Figure 7B*).

4.3.6 *Cryosectioning*

One of the best methods to preserve tissue antigenicity is to use cryo-ultramicrotomy (*Table 9*), thus avoiding both tissue dehydration and any resin embedding prior to immunolabelling. As ice is the embedding medium and the frozen sections are thawed prior to immunolabelling, this procedure gives better accessibility of the antibodies to antigens compared to the resin embedding techniques. Also problems of non-specific binding of antibodies to the embedding resin are obviated, giving lower levels of non-specific background staining. To aid in both the freezing and sectioning procedures, the fixed material is infused with sucrose before freezing in liquid nitrogen, Freon or propane. After immuno-staining, the sections can be post-fixed in osmium tetroxide, stained with uranyl acetate, dehydrated and embedded in a thin layer of resin prior to observation by EM (27). To date there are very few applications of this technique to plant tissues, a fact attributable to the technical difficulties in the cryo-sectioning procedure and the expense of the necessary equipment. However, as the need to probe antigens which are not concentrated in specific areas of the cell increases, it is likely that this technique will come to assume a more important role in plant immunocytochemistry. Before attempting these

Table 9. Basic procedure for immuno-cryoultramicrotomy.

---

1. Fix small (1 mm³) pieces of the tissue in the fixative of choice (containing 0.5 M sucrose).
2. Infuse the tissue with a mixture of 2% paraformaldehyde and 1.75 M sucrose in buffer at 4°C for 2 days.
3. Mount the tissue block on the microtome stub in a small drop of infusion medium.
4. Freeze by plunging into liquid Freon or nitrogen[a] and mount in the chuck of the cryoultramicrotome pre-cooled to −90°C[b].
5. Pick up the ultrathin frozen sections on a small drop of 0.75 M sucrose held in a wire loop and transfer to the surface of a carbon-coated formvar nickel or gold grid.
6. Carry out the immunolabelling procedure with primary and gold-labelled secondary antibodies.
7. After washing with buffer float the grids onto drops of buffered 2% OsO₄.
8. Wash the grids three times for 2 min.
9. Stain with 2% uranyl acetate.
10. Wash and dehydrate in a graded water/ethanol series (see *Table 1*).
11. Infiltrate the sections with 1.1 (v/v) ethanol/LR White resin for 10 min.
12. Infiltrate with LR White twice for 10 min.
13. Gently remove excess resin from the grids by blotting in a filter paper sandwich and polymerize in a vacuum oven (or in a nitrogen atmosphere) at 60°C for 24 h.

---

[a] Other cryogens such as liquid propane can be used.
[b] The exact sectioning temperature will depend on the final section thickness required. Generally the cooler the microtome the harder the block will be, permitting thinner sectioning.

procedures, it is advisable to read the excellent review by Tokuyasu (28) and references quoted therein.

5. LECTINS

Another group of proteins and glycoproteins which can be used as direct probes are lectins. They are non-immune in origin, derived mostly from plants and have the unique ability of binding specifically to carbohydrate groups. As they can be conjugated to fluorochromes and to colloidal gold, they can be used as cytochemical probes at both the light microscope and EM levels. The list of different lectins and their sugar specificities is too long to be given here, but can be found elsewhere in the literature (29,30). It should, however, be emphasized that the simple sugar that may inhibit lectin binding may not be the same as the sugar group which binds the lectin in the tissue. In addition two lectins with the same simple sugar binding site may not bind to identical tissue polysaccharides. Also, a lectin may bind to more than one sugar group, and care has to be taken with the interpretation of cytochemical results of lectin binding. In plant cells, lectins have been used mostly in the study of cell walls or as surface markers on the plasma membrane of protoplasts.

There are two approaches to the use of lectins as cytochemical stains.

(i) They can be conjugated directly to the label of choice (i.e. fluorescein, an enzyme or colloidal gold particles).
(ii) The lectin bound to the tissue or section can be localized with an antibody raised to the lectin.

The advantage of the second method is that it increases the sensitivity of the staining procedure and/or permits the use of more dilute lectin solutions (10 μg/ml). Protocols can be derived from the techniques already outlined for immunofluorescence and immunogold labelling (see Section 4).

Figure 8 shows the fluorescent localization of cell wall polysaccharides in the spores and conidiophores of the deuteromycete state of the fungus *Ceratocystis adiposa* (C.Moreau) (31). α-D-mannopyranosyl and α-D-glucanopyranosyl residues were stained with the jack bean lectin Concanavalin A (Con A) conjugated to FITC (*Figure 8B*) while *N*-acetylglucosamine (cell wall chitin) was probed with wheat germ agglutinin (WGA) (*Figure 8C*).

6. OTHER PROBES

There are a variety of other probes that can be used for the direct localization of macromolecules, especially by fluorescence microscopy. Some useful ones for the study of plant cells are given below.

6.1 Aniline blue

In low concentrations aniline blue can be used as a direct probe for β1–3 linked D-glucans and will stain areas rich in callose such as pit fields and cell plates. Stained specimens absorb in the UV (~370 nm) and fluoresce yellow (peak emission 509 nm). The active fluorochrome, Sirofluor (sodium 4,4-[carbonylbis-(benzen-4,1-diyl)bis(imino)]bisbenzene sulphonate), is a minor component of commercially produced aniline blue and has been isolated and characterized (32). Use of this has the advantage of reduced background fluorescence over aniline blue. Commercial batches of aniline blue may also vary in the amount of fluorochrome, thus causing problems with repeatability of staining.

Staining can be carried out on fresh or aldehyde-fixed material. Specimens may be embedded in water soluble resin such as glycol methacrylate or JB 4 and thick sections stained, or the resin may be removed from epoxy sections (see *Table 4*) prior to staining in aniline blue or Sirafluor:

(i)   Aniline blue. Stain for 30 min in a 0.1% w/v solution of the dye in 0.1 M phosphate buffer at pH 8.5 for 30 min. Wash with buffer before observation.

(ii)   Sirofluor. The stain solution is 0.025 mg/ml Sirofluor. Stain as for aniline blue in step (i).

6.2 Calcofluor White

The optical brightener Calcofluor White binds to β-linked glucans and is excluded from living cell cytoplasm. It is therefore a good probe for the cell wall in both whole living cells and fixed, sectioned material (*Figure 8A*). It will bind to fungal and algal cell walls, as well as to cellulose in higher plants. Calcofluor White is excited by UV light, emitting an intense blue fluorescence and is easily used in conjunction with FITC-labelled antibodies. It may be used as 0.1–0.05% (w/v) solution in the buffer wash of an immunostaining procedure.

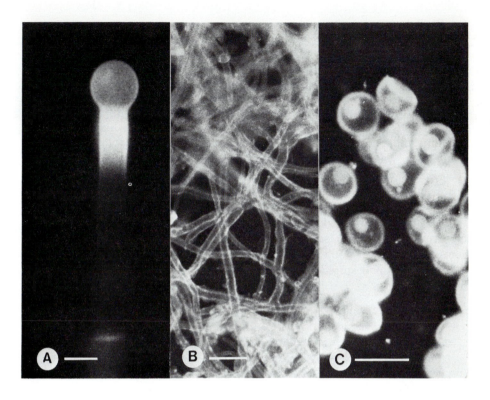

Figure 8. (A) Calcofluor White staining of newly synthesized cell wall in the conidiophore of young conidium of *C.adiposa*. Bar = 6 μm. **(B)** FITC–concanavalin A (1.5 mg/ml in PBS) staining of mycelial cell walls of *C.adiposa*. Bar = 20 μm. **(C)** FITC–wheat germ agglutinin (200 μg/ml in PBS) staining of the inner chitin wall layer of *C.adiposa* conidia. Bar = 30 μm.

6.3 DNA stains

There are a range of fluorescent dyes which are specific for DNA, can be used in low concentration and are easily incorporated into other staining schedules at the final buffer wash step.

(i)   Ethidium bromide. Excite in the UV below 400 nm. Fluorescence is orange/yellow. Stain for 30 sec to 10 min in 0.1% w/v ethidium bromide in buffer. Wash and observe.

(ii) DAPI (4,6-diamidino-2-phenylindole). Excite in the UV. Fluorescence is blue. Use at 1 μg/ml in buffer. Stain for 5 min. As the dye will only fluoresce when bound to DNA, specimens may be mounted in it.

(iii) Hoechst 33258/33342 (a bis-benzimidazole dye). Excite in the UV. Fluorescence is blue. Use as DAPI but at 20 μg/ml. Both DAPI and Hoechst dyes will stain chloroplast and mitochondrial DNA as well as nuclear DNA (*Figure 4B*).

6.4 **Phallotoxins**

The phallotoxins are cyclic peptides which can be isolated from the death cap mushroom (*Amanita phalloides*). They have the property of stabilizing cytoplasmic actin filaments against depolymerizing agents. When conjugated to fluorochromes such as FITC and TRITC, they can be used as specific fluorescent probes for cytoplasmic F-actin. The extent of the distribution of actin filaments in plant cells is only just being realized (33) and this is in part due to the development of techniques to permeabilize the cells to the conjugated phallotoxin, without disrupting the actin filament network. It is now becoming clear that when rhodamine-conjugated phalloidin (RhP) is used to stain highly processed tissue, only the major actin cables and filament bundles are stained and much of actin network is not visualized. The procedures given below have been developed for single cell systems such as hairs, suspension cultures or epidermal strips. Stock solutions of labelled phalloidin should be made to a concentration of 50–500 µg/ml in Pipes microtubule stabilizing buffer (see Section 4.2.1) and stored in the dark. Three alternative protocols are given below.

(i)   Mount the cells on a slide in a 1:1 (v/v) mixture of RhP and weak detergent such as 0.01% Triton X-100 or 0.01–0.05% Nonidet NP40 (in buffer). Observe immediately. 5% (v/v) dimethylsulphoxide may be added to the extraction buffer.

(ii) Mount the cells on a slide in a 1:1 (v/v) mixture of RhP and 3% paraformaldehyde in Pipes MSB.

(iii) Mount the cells on a slide in a 1:1:1 (by vol.) mixture of RhP, detergent and paraformaldehyde.

Staining of the actin cytoskeleton in a suspension cultured carrot cell is demonstrated in *Figure 3*.

7. IN SITU HYBRIDIZATION

By *in situ* hybridization (see also Chapter 1) specific DNA and RNA sequences within chromosomes, nuclei and in the cytoplasm can be localized by both light and electron microscopy. Recombinant DNA technology permits the use of pure cDNA probes, which can either be radiolabelled and visualized by standard autoradiographic techniques (see Section 3) or labelled with biotinylated nucleotide analogues and visualized with labelled avidin, streptavidin or an antibody to biotin (34,35). At the EM level, a compromise has to be made between the conditions which allow successful hybridization and the preparative procedures required for good ultrastructural preservation.

The various procedures for the application of this technology to plant cells are still rapidly evolving and being refined. In this section I shall give an account of the basic procedures alongside a detailed protocol (*Table 10*) for the localization of mRNA by hybridization with cDNA probes.

Table 10. A protocol for *in situ* hybridization.

1. Fix the tissue in 3% paraformaldehyde, 1.25% glutaraldehyde in 0.05 M phosphate buffer, pH 7, for up to 24 h at room temperature.
2. Wash twice for 10 min in phosphate buffer.
3. Dehydrate the tissue in a water/ethanol series for 10 min each change (see *Table 1*).
4. Embed in either wax or PEG mol. wt 1500 [see also Section 4.2.2 (ii)].
5. Cut 10 μm thick sections and place these onto 'subbed' slides (see *Table 2*).
6. Radiolabel the cDNA probe (ideally between 50 and 500 bp in size) by nick translation (use appropriate kit (36) or biotinylate either by nick translation or by using a photoactivated biotin (e.g. using the Vector kit).
7. Remove the embedding medium from the sections and take to water through an ethanol/water series.
7a. Post fix with 4% phosphate-buffered paraformaldehyde (optional).
8. Wash with PBS three times for 10 min.
9. Wash with hybridization buffer three times for 10 min[a].
10. Incubate with cDNA probe in hybridization buffer for 4–24 h at 35–50°C. Use 0.5 μg/ml labelled cDNA plus 250 μg/ml salmon or herring sperm DNA[b].
11. Wash with hybridization buffer overnight.
12. Wash with standard saline citrate (SSC) or PBS[c].
13. Locate the probe.
    (a) Radiolabelled probe. Dip the slides in a suitable emulsion, incubate in dark at 4°C and develop (see Section 3).
    (b) Biotinylated probe. Incubate with streptavidin 5 nm gold (1:200 in PBS for 2 h) or incubate with anti-biotin IgG followed by gold-labelled secondary antibody (see Section 4). Silver enhance (see Section 4.2.7), dehydrate and mount for light microscopy (e.g. in Histosol/Histomount)[d].
    (c) For EM, dehydrate the sections on the slide and embed in a suitable resin, thin section and counterstain with uranyl acetate and lead citrate[e].

[a] Hybridization buffer is 50% formamide, 0.3 M NaCl/0.03 M sodium citrate (SSC), pH 7.
[b] If a double-stranded probe is used, a denaturation step at 78°C for 3.5 min is required prior to hybridization.
[c] The concentration and temperature will govern the stringency of hybridization, i.e. the match between the probe and sample.
[d] For light microscopy the probe can also be visualized with fluorochrome-labelled streptavidin or by fluorescent anti-biotin techniques.
[e] For EM the probe can also be located by streptavidin–peroxidase, streptavidin–ferritin or streptavidin–gold conjugates (35).

7.1 Steps in *in situ* hybridization in plant tissues

(i) Extract mRNA from the tissue of interest and make either single- or double-stranded cDNA probes (see ref. 36 or Chapters 1 and 9).

(ii) Radiolabel or biotinylate the cDNA probes. Standard enzymatic nick translation can be used (35) or the cDNA can be labelled with photoactivated biotin. In this latter procedure, an aryl-azide derivative of biotin with a positively charged arm between the azide group and the biotin is subjected to strong UV light, which converts the azide to a nitrene which covalently binds to the DNA. Incorporation of biotin into the DNA can be tested by a dot blot procedure using peroxidase-labelled avidin (35). If a double-stranded probe is used, this should be denatured before the hybridization step. Due to the ease and speed of use, coupled

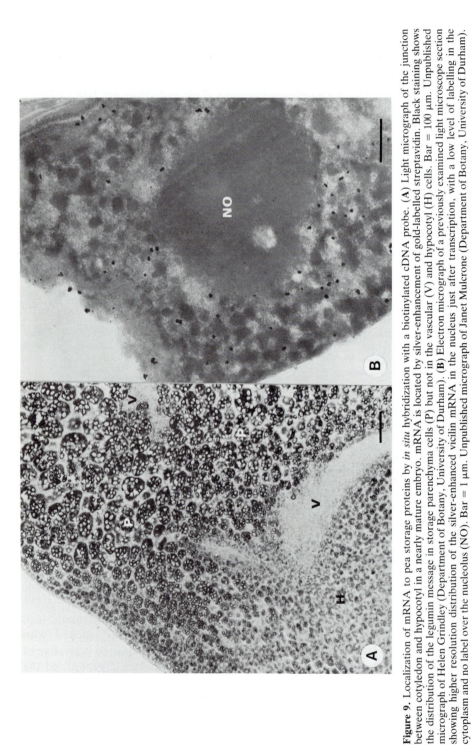

Figure 9. Localization of mRNA to pea storage proteins by *in situ* hybridization with a biotinylated cDNA probe. (**A**) Light micrograph of the junction between cotyledon and hypocotyl in a nearly mature embryo. mRNA is located by silver-enhancement of gold-labelled streptavidin. Black staining shows the distribution of the legumin message in storage parenchyma cells (P) but not in the vascular (V) and hypocotyl (H) cells. Bar = 100 μm. Unpublished micrograph of Helen Grindley (Department of Botany, University of Durham). (**B**) Electron micrograph of a previously examined light microscope section showing higher resolution distribution of the silver-enhanced vicilin mRNA in the nucleus just after transcription, with a low level of labelling in the cytoplasm and no label over the nucleolus (NO). Bar = 1 μm. Unpublished micrograph of Janet Mulcrone (Department of Botany, University of Durham).

with the high resolution and various means of visualization of biotinylated probes, they are likely to supersede radiolabelling procedures.

(iii)   Prepare the tissue for hybridization. Two basic choices can be made here.

(a) Hybridization may be carried out on self-supporting slices of fixed tissue (hybridization to sections supported on slides often leads to higher levels of background labelling). For light microscopy the probe can then be localized by fluorochrome, peroxidase or gold (silver enhanced on the section) conjugated streptavidin or biotin antibodies. For EM the biotinylated cDNA is localized with peroxidase, ferritin or gold-labelled anti-biotin or streptavidin (7), dehydrated, resin-embedded and thin-sectioned.

(b) The tissue is fixed (buffered glutaraldehyde/paraformaldehyde) and embedded in wax or PEG [see Section 4.2.2 (ii)]. Thick sections (5–10 μm) are cut, mounted onto slides and the wax or PEG removed. Hybridization is carried out on the sections, the probe localized and detected as discussed above. *Figure 9* shows the localization of mRNA for the storage proteins, legumin and vicilin, in pea cotyledon parenchyma cells, by silver enhancement of gold, at both the light microscopy and EM levels by the protocol given in *Table 10*.

8. ACKNOWLEDGEMENTS

I wish to thank Brian Wells and Dr Keith Roberts of the John Innes (JI) Institute, Norwich, for information on their Lowicryl embedding procedure and supplying *Figure 6*. Dr Kate VandenBosch (also of JI) kindly gave me micrographs for *Figure 5* and also information on the silver-enhancement procedure. Sue Montgomery of the Plant Sciences Department, Oxford, kindly let me use her micrographs and protocols for SGAT localization. Finally I wish to thank Dr Nick Harris (Botany Department, University of Durham) for invaluable help with information on the *in situ* hybridization technology, and *Figure 9*.

9. REFERENCES

1. Sexton,R. and Hall,J.L. (1978) In *Electron and Microscopy and Cytochemistry of Plant Cells*. Hall,J.L. (ed.), Elsevier/North Holland, Amsterdam, p. 63.
2. Gahan,P.B. (1984) *Plant Histochemistry and Cytochemistry: An Introduction*. Academic Press Inc., New York.
3. Lewis,P.R. and Knight,D.P. (1977) *Staining Methods for Sectioned Material*, Vol. 5 (1), In *Practical Methods in Electron Microscopy*. Glauert,A.M. (ed.), North Holland, Amsterdam, p. 137.
4. Evans,L.V. and Callow,M.E. (1978) In *Electron Microscopy and Cytochemistry of Plant Cells*. Hall,J.L. (ed.), Elsevier/North Holland, Amsterdam, p. 235.
5. Williams,M.A. (1977) *Autoradiography and Immunocytochemistry*, Vol. 6 (1) In *Practical Methods in Electron Microscopy*. Glauert,A.M. (ed.), North Holland, Amsterdam, p. 77.
6. Bullock,G.R. and Petrusz,P. (1982) *Techniques in Immunocytochemistry*, Vol. I. Academic Press Inc., New York.
7. Bullock,G.R. and Petrusz,P. (1983) *Techniques in Immunocytochemistry*, Vol. II. Academic Press Inc., New York.

8. Bullock,G.R. and Petrusz,P. (1985) *Techniques in Immunocytochemistry*, Vol. III. Academic Press Inc., New York.
9. Polak,J.M. and Varndell,I.M. (1984) *Immunolabelling for Electron Microscopy*. Elsevier, Amsterdam.
10. Wang,T. (1986) *Immunology in Plant Science*. Cambridge University Press, London.
11. Catty,D. (ed.) (1988) *Antibodies—A Practical Approach*, Vol. I. IRL Press, Oxford.
12. Goding,J.W. (1983) *Monoclonal Antibodies: Principles and Practice*. Academic Press Inc., New York.
13. Marchalonis,J.J. and Warr,G.W. (1982) *Antibody as a Tool*. J. Wiley & Sons, Chichester.
14. Anderson,A., Sandrin,M.S. and Clarke,A.E. (1984) *Plant Physiol.*, **75**, 1013.
15. Wick,S.M. and Duniec,J. (1986) *Protoplasma*, **133**, 1.
16. Wick,S.M., Seagull,R.W., Osborn,M., Weber,K. and Gunning,B.E.S. (1981) *J. Cell Biol.*, **89**, 685.
17. Van Lammeren,A.A.M., Keijzer,C.J., Willemse,M.T.M. and Kieft,H. (1985) *Planta*, **165**, 1.
18. Wolosewick,J.J., De Mey,J. and Meninger,V. (1983) *Biol. Cell*, **49**, 219.
19. Raikhel,N.V., Mishkind,M. and Palevitz,B.A. (1984) *Protoplasma*, **121**, 25.
20. De Mey,J., Lambert,D.M., Bajer,A.S., Moermans,M. and De Brabander,M. (1982) *Proc. Natl. Acad. Sci. USA*, **79**, 1898.
21. Danscher,G. (1981) *Histochemistry*, **71**, 81.
22. VandenBosch,K.A. (1986) *J. Microsc.*, **143**, 187.
23. Craig,S. and Goodchild,D.J. (1984) *Protoplasma*, **122**, 35.
24. Harris,N. and Croy,R.R.D. (1985) *Planta*, **165**, 522.
25. Wells,B. (1985) *Micron Microscopica Acta*, **16**, 49.
26. De Mey,J. (1983) In *Immunocytochemistry, Practical Applications in Pathology and Biology*. Polak,J.M. and Van Noorden,S. (eds), Wright, Bristol, p. 82.
27. Greenwood,J.S. and Chrispeels,M.J. (1985) *Planta*, **164**, 295.
28. Tokuyasu,K.T. (1986) *J. Microsc.*, **143**, 139.
29. Leathem,A.J.C. and Atkins,N.J. (1983) In *Techniques in Immunocytochemistry*, Vol. 2. Bullock,G.R. and Petrusz,P. (eds), Academic Press, London, p. 39.
30. Horisberger,M. (1985) In *Techniques in Immunocytochemistry*, Vol. 3. Bullock,G.R. and Petrusz,P. (eds), Academic Press, London, p. 155.
31. Hawes,C.R. (1979) *Trans. Br. Mycol. Soc.*, **72**, 177.
32. Stone,B.A., Evans,N.A., Bonig,I. and Clarke,A.E. (1984) *Protoplasma*, **122**, 191.
33. Parthasarathy,M.V. (1985) *Am. J. Bot.*, **72**, 1318.
34. Hutchison,N.J. (1984) In *Immunolabelling for Electron Microscopy*. Polak,J. and Varndell, I.M. (eds), Elsevier, Amsterdam, p. 341.
35. Harris,N. and Croy,R.R.D. (1986) *Protoplasma*, **130**, 57.
36. Hames,B.D. and Higgins,S.J. (1985) *Nucleic Acid Hybridization—A Practical Approach*. IRL Press, Oxford.

CHAPTER 6

Foreign gene expression in plants

LUIS HERRERA-ESTRELLA and JUNE SIMPSON

1. INTRODUCTION

Transfer and expression of foreign genes in plant cells, now routine practice in several laboratories around the world, has become a major tool to carry out gene expression studies and to attempt to obtain improved plant varieties of potential agricultural or commercial interest. Although several plant trans-formation systems have been investigated, the system which has proved most successful is that based on the Ti plasmid of *Agrobacterium tumefaciens*. Whilst vectors based on viral genomes have also been extensively studied, none has yet been developed that is practical for general use in plant transformation. This is mainly due to the pathogenic nature of the virus, the restrictions on genome size and the fact that the virus DNA is not stably transmitted to the progeny of infected plants.

In addition to *Agrobacterium*-mediated transformation, direct DNA trans-formation has also been studied and shown to be successful (1,2). It is now possible to obtain plant transformants at a high frequency using this technique. The main drawback of the direct DNA transformation methods is that plant cell protoplasts must be used (3) involving lengthy and skilful tissue culture. In addition, since for many plant species it is not yet possible to regenerate mature plants from these cells, this technique is of limited use. However, for transient gene expression studies, or for plant species which are not susceptible to *Agrobacterium* infection, direct DNA transfer may prove the method of choice. This technique together with protoplast isolation are discussed in Chapter 7. In this chapter we will concentrate on Ti plasmid-mediated plant transformation, covering three general topics:

(i)   transfer and maintenance in *Agrobacterium* of plasmids carrying sequences to be transferred into plant cells;
(ii)   *Agrobacterium*-mediated transformation of plant cells, selection of transformants and regeneration of mature transgenic plants;
(iii)   reporter genes and methods to study expression of chimaeric genes in plant cells.

2. THE AGROBACTERIUM SYSTEM

The soil bacterium *A.tumefaciens* has the ability to induce tumours on all tested dicotyledonous and some monocotyledonous plants (4,5). A large plasmid called the Ti, or tumour inducing, plasmid has been shown to be responsible for tumour formation, due to a remarkable natural capacity to transfer, insert and express a particular segment of DNA in the plant cell genome (6–8). The segment of the Ti plasmid DNA which is transferred from the bacterium and becomes stably integrated into the plant genome has been called the transferred or T-DNA (9,10). Genetic analysis of the Ti plasmid reveals that two regions of this plasmid are essential for tumorigenesis, the T-DNA and the *vir* or virulence region. Mutations in the T-DNA produce tumours with altered morphology, whereas most of the mutations in the *vir*-region abolish or decrease tumour formation.

Upon integration, the T-DNA encoded genes are expressed in the plant cell nucleus. It has been shown that the gene products of the so-called *onc* genes of the T-DNA, genes 1, 2 and 4 are involved in the synthesis of phytohormones such as the auxin, indoleacetic acid and the cytokinin, isopentenyladenine (11,12). The over-production of these hormones is the direct cause of the proliferation of the plant cells and prevents differentiation of the transformed cells. Other T-DNA-encoded genes are responsible for the synthesis of novel amino acid or sugar derivatives, termed opines, or the secretion of these opines from the plant cell. The opine encoded by the *Agrobacterium* strain inciting the tumour can be used exclusively by that strain as a carbon and nitrogen source (13). *Agrobacterium* Ti plasmids are classified by the type of opine they induce the plant to produce. The two most widely studied are the octopine and nopaline Ti plasmids, pTiAch5 and pTiC58 respectively.

Extensive analysis shows that the T-DNA-encoded genes are responsible for changing the regulatory pattern of the plant cells and causing the tumour to form, but are not involved in the mechanism of transfer of the T-DNA. The *vir*-region however, is essential for the transfer and integration of the T-DNA in plant cells and is functional even when the T-DNA is on a separate replicon. The flexibility of the mechanism by which *Agrobacterium* can transfer DNA into plant cells can be readily observed from the structure of the natural T-DNA in different Ti plasmids: the T-DNA of the nopaline-type Ti plasmid pTiC58 is transferred as a single 23 kb fragment which encodes approximately 13 genes (14,15); whereas the octopine plasmid pTiAch5 can be transferred as a composite segment including two DNA segments of 13.2 kb and 7.9 kb, or a single T-DNA comprising either of these two segments (16,17).

The mode of transfer of the T-DNA to plant cells is not as yet completely understood. However, directly repeated 25 bp sequences known as border sequences have been shown to be essential for T-DNA transfer and integration into the plant genome (18–20). The transferred segment normally ends within or very close to these 25 bp sequences and any DNA placed between these borders may be efficiently transferred and stably integrated into the plant genome (21). The 25 bp border repeats, in the correct orientation, are sufficient to promote DNA transfer when complemented by a functional *vir*-region (22). This property forms the basis for producing vectors for plant cell trans-formation.

2.1 Introduction of gene constructs into *Agrobacterium*

The general process for manipulating genes to be transferred into the genome of plant cells is carried out in two phases. First, all the cloning and DNA modification steps are done in *Escherichia coli*, and then the plasmid containing the gene construct of interest is transferred by conjugation into *Agrobacterium*. The resulting *Agrobacterium* strain is finally used to transform plant cells. In some cases, as for binary vectors which contain a broad host range origin of replication, direct DNA transformation can be used instead of conjugation to introduce the corresponding plasmid into *Agrobacterium* (23). A general outline of the transformation procedure is shown in *Figure 1*.

Many of the widely used cloning vectors (e.g. pBR322) cannot be directly conjugated into *Agrobacterium*, since they do not normally encode the transfer (*tra*) and mobility (*mob*) functions. However, they usually retain the *bom* site where conjugative transfer is initiated by the *mob* functions. Therefore, if the *mob* and *tra* functions are provided in *trans* by helper plasmids (such as R64drd11, pGJ28 or pRK2013), *mob*⁻ plasmids (e.g. pBR322) can be mobilized to *Agrobacterium* or any other Gram-negative bacteria (24).

Conjugative transfer of pBR322-like vectors from *E.coli* to *Agrobacterium* can be achieved in two different ways.

(i)   A two-step conjugation, where the helper plasmids are first conjugated into the cloning vector-containing strain. The strain containing both sets of plasmids is then isolated and used to mate with the *Agrobacterium* acceptor strain.

(ii)   A 'triparental' mating may be carried out where the strain containing the cloning vector, that containing the helper plasmid and the *Agrobacterium* strain are mixed together and *E.coli–E.coli–Agrobacterium* conjugations occur simultaneously.

The protocol for two-step conjugation is as follows.

(i)   Grow fresh agar plates of donor (helper) and recipient strain (intermediate vector) on LB (*Table 1*) plus antibiotics (overnight). Also grow a fresh plate of the *Agrobacterium* vector strain on YEB (*Table 1*) plus antibiotics.

(ii)   On the next day take one isolated colony from the donor and streak in one direction on an LB plate containing no antibiotics.

(iii) Take one isolated colony of the recipient strain and streak on the same LB plate at 90° to the original strain, to form a cross.

(iv)   Leave to grow at 37°C for at least 3–4 h or overnight.

(v)   Take a loopful of bacteria from the cross junction and plate out to single colonies on one half of an LB plate containing antibiotics to select for the transconjugants.

(vi)   In one remaining quarter of the plate streak the donor alone and in the other quarter streak the recipient.

(vii) Grow overnight at 37°C. Only the transconjugants should grow on selective media, the donor and recipient alone should not grow.

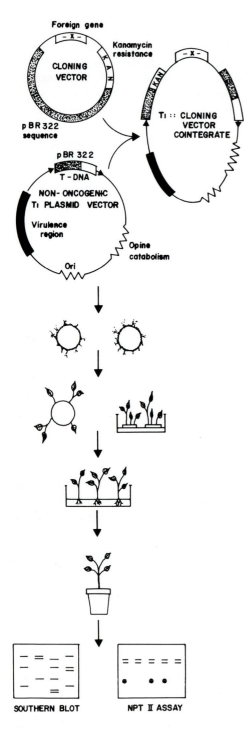

1) Cloning of foreign gene of interest into a pBR 322 derivative containing an antibiotic resistance marker for selection in plant cells.

2) Transfer of cloning vector from E. coli to agrobacterium and selection of cointegrates.

3) Infection of leaf discs.

4) Shoot regeneration on selective media.

5) Root formation and second round of selection.

6) Transfer of transformed plantlets to soil for further analysis.

7) Analysis of gene stucture (southern) and gene expression (RNA or enzymatic assay).

Figure 1. Plant transformation procedure.

Table 1. Bacterial media.

Liquid Medium	
LB	
Tryptone	1%
Yeast extract	0.5%
NaCl	1%
pH	7.2
YEB	
Beef extract	0.5%
Yeast extract	0.1%
Peptone	0.5%
Sucrose	0.5%
$MgSO_4$	2 mM
pH	7.2

Solid media contain 1.5% agar

λ *buffer*
10 mM Tris–HCl pH 7.2
10 mM $MgSO_4$

Minimal A medium (MinA)	
K_2HPO_4	10.5 g/l
KH_2PO_4	4.5 g/l
$(NH_4)_2SO_4$	1 g/l
Sodium citrate.$2H_2O$	0.5 g/l
$MgSO_4·7H_2O$	0.2 g/l
Glucose	2 g/l

(viii) Inoculate an isolated colony of the *Agrobacterium* vector strain in 3 ml of YEB plus antibiotics and grow overnight.

(ix) Grow three isolated colonies of the transconjugants in 1.5 ml of LB for 4–5 h.

(x) Take 100 μl of *Agrobacterium* culture and 150 μl of *E.coli* culture, mix in a sterile Eppendorf tube by vortexing, plate on a plain YEB plate and grow overnight at 28°C.

(xi) Resuspend the bacteria from the plate in 5 ml of λ buffer (*Table 1*).

(xii) Transfer to a clean test tube, mix by vortexing and plate aliquots (e.g. 200, 100 and 20 μl) on selective YEB plates.

(xiii) Leave for 2–4 days at 28°C.

(xiv) Streak out several individual *Agrobacterium* colonies on selective media. *E.coli* mutate at a relatively high frequency and can grow on the selective YEB plates. As a rough guide, colonies appearing in 1–2 days are usually *E.coli*. *Agrobacterium* colonies are generally round and smooth and difficult to pick up with a wire loop, whereas *E.coli* are not rounded and smooth and are easily picked up.

The corresponding protocol for triparental mating is given below.

(i)   From fresh plates inoculate separately: *E.coli* intermediate vector in 5 ml of LB plus antibiotics; *E.coli* helper in 5 ml of LB plus antibiotics; *Agrobacterium* strain in 5 ml of YEB plus antibiotics. Grow overnight at 37°C (*E.coli*) and 28°C (*Agrobacterium*).

(ii)   Combine 1 ml of each overnight culture in a single tube, centrifuge (2 min) to pellet the bacteria. Resuspend the bacteria in 1 ml of λ buffer.

(iii) Filter the solution through a 0.22 micron filter using a syringe and transfer the filter to a plain YEB plate OR resuspend the cells in 10 ml of λ buffer and spot on a YEB plate.

(iv) Incubate overnight at 28°C.

(v)   Resuspend the bacteria from the filter or plate, in 1 ml of λ buffer and plate dilutions on selective YEB plates.

(vi) In 2–3 days colonies appear. Take care as in the two-step conjugation for the appearance of *E.coli* background colonies.

The most important step in the process of transferring foreign genes to plant cells is to place the desired gene construct between the 25 bp border repeats of the T-DNA. Two different vector systems have been developed in order to accomplish this (*Figure 2*).

(i)   The binary vector system, where the sequence of interest is directly cloned between the T-DNA border repeats present in a specially designed cloning vector.

(ii)   The co-integrative vector system, where the sequence of interest is inserted between T-DNA borders by co-integrating the cloning vector with the Ti plasmid, via a single homologous recombination event.

2.2 Binary vector systems

This system is based on the fact that the *vir*-region and the T-DNA need not be linked on the same replicon for functional T-DNA transfer to occur. The *vir*-region acts in *trans* and recent reports have shown that *vir*-region encoded proteins bind at the T-DNA border repeats, during the transfer process (33). The binary vector system consists of two components (*Table 2*).

(i)   A cloning vector consisting of an origin of replication active both in *E.coli* and *Agrobacterium*, a selectable marker also functional in both bacteria, and a small T-DNA. This latter is composed of the 25 bp border sequences flanking multiple restriction sites for cloning the desired sequences and a dominant selectable marker gene functional in plant cells (*Figure 2*).

(ii)   An acceptor *Agrobacterium* strain which contains a Ti-plasmid depleted of its T-DNA, and which provides the necessary functions to complement the T-DNA present in the cloning vector.

Although not the optimal situation, any *Agrobacterium* strain containing a wild-type Ti-plasmid can be used as an acceptor strain for binary vectors. The

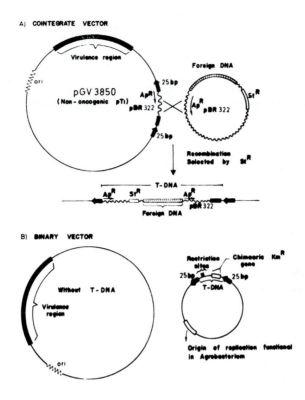

Figure 2. Ti plasmid-derived plant transformation vectors. **A** shows the co-integrate vector system and **B** shows the binary vector system. The *vir*-region is indicated by the stippled box; the solid black arrows indicate the T-DNA border repeats; the wavy line, pBR322 sequences; the solid black box, the nopaline synthase gene; and the white boxes antibiotic resistance markers: St^R, streptomycin; Km^R, kanamycin; and Ap^R, ampicillin; for selection of conjugation/co-integration in *Agrobacterium*.

main disadvantage of using wild-type Ti plasmids is that around 50% of the transformants containing the target sequence will also contain the wild-type T-DNA and therefore still produce tumour tissue. However, since only limited amounts of tissue or plantlets are needed for further work, and in most cases the transformation frequency is high, this is not really a limiting factor for the use of wild-type *Agrobacterium* strains.

2.3 Co-integrative vector systems

In this system both the T-DNA and the *vir*-region are present on the same Ti plasmid (*Table 3*). Insertion of the gene of interest in between T-DNA borders depends on a co-integration event between homologous sequences present in the cloning vector and the Ti plasmid T-DNA. The quintessential example of this system is the vector pGV3850 (19). This vector comprises a Ti plasmid containing a T-DNA from which all the oncogenic genes have been deleted and

137

Table 2. Binary vectors.

E.coli plasmid	Description	Agrobacterium strain	Description	Reference
Bin 19	pRK252 derivative harbouring Km^R-resistant gene for selection in *Agrobacterium*. Carries left and right T-DNA borders encompassing intact nopaline synthase gene and multiple restriction sites for cloning: *Eco*RI; *Bam*HI; *Hind*III; *Sst*I; *Kpn*I; *Sma*I; *Xba*I; *Sal*I.	LBA4404 (pAL4404)	Derivative of wild-type strain Ach5 containing octopine type Ti-plasmid pTi Ach5 deleted of the T-DNA functions.	25
pRAL3940	pR29 (pBR325) derivative carrying erythromycin resistance marker for selection in *Agrobacterium*, carries left and right T-DNA borders encompassing the octopine synthase gene and gene 5 of the octopine T-DNA. Restriction sites available for cloning between T-DNA borders are *Bam*HI, *Bgl*II and *Sma*I.	LBA4404 (pAL4404)	As above.	26
pGA492	Carries a left and a right T-DNA border. Between the T-DNA borders are a Nos–NPT (II)–ocs chimaeric cassette and a promoter probe chimaeric cassette consisting of the CAT gene fused to the termination signals of the octopine T-DNA. Upstream to the CAT coding sequence are multiple restriction sites to facilitate cloning: *Hind*III; *Xba*I; *Sst*I; *Hpa*I; *Kpn*I; *Cla*I; *Bgl*II.	LBA4404 (pAL4404)	As above.	27

Table 2. Continued.

E.coli plasmid	Description	Agrobacterium strain	Description	Reference
pL22	Plasmid carrying left and right T-DNA borders. Also carries the replication origins of pSa and RI, functional in *Agrobacterium*. Between T-DNA borders are a cos site, a chimaeric nos–NPT (II)–ocs cassette and multiple restriction sites for cloning: *Apa*I; *Sma*I; *Eco*RI; *Xba*I.	pGV2260	pTiB6S3 (octopine type Ti plasmid) derivative from which the T-DNA is deleted and replaced by pBR322 sequences.	28
pAGS111	pUC8 derivative carrying left and right T-DNA borders flanking a nos–NPT (II)–ocs chimaeric construct functional in plant cells.	LBA4404 (pAL4404)	As Bin19	29
pAGS112	As above but derived from pLAFR1.	As above.		
pAGS113	As pAGS111 but pRK404 derivative.	As above.		
pAGS127	Equivalent to pAGS111 but cloned into pRK252 cos containing the *cos* site. This vector also carries the *lacZ* multi-linker downstream to the nos–NPT (II)–ocs chimaeric construct and between T-DNA borders.	As above.		

replaced by the cloning vehicle pBR322. Therefore, any sequence cloned in pBR322 derivatives can be conjugated into *Agrobacterium* and inserted in between T-DNA borders by a single homologous recombination event between the two plasmids (pGV3850 and an example of a co-integration event is shown in *Figure 2*). Since pBR322-like vectors cannot replicate in *Agrobacterium*, exconjugants harbouring the co-integrates can be easily isolated by selecting for an antibiotic resistance marker present in the pBR322 derivative, but which is absent from the Ti plasmid.

Table 3. Co-integrate vectors.

Agrobacterium Ti plasmid	Description	E.coli plasmids	Reference
pGV3850	pGV3839 (30) derivative carrying left and right T-DNA borders, flanking an intact nopaline synthase gene and pBR322 sequences. Plasmid also carries intact *vir*-region.	Any pBR322-derived plasmid.	19
pTiB6S3-SE	pT1B6S3 derivative carrying the TL DNA left border, the nopaline synthase gene and the Km^R gene from Tn903.	pMon 120 type intermediate vectors carrying the T-DNA right border.	31
pGV2260	pT1B6S3-derived vector. Deleted completely of T-DNA functions but carrying pBR322 sequences and the Km^R gene from Tn903.	pGV831 or other pBR322-derived vector carrying T-DNA borders.	32

Although the great majority of these co-integration events occur as expected, it is always useful to confirm that the expected co-integration has occurred. Due to the large size of the Ti plasmid, it is impractical to isolate it and examine the co-integrates by conventional restriction enzyme analysis. Therefore the most convenient method is to isolate total *Agrobacterium* DNA and analyse the co-integrates by Southern blot hybridization. A straightforward method to isolate total *Agrobacterium* DNA is presented in *Table 4*.

Both vector systems have advantages but also some limitations. Co-integrative vectors have the disadvantage that their efficiency relies on the frequencies of both conjugal transfer and homologous recombination. The number of independent fragments which can be co-integrated with the Ti plasmid in a single conjugation event is limited, because the frequency of co-integration is relatively low. This is important if one wishes to transfer a gene library to plant cells, but for most purposes is not a limiting factor. The major disadvantages of binary vectors are that most of them are still rather large and carry a limited number of cloning sites.

In addition, stability of these plasmids is not always optimal in *Agro-bacterium*. Possibly the most important drawback is that recent reports have shown that for some plant species such as tomato, the efficiency of transformation by binary vectors is significantly lower than that obtained using co-integrative vectors (34).

3. PLANT CELL TRANSFORMATION TECHNIQUES

The use of *Agrobacterium*-mediated plant transformation systems is now firmly established, and is becoming routine for model systems such as tobacco and petunia. However, the lack of tissue culture techniques for the regeneration of plants from explants or protoplasts has become the main limiting factor in

Table 4. Isolation of total *Agrobacterium* DNA.

---

1. Inoculate 3 ml of YEB (*Table 1*) with the *Agrobacterium* strains to be characterized and include the co-integrate vector alone as a control. Grow at 28°C overnight.
2. Transfer 1.5 ml of culture to an Eppendorf tube and centrifuge for 1.5 min.
3. Resuspend the pellets in 300 μl of TE[a].
4. Add 100 μl of 5% Sarkosyl in TE.
5. Add 100 μl of 2.5 mg/ml pronase in TE.
6. Incubate at 37°C for at least 1 h to lyse the cells.
7. Shear the lysate by passing once or twice through a syringe with 20-gauge needle.
8. Extract aqueous phase twice with equal volumes of phenol saturated in TE (some patience and care is required since the interphase is large and the aqueous phase difficult to remove without contamination).
9. Extract the aqueous phase 2–3 times with chloroform.
10. Add 25 μl of 5 M NaCl (0.25 M final) and 2 vols of ethanol.
11. Precipitate at −20°C for 1–2 h.
12. Pellet the DNA in a microcentrifuge for 5 min, dry the pellet and dissolve in 100 μl of H_2O.
13. Determine the DNA concentration in a spectrophotometer. The yield obtained is normally 80–100 μg.
14. Digest 4 μg of DNA with appropriate enzymes for use in Southern hybridization.

---

[a]TE is 40 mM Tris–HCl, pH 7.5, 10 mM EDTA.

extending this methodology to other species, especially for those which are economically important. Nevertheless, recent reports have shown that several different crop species can be transformed and regenerated. These include tomato, potato and alfalfa (34–36). In this section we describe the most efficient methods for the transformation of plant cells focusing on the model systems of tobacco and petunia, and report modifications to these methods employed for other plant species.

The plant growth media used are given in *Table 5*.

3.1 Growth of sterile plant tissues

To transform plant cells and obtain mature plants from them it is necessary to start with axenic plant tissues. Sterile plant material may be obtained by sterilizing seeds and germinating plants under sterile conditions. The following protocol should be carried out in a sterile flow hood.

(i)   Cover the seeds in 75% ethanol for 2 min stirring with a sterile pipette or glass rod.
(ii) Remove the ethanol with a sterile pipette and add ¹⁄₁₀ strength commercial bleach.
(iii) Stir occasionally for 10 min.
(iv) Remove the bleach and wash 3–5 times in sterile water.

Some seeds such as certain types of *Phaseolus* or rice require a stronger sterilization utilizing mercuric chloride. For these cases, sterilize for 1 min in absolute ethanol followed by 5–6 min in 0.1% mercuric chloride dissolved in absolute ethanol.

Alternatively, plants may be grown under greenhouse conditions, and mature material sterilized immediately before use. For this route, use the protocol above but if the tissue is difficult to sterilize, use cotton soaked in ¹⁄₁₀ strength bleach, 0.1% SDS and gently clean the tissue wearing gloves. Wash twice in sterile water then proceed with the protocol above.

Table 5. Plant growth media.

Murashige and Skoog medium (37)

Macro-element	NH_4NO_3	1650	mg/l
	KNO_3	1900	mg/l
	$CaCl_2 \cdot 2H_2O$	440	mg/l
	$MgSO_4 \cdot 7H_2O$	370	mg/l
	KH_2PO_4	170	mg/l
Micro-elements	H_3BO_3	6.2	mg/l
	$MnSO_4 \cdot 1H_2O$	22.3	mg/l
	$ZnSO_4 \cdot 4H_2O$	8.6	mg/l
	KI	0.83	mg/l
	$Na_2MoO_4 \cdot 2H_2O$	0.25	mg/l
	$CuSO_4 \cdot 5H_2O$	0.025	mg/l
	$CoCl_2 \cdot 6H_2O$	0.025	mg/l
Fe-EDTA	Na_2EDTA	37.2	mg/l
	$FeSO_4 \cdot 7H_2O$	27.8	mg/l

pH 5.7 before autoclaving the medium, with KOH.

K_3 Media (38)

Macro-elements	NH_4NO_3	250	mg/l
	KNO_3	2500	mg/l
	$CaCl_2 \cdot 2H_2O$	900	mg/l
	$MgSO_4 \cdot 2H_2O$	250	mg/l
	$NaH_2PO_4.1H_2O$	150	mg/l
	$(NH_4)_2SO_4$	134	mg/l
	$CaHPO_4 \cdot 1H_2O$	50	mg/l
Micro-elements	H_3BO_3	3	mg/l
	$MnSO_4 \cdot 1H_2O$	10	mg/l
	$ZnSO_4 \cdot 4H_2O$	2	mg/l
	KI	0.75	mg/l
	$Na_2MoO_4 \cdot 2H_2O$	0.25	mg/l
	$CuSO_4 \cdot 5H_2O$	0.25	mg/l
	$CoCl_2 \cdot 6H_2O$	0.025	mg/l
Fe-EDTA	Na_2EDTA	37.2	mg/l
	$FeSO_4 \cdot 7H_2O$	27.8	mg/l
Inositol		100	mg/l
Sucrose		137	g/l (= 0.4 M)
Xylose		250	mg/l
Vitamins	Nicotinic acid	1	mg/l
	Pyridoxine	1	mg/l
	Thiamine	10	mg/l
Hormones	NAA	0.1	mg/l
	Kinetin	0.2	mg/l

Adjust the pH to 5.6 with KOH.

Table 5. Continued.

Media for callus and shoot induction: Murashige and Skoog salts plus vitamins (39)

Vitamins			
	Glycine	2	mg/l
	Nicotinic acid	0.5	mg/l
	Pyridoxine–HCl	0.5	mg/l
	Myo-inositol	100	mg/l
	Thiamine	0.4	mg/l

Callus media contains 3% sucrose
Shoot inducing media 1% sucrose
Both contain 0.8% agar
pH is adjusted to 5.7 with KOH

Having the necessary facilities, we find it more convenient to establish and maintain sterile plants, to avoid contamination problems during the development of an important experiment. However, problems of storage in a suitable growth chamber may arise when very large numbers of plants are required to provide enough material for the transformation experiments. In addition, certain types of seeds (e.g. rice and beans) may prove much more difficult to sterilize than the different tissues obtained from greenhouse-grown plants.

3.2 *In planta Agrobacterium* transformation

In vivo inoculation of seedlings (*Table 6*) or well established plants that have been propagated *in vitro* under aseptic conditions is the traditional way to obtain *Agrobacterium*-transformed cells. This method has been the most successful in the past, but has now been superseded by the leaf-disc infection technique (see Section 3.3). However, in certain cases it may still prove useful. One case is where the infectivity of *Agrobacterium* on a particular plant is to be tested. This can be done using wild-type *Agrobacterium* or an engineered strain retaining at least part of the oncogenic functions. Strains are inoculated using a tooth pick or syringe needle at the internodes of the plant. If the plant is susceptible to the bacterium, a typical crown gall tumour will be produced in 2–3 weeks. If necessary, this tumour tissue can be excised and grown axenically on hormone-free medium.

In certain plants it may prove very difficult to regenerate shoots from explants grown in tissue culture, thus limiting the possibility of transforming the plant. Although it gives transformed shoots at low frequencies one technique which may be employed in such cases is the co-infection technique.

Co-infection involves inoculating two different *Agrobacterium* strains on the same plantlet. The shoot apex of the plantlet is removed and a mixture of the two strains applied with a toothpick. One strain should be of the oncogenic 'shooter' mutant type (40) and the other should harbour the genes of interest to be transferred to the plant. On infection, the shooter mutant induces shoot formation. Some of these shoots will be transformed at low frequency by the co-infecting non-oncogenic strain. These shoots can then be excised and used to

Table 6. *In vivo* inoculations.

---

1. Inoculate 3 ml of MinA medium (*Table 1*) with *Agrobacterium* strain and grow overnight at 28°C.
2. Centrifuge 1.5 ml in a microcentrifuge for 5 min.
3. Remove the supernatant.
4. With a toothpick or syringe needle held in sterile tweezers, take some of the pellet and either prick the plant at the internodes (for oncogenic strains) or inoculate at the apex of a decapitated plant (for non-oncogenic strains).
5. Leave the plants for 3–4 weeks, then remove either tumour tissue or wound callus formed to solid growth media. Tumour tissue is selected by culture in the presence of the appropriate antibiotic and the absence of hormones. Wound callus is selected in the presence of both antibiotics and hormones. Both media should contain antibiotics to eliminate *Agrobacterium*.
6. Grow for several weeks in selective media, changing the media every 2–3 weeks until the tissue is large enough to analyse. Wound callus from non-oncogenic infections may be further selected on hormones to induce shoots.
7. After a further 3–4 weeks small shoots can be removed and grown in the absence of hormones to allow root formation to occur.

---

generate mature plants. This system is not widely used since large numbers of shoots must be screened to obtain a single transformant. However, if there is no alternative, this method may be the only choice.

3.3 The leaf-disc technique

A technique which is rapidly becoming the method of choice for the production of transformed or transgenic plants from many plant species, due to its simplicity and efficiency, is the leaf-disc infection technique (41). This technique (*Table 7*), involves the co-culture of leaf discs with a suspension of *Agrobacterium*. After two days of co-culture, the leaf discs are washed to eliminate most of the bacteria, and directly placed on regeneration media containing an antibiotic or herbicide, which allows the selection of transformed tissue. Within 2–6 weeks, depending upon the species, small shoots appear at the edges of the discs, which can be regenerated to form mature plants when placed on the appropriate medium. Two alternatives (*Table 7*) have been described for the co-culture step of this technique. In the original method, the explants are dipped in an *Agrobacterium* culture in a test tube, and gently agitated (taking care not to damage the leaf discs) for 1–2 min. Explants are then transferred to solid medium containing a layer of a nurse suspension culture. The alternative method, which we find works almost as efficiently and avoids the need of keeping the nurse suspension culture, is to co-culture the explants with the *Agrobacterium* in liquid medium for 2 days in a Petri dish to allow the infection process to occur, then directly transfer the explants to selectable solid media.

The leaf-disc technique can be easily modified for use with explants other than leaves, such as stem sections, midrib sections, cotyledon sections etc., depending upon the organ which is most responsive for regeneration of plants via organogenesis or embryogenesis. In some cases shoots can be directly induced to form from the explant, but in other cases callus tissue has first to be induced.

Table 7. Leaf-disc infections.

1. Inoculate 3 ml of MinA medium (*Table 1*) with *Agrobacterium* strain and grow overnight at 28°C.
2. To a sterile Petri dish in a flow bench add 20 ml of K₃ media (*Table 5*) containing 0.2 M sucrose, no antibiotics or hormones.
3. Place one leaf upside down in the medium, without wetting the normal basal side of the leaf. From this leaf cut either 1 cm diameter discs using a sterile cork borer or a sharp paper punch, or alternatively first cut around the edge of the leaf and discard the outer edge, then slice into 1 cm squares. Make about 10 squares per *Agrobacterium* strain.
4. Add 3–4 ml of K₃ media (*Table 5*) to a small (4 ml) Petri dish and place five discs or slices in each dish, keeping the leaf pieces upside down at all times.
5. To each dish add 50 µl of overnight *Agrobacterium* culture. The optimal amount of bacteria to be added has to be experimentally determined.
6. Seal the dishes and leave to incubate in a plant growth chamber at 25°C for 2 days.
7. (a) Either add claforan (cefotaxime; Hoechst) or another suitable antibiotic to the culture to kill *Agrobacteria* and incubate for one more day, and then plate on shoot-inducing media or
   (b) wash the leaf discs, taking care not to wet the lower surface, in K₃ containing claforan and plate directly on growth media containing hormones to stimulate shoots.

Leaf-disc infections using feeder layer technique
1. Carry out steps 1–3 above.
2. Dip the leaf pieces in an overnight culture of *Agrobacterium* for 2–3 min and transfer to feeder plates. In many cases explants can be directly transferred to shoot-inducing media.
3. Prepare the solid plant media plus necessary hormones.
4. Cover the media with a layer of tobacco suspension culture cells.
5. Place on top an 8.5 cm Whatman filter circle to cover the cells completely.
6. Place the leaf discs still upside down onto the filter paper.
7. Leave the leaf pieces on feeder plates for two days, then remove them to selective media containing hormones and an antibiotic to kill the bacteria.
8. Place the leaf pieces on solid selective media containing the appropriate hormones.

From this callus, embryos or shoots can be obtained from which mature plants can in turn be regenerated.

Another important factor in these infections is to determine the optimal concentration of bacteria needed to infect the explant. In some cases too high a concentration of bacteria might result in severe damage of the plant tissue. Therefore it is recommended to use dilutions from 10^{-1} to 10^{-4} of a saturated *Agrobacterium* culture in the initial transformation experiments.

In plant species for which transformation conditions have not been determined previously, and where binary vectors are intended to be used, it is also recommended to include as a control a co-integrative vector, since it has been reported that for some species co-integrative vectors have a significantly higher frequency of transformation.

3.4 Regeneration of transformed plants

Determining the optimal conditions to regenerate plants from protoplasts or explants is largely a trial and error process. For some plants, such as petunia and tobacco, the culture conditions for efficient *in vitro* regeneration are well established and this has greatly facilitated the development of transformation methods. In other plant species, such as potato and tomato, even individual varieties require different hormone combinations to induce them to undergo regeneration. In the majority of the cases it will be necessary to establish the conditions which allow a frequency of plant regeneration sufficient to allow transformation experiments to be attempted. Reported hormone concentrations for various plant types used in transformation experiments are described in *Table 8*.

Once the tissue culture conditions for transformation experiments are determined and small shoots are obtained in selective media, these shoots should be excised from the callus or explant and transferred to selective media which allows root formation. This latter step helps to eliminate those shoots which escape the first round of selection. When roots have formed, the plantlets can be removed from the culture medium and placed in earth or vermiculite to be grown in greenhouse conditions, where growth is normally much faster.

(i)   Gently remove the plantlet from agar with tweezers, trying not to break the roots.
(ii)   On an absorbent paper towel clean off any excess agar sticking to the roots.
(iii) Place the plantlet in a small pot containing earth or vermiculite which is already soaked with water.
(iv)   Cover with cling film to prevent dehydration.
(v)   After a few days if the plant looks comfortable in soil, pierce the cling film in several places to allow a little air to enter.
(vi)   Every few days allow more air to enter until the plant can be grown without a plastic covering.

4. GENE EXPRESSION ANALYSIS IN TRANSGENIC PLANTS

Before analysis of gene expression of transferred genes in transgenic plants is carried out, the structure and copy number of the DNA of interest should be determined by Southern blot hybridization analysis. This can be carried out

Table 8. Hormone concentrations used in plant transformation experiments.

Hormone concentration for tissue type			
Plant	*Callus mg/l*	*Shoots mg/l*	*Reference*
Tobacco	0.5 BA + 2 mg NAA	0.5 BA	35
Potato	0.5 BA + 15 ml Glu + 2 mg 2,4-D	0.5 BA + 15 mg Glu	35
Tomato	2 mg 2,4-D	2.5 mg BA + 1 mg IAA	35
		or	
		1 mg BA + 0.2 mg IAA)	34
Arabidopsis	0.5 mg 2,4-D + 0.05 mg K		35
Petunia	0.1 mg NAA + 1 mg BA		41
Alfalfa	0.5 mg 2,4-D + 0.2 mg BA[a]	10 mg 2,4-D + 1 mg BA	36

BA, benzyl amino purine; 2,4-D, 2,4-dichlorophenoxy acetic acid; Glu, glutamine; IAA, indole acetic acid; K, kinetin; NAA, naphthalene acetic acid.
[a] Reported for induction of embryos.

using crude total plant DNA, easily prepared by a miniprep method (42) in which as little as 100 mg of tissue can be used. *Table 9* gives a step by step guide to this technique.

4.1 Dominant selectable marker genes

The establishment of transformation systems for any organism requires the development of dominant selectable markers which allow the selection of the cells that have acquired new genetic information. For plant cells several dominant selectable marker genes have been developed. At present, the most widely used selectable marker genes are based on the neomycin phosphotransferase (II) gene [NPT (II)] from the bacterial transposon Tn5. This gene confers

Table 9. Dellaporta plant DNA miniprep technique (42).

Reagents
1. Extraction buffer
       100 mM Tris–HCl, pH 8
       50 mM EDTA, pH 8
       500 mM NaCl
       10 mM mercaptoethanol
2. 20% SDS
3. 5 M KAc
4. Miracloth filter
5. Resuspension buffer I
       50 mM Tris–HCl, pH 8
       10 mM EDTA, pH 8
6. Resuspension buffer II
       10 mM Tris–HCl, pH 8
       1 mM EDTA, pH 8
7. 3 M NaAc
8. Isopropanol

Protocol
1. Weigh out at least 0.5–0.75 g of leaf tissue.
2. Freeze the tissue in liquid N_2 and grind to a fine powder in a small mortar and pestle. Transfer to a 30 ml Corex tube. Do not allow to thaw and do not cap tubes.
3. Add 15 ml of extraction buffer (at this stage if desired tissue may be further homogenized in a polytron). Add 1 ml of 20% SDS, mix thoroughly, shaking vigorously and incubate at 65°C for 10 min.
4. Add 5 ml of 5 M KAc, shake vigorously and incubate at 0°C for 20 min.
5. Spin at 25 000 g for 20 min.
6. Pour the supernatant through a Miracloth filter into a clean 30 ml tube containing 10 ml of isopropanol. Mix and incubate tubes at −20°C for 30 min.
7. Pellet DNA at 20 000 g for 15 min, remove the supernatant and invert the tubes on a clean paper towel to dry for 10 min.
8. Resuspend the pellets in 0.7 ml of resuspension buffer I and transfer to an Eppendorf tube. Spin in a microcentrifuge for 10 min.
9. Transfer the supernatant to a clean Eppendorf tube, add 75 μl of 3 M NaAc and 500 μl of isopropanol. Mix well and pellet the DNA in a microcentrifuge for 30 sec.
10. Wash the pellet in 80% ethanol, dry and re-dissolve in resuspension buffer II.

In our hands 1–2 g of leaf tissue yields 100–200 μg of DNA.

resistance in plant cells to aminoglycoside antibiotics, such as kanamycin and G418. The hygromycin phosphotransferase (HP) gene from *E.coli* has also been used for the construction of selectable marker genes for plant cells and although not as widely used as the NPT (II) gene, has also been successfully used with several plant species (43). A recent report (44) discusses the use of a bleomycin resistance marker in plant cell cultures. Bleomycin is a glycopeptide which interacts with DNA producing single- and double-strand breaks. The bleomycin resistance gene was also found on Tn5 and has been used successfully as a selectable marker gene in *Nicotiana plumbaginifolia*. Other selectable markers using genes such as the bacterial or mammalian dihydrofolate-reductase, have been constructed and used in some cases, but are not as generally useful as those based on the NPT (II) or HP genes. When developing the transformation system for a given species, before deciding which selectable marker to use it is very important to determine the susceptibility of the plant to the different selectable agents, and the minimal concentration of this agent which kills effectively most if not all of the non-transformed cells. Concentrations of antibiotics currently in use are shown in *Table 10*. New types of selectable markers are at the moment being developed which involve conferring herbicide resistance on a plant, such as glyphosate resistance (45). Resistant strains are selected by growth in the presence of the herbicide. Due to the commercial interest in such vectors and plants, these gene constructs are not widely available. A demonstration of selection of the R_1 progeny of a transformed plant on kanamycin containing media is shown in *Figure 3*.

4.2 mRNA analysis

There are several alternative and complementary ways to examine gene expression in transgenic plants. When intact genes are transferred from one plant species to another, gene expression can be analysed at the RNA level by hybridization techniques, or at the protein level by the use of specific antibodies. Several techniques of different degrees of complexity are used to determine the level of mRNA present in the different organs of a plant. One of the easiest and more sensitive techniques which requires small amounts of RNA is the dot blot hybridization technique.

The protocol below is a simple method provided by L.Willmitzer (Berlin) to isolate RNA from small amounts of tissue (~1 g) which can be used in a hybridization technique utilizing a dot blot device.

All glassware and solutions (except buffer A) should be sterilized and preferably treated with diethyl pyrocarbonate (DEPC).

Table 10. Concentrations of antibiotics used in tobacco transformation experiments.

Antibiotic	Concentration µg/ml
Kanamycin	75–100
Hygromycin	20–50
Methotrexate	0.5–1
Bleomycin	5–10

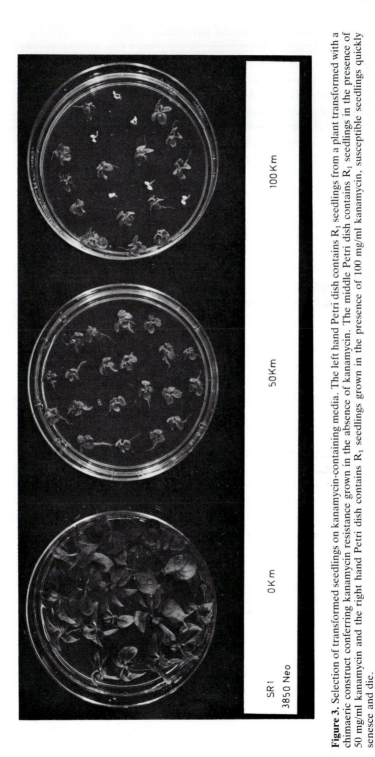

Figure 3. Selection of transformed seedlings on kanamycin-containing media. The left hand Petri dish contains R₁ seedlings from a plant transformed with a chimaeric construct conferring kanamycin resistance grown in the absence of kanamycin. The middle Petri dish contains R₁ seedlings in the presence of 50 mg/ml kanamycin and the right hand Petri dish contains R₁ seedlings grown in the presence of 100 mg/ml kanamycin, susceptible seedlings quickly senesce and die.

(i)   Freeze 1 g of tissue in liquid nitrogen.
(ii)   Homogenize with a mortar and pestle until a fine powder is obtained. Be careful never to allow the powder to thaw.
(iii)   Add 500 μl of buffer A (8 M guanidium hydrochloride, 20 mM MES, pH 7.0, 20 mM EDTA, 50 mM β-mercaptoethanol).
(iv)   Add more liquid nitrogen, and homogenize to a very fine paste.
(v)   Add 3 ml of buffer A.
(vi)   Transfer completely into a Corex tube.
(vii)   Extract with an equal amount of phenol/chloroform/iso-amyl alcohol in a 25:24:1 (by vol.) proportion.
(viii) Centrifuge for 30 min at 10 000 r.p.m. in a SS34 Sorvall rotor or equivalent.
(ix)   Transfer the upper phase to a clean tube.
(x)   Add $\frac{1}{20}$ vol. of 1 M CH_3COOH and 0.7 vol. of ethanol.
(xi)   Incubate at $-70°C$ for 30 min.
(xii)   Pellet the RNA by centrifuging for 5 min at 5000 r.p.m. in a SS34 Sorvall rotor or equivalent.
(xiii) Wash the pellet twice with 3 ml of 70% ethanol (room temperature).
(xiv)   Resuspend the pellet in 1 ml of cold buffer B (4 M potassium acetate, pH 4.8).
(xv)   Pellet the RNA (5 min at 5000 r.p.m. in a SS34 or equivalent).
(xvi)   Wash the pellet twice with 3 ml of 70% ethanol.
(xvii) Dry the pellet by incubating the tube upside down for 30 min.
(xviii) Dissolve the RNA in 200 μl of sterile water (if necessary heat up to 60°C).
(xix)   Take 10 ml for a spectrophotometric analysis.

When extracting from tissues other than leaves, it is sometimes necessary to include a precipitation step with 4 M LiCl after step (xii). Resuspend the pellet in 3 ml of water and add an equal volume of 3 M LiCl.

Techniques which are more complicated but that produce more precise data about the size or structure of the mRNA are the Northern blot and nuclease S1 mapping techniques discussed in the first chapter of this book.

4.3 Reporter genes

The use of intact genes can provide very useful and detailed information about the expression of a particular gene. However, this strategy has two major drawbacks.

(i)   Unless nuclear run-off transcriptional analysis is used, no difference between transcriptional regulation and mRNA stability can be made.
(ii)   Often it is not easy to find a DNA-specific probe which does not cross hybridize with mRNA from the host plant.

Therefore, for more detailed analyses of the transcriptional regulatory signals of a given gene, it is more straightforward to link this sequence to a reporter gene. Reporter genes are those which are well characterized both genetically and biochemically, and have a coding region which can be easily fused to the regulatory sequences of other genes. The enzymatic activity of most reporter genes is not normally present in the host plant, or is easily distinguishable from other endogenous gene activities. This allows the study of the regulation of a gene under different environmental conditions or in different organs of a plant. The methods for the evaluation of enzymatic activity of the most commonly used reporter genes in plant gene expression studies are described below.

4.3.1 *Octopine synthase*

The first reporter gene utilized in the development of expression vectors for plant cells was the octopine synthase gene. Octopine synthase is one of the enzymes encoded by the T-DNA of the Ti-plasmid and is responsible for the production of the opine, octopine. Cassette constructs containing the octopine synthase coding sequence flanked by restriction sites have been constructed previously (46,47). A simple protocol (39,48) to determine the enzymatic activity of this enzyme is presented here.

(i)   Grind 10–20 mg of plant tissue in extraction buffer (*Table 11*; buffer A; 1 ml/g of tissue) in an Eppendorf tube.
(ii)   Centrifuge briefly.
(iii)   Add 7 µl of enzyme substrate (*Table 11*, buffer 2).
(iv)   Incubate at room temperature for 1–10 h.
(v)   Prepare Whatman 540 paper for descending chromatography. Number the samples and space them 1.5 cm apart.
(vi)   Spot the following colour markers at each edge: 10 mg/ml Methyl Green, 10 mg/ml Orange G, 10 mg/ml Xylenecyanol (make up as a single solution).

Table 11. Reagents for octopine synthase assay.

---

1. Extraction buffer
      0.2 M Tris–HCl, pH 8
      0.5 M sucrose
      0.1% ascorbic acid
      0.1% cysteine–HCl
2. Enzyme substrate
      6.3 mg L-arginine–HCl
      8.2 mg sodium pyruvate
      0.5 ml H_2O
      0.5 ml 0.4 M sodium phosphate pH 6.8
      pH should be approximately 7

Dissolve 1 mg of NADH in above to a concentration of 14.2 mg/ml.

Prepare solutions fresh and keep on ice.

NOTE: The presence of octopine or nopaline in plant tissue can interfere with the enzymatic assay and should be removed by filtration of the plant extracts through a column of Sephadex G-100 before proceeding with the assay.

(vii)   Spot separately, the following control markers at each edge: 500 mg/ml octopine; 500 mg/ml nopaline, 500 mg/ml arginine.

(viii) Place the chromatogram in a tank for descending chromatography, using a buffer containing: *n*-propanol:ammonia (3:2). Leave to run overnight.

(ix)   Allow the chromatogram to dry then stain with a 1:1 mixture of 0.02% phenanthrequinone in 60% ethanol and 10% NaOH in 60% ethanol.

(x)   Dry and examine under UV illumination.

(xi)   Alternatively, samples may be spotted on a 20 × 30 cm sheet of Whatman 3 MM paper 4.5 cm from one edge. The paper is then placed in an electrophoresis apparatus with the samples at the anodal side. The running buffer is 5% formic acid, 15% acetic acid, in H_2O (pH 1.8). Samples are run at 20 V/cm until the green marker reaches the edge (~50 min). The paper is then dried and treated as in steps (ix) and (x).

The corresponding opine synthase gene from the nopaline T-DNA, nopaline synthase, has also been used as a reporter gene (49,50).

4.3.2 *Chloramphenicol acetyl transferase*

To date, the most frequently used reporter gene in prokaryote, mammalian and plant cells is the chloramphenicol acetyl transferase (CAT) gene (51,52). The CAT enzyme specifically acetylates chloramphenicol to its 1, 3 and 1–3 derivatives, and is used by bacteria as a mechanism of resistance to the antibiotic. The enzyme assay reaction mix (*Table 12*) contains the substrates [^{14}C]chloramphenicol and acetyl CoA. After incubation with a plant extract the chloramphenicol and its derivatives are extracted with an organic solvent and then resolved by TLC on silica gel plates. For plant extracts which contain large amounts of polyphenoloxidases it is recommended that ascorbic acid and L-cysteine are included in the reaction mix to avoid the formation of certain compounds which inhibit CAT activity. Before going ahead with transformation experiments using the CAT gene in a new plant species, it is wise to determine that there are no non-specific acetylases that can use chloramphenicol as a substrate, to avoid the presence of high levels of background activity.

4.3.3 *Neomycin phosphotransferase (II)*

This enzyme specifically phosphorylates aminoglycoside antibiotics of the neomycin family such as kanamycin and G418. The gene, originally found on transposon Tn5, has also been used previously in mammalian and prokaryote systems. In these systems the assays are carried out by incubating cell extracts with kanamycin and ^{32}P, then spotting the mixture on phosphocellulose paper (P81 paper) and counting in a scintillation counter. Plants however, have non-specific phosphorylases which can also phosphorylate kanamycin and mask the NPT II effects. Therefore it is necessary to use a gel assay system. This was first developed for mammalian cells (53) and later adapted for use in plant cell systems (54). The gel assay protocol is outlined below.

Steps ii–xii (with the exception of the protein determination) are all carried out in a cold room. Reagents for this assay are described in *Table 13*.

Table 12. CAT assay.

Reagents
1. Extraction buffer
        0.5 M   sucrose[a]
        0.25 M   Tris-HCl
        1%   ascorbic acid[a]
        0.5 mM   leupeptin
        10 mM   EDTA
        1%   cysteine–HCl pH 7.5[a]
2. [^{14}C]chloramphenicol
        1.5 mCi/nmol
3. Acetyl coenzyme A
        4 mg/ml in H_2O (always prepare fresh solution)
        Keep reagents on ice until needed.

Protocol
1. Take 100 mg of tissue in an Eppendorf tube. Include an untransformed negative control and a sonicated extract of *E.coli* containing pBR325 (harbouring the CAT gene) as a positive control.
2. Add 100 μl of extraction buffer to the tissue.
3. Homogenize the tissue with a glass rod.
4. Heat for 3 min at 65°C and centrifuge to eliminate the precipitate.
5. Add 10 μl of [^{14}C]chloramphenicol.
6. Heat the reaction mix to 37°C for 5 min, then add acetyl CoA to 0.3 mM.
7. Incubate at 37°C for 45 min.
8. Add 0.5 ml of ethyl acetate, vortex, centrifuge for 2–3 min and remove supernatant to a clean Eppendorf tube. Repeat the extraction twice and pool the supernatants.
9. Dry in a rotary evaporator.
10. Resuspend in 10 μl of ethyl acetate.
11. Run ascendant thin layer chromatogram on 20 × 20 silica gel plates. Spot the samples 2 cm from edge and 1.5 cm apart. Spot 1 μl at a time and dry with a hairdryer. Run the chromatogram in chamber saturated with and containing chloroform:methanol, 95:5.
12. Expose to autoradiography for 24–48 h.

[a] For most plants a 0.25 M Tris–HCl solution is sufficient as extraction buffer. Ascorbic acid and cysteine are most useful for plant extracts which easily oxidize.

(i)   First prepare a polyacrylamide gel and leave at 4°C to cool.
(ii)   Prepare three sets (a, b, c) of Eppendorf tubes for each sample and keep on ice:
         (a) containing 50 μl of extraction buffer (*Table 13*);
         (b) containing sterile water for protein determination (Biorad);
         (c) containing 5 μl of loading buffer (*Table 13*).
(iii)   Add 70–100 mg of plant tissue to the tubes containing extraction buffer.
(iv)   Grind the tissue in extraction buffer at 4°C with a glass rod.
(v)   Spin the samples for 2–3 min in an Eppendorf centrifuge.
(vi)   Transfer the samples to tubes (b) for protein determination (Biorad).
(vii)   Transfer the rest of the supernatant to the tubes containing loading buffer.
(viii) Determine the protein content in the samples and load equivalent amounts of each sample on gel. 20–100 μg protein is optimal.

Table 13. Reagents for neomycin phosphotransferase assay (NPT II)

1. Extraction buffer (2×)
    1% β-mercaptoethanol
    50 mM Tris–HCl, pH 6.8
    0.13 mg/ml leupeptin
2. Loading buffer (10×)
    50% glycerol
    0.5% SDS
    10% β-mercaptoethanol
    0.005% Bromophenol blue
3. Running buffer (1×)
    60 g Tris
    144 g glycine in 10 litres
4. Reaction buffer (2×)
    100 mM Tris–HCl, pH 7.5
    50 mM $MgCl_2$
    400 mM NH_4Cl
    1 mM DTT
5. Washing solution (100×)
    0.5 M Na_2HPO_4
    Dilute to 1× before use
6. 10% Polyacrylamide gel

		Separation	Stacking
Acrylamide	30%	10.00 ml	1.67 ml
Bisacrylamide	2%	3.90 ml	1.30 ml
1 M Tris–HCl pH 8.7		11.20 ml pH 6.8	1.25 ml
H_2O		4.66 ml	5.67 ml
Ammonium persulphate	10%	0.20 ml	0.10 ml
Temed		0.04 ml	0.01 ml
Total		30.00 ml	10.00 ml

7. Agarose gel (25 ml is enough for 140 cm gel)
    12.5 ml 2× reaction buffer
    12.5 ml water
    0.25 g agarose

Melt agarose and allow to cool to 45°C.

Add 25 μl of 100 mg/ml kanamycin sulphate
    10 μl of [^{32}P]γ-ATP, sp. act. 3000 mCi/mM

(ix)   Run the gel at 85 V for 8–10 h at 4°C, or until blue dye front is at least 10 cm from the bottom of the stacking gel.

(x)   Take the gel and remove the stacking gel and anything lower than the dye front. Remember to mark the orientation of the gel.

(xi)   Wash twice in cold water for 5 min at 4°C.

(xii)   Equilibrate for 30 min in reaction buffer (*Table 13*) at 4°C with shaking.

(xiii)   While the gel is equilibrating prepare the agarose gel solution omitting the γ-[^{32}P]ATP.

(xiv)   Place the gel in a shallow tray which fits neatly (e.g. a sandwich box lid).

(xv)·   Add the γ-[^{32}P]ATP to the agarose and pour over the polyacrylamide gel taking care to cover it completely.

(xvi)   Allow the agarose to dry and overlay with one piece of P81 (Whatman) paper, wetted in reaction buffer and cut to gel size.

(xvii) Leave to incubate for 30 min at 37°C.

(xviii) Add three layers of 3 MM paper and 3–4 cm of paper towels, cover with a glass plate and add a 1 kg weight. Blot for 2–3 h.

(xix)   Remove the P81 paper and wash 4–5 times for 10 min each in hot (80°C) washing solution.

(xx)   Dry the P81 partially and expose to autoradiography overnight.

4.3.4 β-Galactosidase

The reporter enzyme used in the very earliest cloning experiments in bacteria was β-galactosidase and this system too has been adapted for use in plant cells. A gel system similar to that for the NPT II is used to separate the proteins and the gel is then stained in 4-methylumbelliferyl-β-D-galactoside (4-MUG) and analysed under UV light. β-Galactosidases also exist in plants but these are easily separated from the *E.coli* β-galactosidase in the gel. A protocol for the β-galactosidase assay is presented below with kind permission of T.Teeri (University of Helsinki).

Buffers and solutions for this assay are described in *Table 14*.

(i)   Prepare an 8% SDS–polyacrylamide gel (*Table 13*) and cool to 4°C before running.

(ii)   Take 70–100 mg of plant tissue.

(iii)   Add to 50 μl extraction buffer (*Table 14*) and homogenize with a glass rod.

(iv)   Spin for 2–3 min (microcentrifuge) and remove the supernatant to a clean Eppendorf tube.

(v)   Just before loading add ⅓–½ volume of sample buffer to each sample. DO NOT BOIL. Add directly to the gel (SDS will often precipitate but dissolves as samples enter the gel).

(vi)   Run the gel at 100 V for 12 h (for 15 cm gel). The dye front leaves the gel but β-galactosidase is a large protein and is retained.

(vii)   Cut off the separation gel, marking the orientation and soak the gel for 5, 10 and 15 min in changes of Z-buffer (*Table 14*). This can be done at room temperature.

(viii) Prepare fresh 4-MUG solution (20 mg of 4-MUG in 1 ml of DMSO). Add this to 1 litre of Z-buffer.

(ix)   Pour this immediately onto the drained gel.

(x)   Shake for 5–10 min.

(xi)   Wash the gel with water.

(xii)   Examine the gel on a UV transilluminator and photograph as for a normal agarose gel. Pale yellow filters are useful to block out the UV (Kodak Wrattan 2 E).

Table 14. Reagents for β-galactosidase assay and preparation of a positive control.

Reagents
1. Extraction buffer (2×)
    1% β-mercaptoethanol
    50 mM Tris–HCl, pH 6.8
    0.13 mg/ml leupetin
2. STET buffer
    8% sucrose
    5% Triton X-100
    50 mM EDTA
    50 mM Tris–HCl, pH 8
3. Z-buffer
    100 mM sodium phosphate (pH 7)
    (pH 7 = 60 mM Na_2HPO_4/40 mM NaH_2PO_4)
    10 mM KCl
    1 mM $MgSO_4$
    40 mM β-mercaptoethanol
4. Sample buffer
    (for 3 ml)
    1.6 ml 2× Upper gel buffer (*Table 13*)
    1.2 ml 10% SDS
    0.15 ml 0.1% Bromophenol blue
    0.15 ml β-mercaptoethanol

Preparation of E.coli extract for use as positive control
1. Grow an *E.coli* strain with wild-type *LacZ* operon in rich medium to late log phase.
2. For induction add 0.4% lactose (no glucose) or 5 mM IPTG (isopropyl-1-thio-β-D-galactoside) for several hours.
3. Pellet 1 ml of the culture in an Eppendorf tube in a microcentrifuge.
4. Resuspend the pellet in 50 μl of STET buffer.
5. Add 0.5 ml of extraction buffer and vortex well.
6. Spin down for 10 min in a microcentrifuge.
7. Mix the supernatant with 1 vol. of glycerol.
8. Store at −20°C.

The gel may also be stained in X-Gal (5-bromo-4-chloro-3-indolyl-β-D-galactoside; 200 μl of 2% solution in 30 ml buffer) overnight, although the sensitivity is considerably lower.

The gel assay systems of the NPT II and β-galactosidase are also useful for examining fusion proteins produced by constructing in frame chimaeric gene constructs. The fusion proteins can be detected by their differential mobility within the gel in comparison to the normal NPT II or β-galactosidase protein. Both of these proteins can tolerate fused protein sequences with very little effect on their enzymatic activity.

4.3.5 *Luciferase*

Reporter enzyme systems are constantly being improved and new markers developed. One of the most spectacular systems recently published (55) involves the luciferase gene which was transformed into tobacco plants. When the plants are watered with the enzyme substrate luciferin they luminesce, making a quite

Table 15. Luciferase assay.

Reagents
1. Extraction buffer
        100 mM potassium phosphate, pH 7.5
           1 mM dithiothreitol
2. Assay buffer
        14 mM glycylglycine buffer, pH 7.8
        14 mM $MgCl_2$
         6 mM ATP

Procedure
1. Grind 50–200 mg of tissue in an equal amount of extraction buffer in a 1.5 ml microcentrifuge tube in the presence of liquid N_2.
2. Centrifuge in a microcentrifuge for 5 min at 4°C.
3. Transfer the supernatant to a clean tube.
4. Take an aliquot and determine the protein content (Biorad).
5. Mix 50 µl of the extract with 450 µl of assay buffer.
6. Add 100 µl of 1 mM luciferin.
7. Measure the peak intensity with a luminometer (e.g. LKB, model 1250).

beautiful effect! A protocol for the assay of luciferase (55) is presented in *Table 15*.

Note added in proof. The recently developed *E.coli* β-glucuronidase (GUS) reporter gene (57) is described in the Appendix (p.301).

5. FACTORS INFLUENCING GENE EXPRESSION IN TRANSFORMED PLANTS

Many factors may influence gene expression in transformed plants. Two of the most important factors are the position and the number of integrations of the foreign sequence within the plant genome. Several authors (e.g. 27,47) have reported significant differences in the levels of expression between different transformation events utilizing the same foreign sequences. This has been observed in both callus tissue and in mature transformed plants. Up to a 200-fold difference in expression may be observed between individual calli and there seems to be little relationship between expression levels of two distinct genes transferred in the same T-DNA (27).

Simpson *et al.* (56) have shown that not only quantitative, but also qualitative changes may occur, resulting in altered organ-specific patterns of expression, albeit at a lower frequency. This arose from studying the organ-specific expression of a chimaeric gene under the control of the upstream (5′) sequence of the small subunit ribulose bisphosphate carboxylase gene (ss-rubisco), which is normally expressed only in green tissues (i.e. leaves and stem). It was found that in one transgenic plant this gene was expressed at high levels in petals, an organ in which rubisco genes are not normally expressed at significant levels. These effects are presumably due to regulatory sequences around the site of insertion of the foreign sequence which can influence genes over long stretches of DNA.

The copy number of the transferred DNA may also be a factor affecting the level of expression, although no correlation between copy number and level of expression has been found so far. Nevertheless, in most *Agrobacterium* transformed plants it has been shown that only one functional locus of the T-DNA is present in the genome of the plant cells, as determined by Southern blot hybridization and analysis of T-DNA segregation in the progeny of transgenic plants. This latter analysis however does not distinguish between a single copy or tandem copies of the T-DNA in the same chromosomal location.

Other factors that have to be taken into consideration when studying gene expression are the developmental stage of the plant, light intensities, and additives such as sucrose and hormones included in the culture media. Although not very practical, it is recommended to perform the final analysis of the expression of genes in the progeny of transgenic plants, to eliminate the effects of passing through tissue culture on the original transformed plant.

6. CONCLUSIONS

The development of 'reverse-genetic' techniques in plants has allowed us to begin to study the regulatory systems involved in light-inducible and develop-mentally regulated expression in plant cells. These techniques have also been used to target foreign proteins to plant cell organelles *in vivo*, and will be useful in elucidating the signal sequences required for proteins to enter and pass through the various membrane systems of plant cells. Perhaps the greatest interest currently is in the potential for improvement of agriculturally important plant species. Transgenic plants resistant to viruses or certain insects have already been produced and within the next few years much research will be concentrated on exploiting this potential to the full.

7. ACKNOWLEDGEMENTS

Grateful thanks to Drs Robert Horsch and Robert Fraley of Monsanto and Dr Peter van den Elzen of Mogen for supplying articles and protocols used in this chapter, Leticia Anguiano J., and Gabriela Padilla N., for preparation of the manuscript, Luis Alberto Tinoco for figures and Alba Jofre and Diego González de León for preparation of photographs.

8. REFERENCES

1. Krens,F.A., Molendijk,L., Wullems,G.J. and Schilperoort,R.A. (1982) *Nature*, **296**, 72.
2. Fromm,M., Taylor,L. and Walbot,V. (1985) *Proc. Natl. Acad. Sci. USA*, **82**, 5824.
3. Márton,L., Wullems,G.J., Molendijk,L. and Schilperoort,R.A. (1979) *Nature*, **277**, 129.
4. Braun,A.C. (1978) *Biochim. Biophys. Acta*, **516**, 167.
5. Braun,A.C. (1982) In *Molecular Biology of Plant Tumours*. Kahl,G. and Schell,J. (eds), Academic Press, New York, p. 155.
6. Zaenen,I., Van Larebeke,N., Teuchy,H., Van Montagu,M. and Schell,J. (1974) *J. Mol. Biol.*, **86**, 109.
7. Van Larebeke,N., Engler,G., Holsters,M., Van den Elsacker,S., Zaenen,I., Schilperoort, R.A. and Schell,J. (1974) *Nature*, **252**, 169.
8. Watson,B., Currier,T.C., Gordon,M.P., Chilton,D.-D. and Nester,E.W. (1975) *J. Bacteriol.*, **123**, 255.

9. Chilton,D.-D., Drummond,M.H., Merlo,D.J., Sciaky,D., Montoya,A.L., Gordon,M.P. and Nester,E.W. (1977) *Cell*, **11**, 263.
10. Willmitzer,L., De Beuckeleer,M., Lemmers,M., Van Montagu,M. and Schell,J. (1980) *Nature*, **287**, 359.
11. Schröder,G., Waffeneschmidt,S., Wiler,E.W. and Schröder,J. (1984) *Eur. J. Biochem.*, **138**, 387.
12. Inzé,D., Follin,A., Van Lijsebettens,M., Simoens,C., Genetello,C., Van Montagu,M. and Schell,J. (1984) *Mol. Gen. Genet.*, **194**, 265.
13. Tempé,J. and Petit,A. (1982) In *Molecular Biology of Plant Tumours*. Kahl,G. and Schell,J. (eds), Academic Press, New York, p. 451.
14. Lemmers,M., De Beuckeleer,M., Holsters,M., Zambryski,P., Depicker,A., Hernalsteens, J.P., Van Montagu,M. and Schell,J. (1980) *J. Mol. Biol.*, **144**, 353.
15. Ursic,D., Slightom,J.L. and Kemp, J.D. (1983) *Mol. Gen. Genet.*, **190**, 494
16. Thomashow,M.F., Nutter,R., Montoya,A.L., Gordon,M.P. and Nester,E.W. (1980) *Cell*, **19**, 729.
17. De Beuckeleer,M., Lemmers,M., De Vos,G., Willmitzer,L., Van Montagu,M. and Schell,J. (1981) *Mol. Gen. Genet.*, **183**, 283.
18. Zambryski,P., Depicker,A., Kriger,K. and Goodman,H. (1982) *J. Mol. Appl. Genet.*, **1**, 361.
19. Zambryski,P., Joos,H., Genetello,C., Leemans,J., Van Montagu,M. and Schell,J. (1983) *EMBO J.*, **2**, 2143.
20. Shaw,C.H., Watson,M.D., Carter,G.M. and Shaw,C.H. (1984) *Nucleic Acids Res.*, **12**, 6031.
21. Hernalsteens,J.P., Van Vliet,F., De Beuckeleer,M., Depicker,A., Engler,G. Lemmers,M., Holsters,M., Van Montagu,M. and Schell,J. (1980) *Nature*, **287**, 654.
22. Wang,K., Herrera-Estrella,L., Van Montagu,M. and Zambryski,P. (1984) *Cell*, **38**, 455.
23. Lichtenstein,C. and Draper,J. (1985) In *DNA Cloning—A Practical Approach*. Glover,D.M. (ed.), IRL Press, Oxford, Vol. II, p. 67.
24. Van Haute,E., Joos,M., Maes,M., Warren,G., Van Montagu,M. and Schell,J. (1983) *EMBO J.*, **2**, 411.
25. Bevan,M. (1984) *Nucleic Acids Res.*, **12**, 8177.
26. Hoekema,A., Van Haaren,M.J.J., Fellinger,A.J., Hooykas,P.J.J. and Schilperoort,R.A. (1985) *Plant Mol. Biol.*, **5**, 85.
27. An,G. (1986) *Plant Physiol.*, **81**, 86.
28. Simoens,C., Alliotte,Th., Mendel,R., Müller,A., Schiemann,J., Van Lijsebettens,M., Schell, J., Van Montagu,M. and Inzé,D. (1986) *Nucleic Acids Res.*, **14**, 8075.
29. Van den Elzen,P., Lee,K.Y., Townsend,J. and Bedbrook,J. (1985) *Plant Mol. Biol.*, **5**, 149.
30. Joos,H., Inzé,D., Caplan,A., Sormann,M., Van Montagu,M. and Schell,J. (1983) *Cell*, **32**, 1057.
31. Fraley,R.T., Rogers,S.G., Horsch,R.B., Eichholtz,D.A., Flick,J.S., Fink,C.L., Hoffmann, N.L. and Sanders,P.R. (1985) *Bio/technology*, **3**, 629.
32. De Blaere,R., Bytebier,B., De Greve,H., Deboek,F., Schell,J., Van Montagu,M. and Leemans,J. (1985) *Nucleic Acids Res.*, **13**, 4777.
33. Stachel,S.E., Timmerman,B. and Zambryski,P. (1987) *Nature*, in press.
34. McCormick,S., Niedermeyer,J., Fry,J., Banason,A., Horsch,R. and Fraley,R. (1986) *Plant Cell Rep.*, **5**, 81.
35. An,G., Watson,B.D. and Chiang,C.C. (1986) *Plant Physiol.*, **81**, 301.
36. Deak,M., Kiss,G.B., Koncz,C. and Duclits,D. (1986) *Plant Cell Rep.*, **5**, 97.
37. Murashige,T. and Skoog,G. (1962) *Physiol. Plant*, **15**, 473.
38. Nagy,J. and Maliga,P. (1976) *Z. Pflanzen Physiol.*, **78**, 453.
39. Aerts,M., Jacobs,M., Hernalsteens,J.P., Van Montagu,M. and Schell,J. (1979) *Plant Sci. Lett.*, **17**, 43.
40. Leemans,J., Shaw,C., Deblaere,R., OcGreve,H., Hernalsteens,J.P., Maes,M., Van Montagu, M. and Schell,J. (1981) *J. Mol. Appl. Genet.*, **1**, 149.
41. Horsch,R.B., Fry,J.E., Hoffmann,N.L., Eichholtz,D., Rogers,S.G. and Fraley,R.T. (1985) *Science*, **227**, 1229.
42. Dellaporta,S.L., Wood,J. and Hicks,J.B. (1983) *Plant Mol. Biol. Rep.*, **1**, 19.
43. Van den Elzen,P.J.M., Townsend,J., Lee,K.Y. and Bedbrook,J.R. (1985) *Plant Mol. Biol.*, **5**, 299.
44. Hille,J., Verheggen,F., Rocluink,P., Franssen,H., Van Kammen,A. and Zabel,P. (1986) *Plant Mol. Biol.*, **7**, 171.
45. Comai,L., Sen,L.C. and Stalker,D.M. (1983) *Science*, **221**, 370.
46. Herrera-Estrella,L., Depicker,A., Van Montagu,M. and Schell,J. (1983) *Nature*, **303**, 209.

47. Jones,J.D.G., Dunsmuir,P. and Bedbrook,J. (1985) *EMBO J.*, **4**, 2411.
48. Otten,L.A. and Schilperoort,R.A. (1978) *Biochim. Biophys. Acta*, **527**, 497.
49. Shaw,C.H., Carter,G.H., Watson,M.D. and Shaw,C.H. (1984) *Nucleic Acids Res.*, **12**, 7831.
50. Shaw,C.H., Sanders,D.M., Bates,M.R. and Shaw,C.H. (1986) *Nucleic Acids Res.*, **14**, 6603.
51. Shaw,W.V. (1983) *Crit. Rev. Biochem.*, **14**, 1.
52. Gorman,C.M., Moffat, L.F. and Howard,B.H. (1982) *Mol. Cell. Biol.*, **2**, 1044.
53. Reiss,B., Sprengel,R., Will,H. and Schaller,H. (1984) *Gene*, **30**, 211.
54. Van den Broek,G., Timko,M.P., Kausch,A.P., Cashmore,A.R., Van Montagu,M. and Herrera-Estrella,L. (1985) *Nature*, **313**, 358.
55. Ow,D.W., Wood,L.V., De Luca,M., De Wet,J.R., Helinski,D.R. and Howell,S.H. (1986) *Science*, **234**, 856.
56. Simpson,J., Van Montagu,M. and Herrera-Estrella,L. (1986) *Science*, **233**, 34.
57. Jefferson,R.A. (1987) *Plant Mol. Biol. Rep.*, **5**, 387.

CHAPTER 7

Protoplast isolation and transformation

R.D.SHILLITO and M.W.SAUL

1. INTRODUCTION

The transformation of isolated plant protoplasts by naked DNA and other methods has recently become a routine procedure. There are a number of ways in which protoplasts can be induced to take up and integrate DNA, including the use of polyethylene glycol (PEG) and electroporation. These methods using naked DNA have become known by the general term of Direct Gene Transfer. They can be applied to both monocots (1,2) and dicots (3,4), but have the limitation that DNA can only be transferred into protoplasts, and these protoplasts must be able to divide and form colonies. For this to be of use, one must then be able to regenerate a normal plant from the tissue produced by this process. While this is the case for many solanaceous plants, such as tobacco, petunia and potato, this is not true for the majority of the important graminaceous monocot plants which make up most of the world's food supply.

The techniques of spheroplast and liposome fusion (5,6) may also be used to transfer DNA, but there is little to be gained over naked DNA transformation from using these techniques. The use of microinjection (7,8) can give high tranformation frequencies, but only a limited number of protoplast derived cells can be treated. However, if this could be applied directly to cell transformation, then it might afford a generally useful method of genetic modification.

It is not possible within the scope of this article to review all of the factors involved in protoplast isolation. We therefore will discuss some of these factors in a general sense, and refer the reader to the copious literature available in the field (9). Protocols are given for the isolation and transformation of protoplasts, and the regeneration and characterization of transformed plants of *Nicotiana tabacum* strain SR1, which can easily be applied to other tobacco lines and to other protoplast systems.

2. PROTOPLAST ISOLATION

2.1 Choice of source tissue

The plant material which is often chosen as a source of protoplasts in dicots is leaf mesophyll. This is because this tissue is available in large amounts and yields fairly uniform protoplasts with a good yield. In some cases, the leaf material is taken from shoot cultures maintained axenically in a sterile culture system (10,11). This method obviates the need to sterilize tissue, and provides stable and reproducible material which is free from the vagaries of the weather, and has a reduced seasonal response compared to greenhouse material.

Protoplasts can also be isolated from cell cultures (normally suspension) or from other portions of the whole plant. The use of roots and seedlings as a source of protoplasts has been particularly successful in the case of forage legumes (12,13).

In the case of graminaceous monocots, protoplasts isolated from the whole plant rarely divide, and certainly have not yet been shown to be able to form cultures and regenerate into plants. In these cases the protoplasts normally come from non-morphogenic suspension cultures (14,15) or, rarely, from embryogenic suspensions (16,17). Only protoplasts from embryogenic rice suspensions have been reported to be regenerable to plants in a repeatable way (18,19).

2.2 Preparation of protoplasts

Protoplasts, being devoid of their cell walls, are osmotically fragile. Their internal pressure has therefore to be balanced by the use of an osmotically active suspending buffer solution with an osmotic potential of 400–700 mOs/kg H_2O. Either a salt or sugar (or sugar alcohol) solution or a mixture of the two is normally used for this purpose. For the enzyme digestion step, this can also be a salt solution. The choice of osmoticum is usually determined by whether one wishes to sediment or float the protoplasts (see Section 2.4). Typical solutions consist of mannitol (e.g. 0.4 M) with a low concentration of calcium chloride to stabilize the membranes, or a mixture of calcium, magnesium and potassium chlorides. The experimenter should determine the best mixture for their particular application.

The enzymes used in the preparation of protoplasts are mostly derived from fungi. The preparations are crude and contain a cocktail of enzymes, including, in most cases, some protease, DNase and RNase activity. Historically, a macerase was first added to separate the cells, the cell walls of which were then removed by addition of a cellulase (20). Most researchers now employ a one-step technique in which both enzymes are added simultaneously. There are some enzyme preparations which contain both activities, as well as some tissues which require only one enzyme type in order for protoplasts to be released (21).

In order for the enzymes to penetrate the tissue, it is usually necessary to slice the tissue in the enzyme solution and/or to vacuum infiltrate the enzyme into the tissue, neither of which are required for suspension culture cells.

2.3 Protoplast isolation (cleaning)

After digestion of the tissue, the protoplasts have to be separated from undigested material, the enzymes themselves, and any toxic materials produced during the digestion. This is achieved by a two-step process, consisting of filtration through a 50–100 μm stainless steel sieve, followed by a series of washing steps. Sieves suitable for such work obtained from Saulas et Cié, Paris, France are shown in *Figure 1a*. A flotation step may also be included, in order to separate the heavier undigested cells and cell wall material from protoplasts.

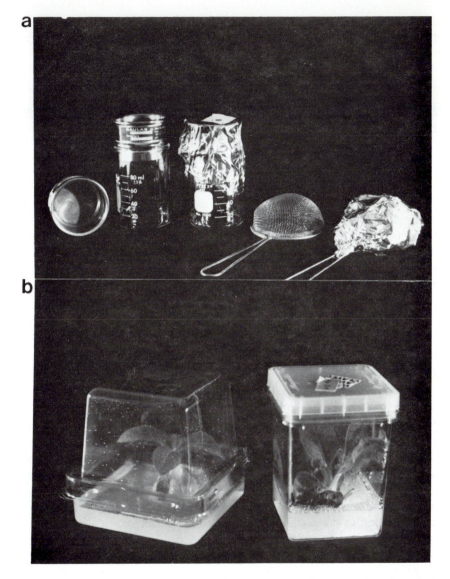

Figure 1. (a) Sieves used for filtration of protoplast preparations. Shown are sieves from Saulas, packaged for sterilization, and a wire mesh tea strainer. (b) Containers used to grow sterile shoots of *N.tabacum* SR1. Left: PlantCon (Flow Labs); Right: GA7 (Magenta Corp.).

Washing is normally carried out using salt and sugar, or sugar alcohol solutions similar or identical to those used for the digestion step, in which the protoplasts sediment, and sucrose solutions (see protocols in Section 7), or salt solutions containing Percoll (22), in which they float. It is not always possible to carry out this separation cleanly, and many preparations, especially those from suspension cultures, contain some undigested cells. In most cases these cells degenerate and die, but the presence of such clumps of cells, which may act as a

163

source of meristematic tissue, has been a major source of controversy in the field of protoplast culture from graminaceous monocots, where the protoplast preparations have often not been cleaned by flotation (23).

2.4 **Culture of protoplasts**

The protoplasts which are obtained after the cleaning steps have to be suspended in a suitable medium in order to allow them to reform a cell wall and initiate divisions (*Figure 2*). There are very many media formulations used, and the formulation will have to be modified to suit the particular species and even the variety of the species used. Most media are based on variations of existing cell culture media with an osmotic stabilizing agent added. Undefined additions such as coconut milk are often made. Some media (24) contain a wide range of vitamin and other supplements, and have been found to be generally useful for protoplast culture (25). The formulations used in the protocols described here are given in *Table 1*. The experimenter is referred to the literature for the medium which best suits their particular species or cell type. However, there are a number of guidelines which can be followed when setting up a culture system.

2.4.1 *Growth regulators*

A protoplast culture will in general require higher levels of plant growth regulators, in the initial stages, than the corresponding suspension or callus culture. A reduction in the levels of plant growth regulators may be required within 1 or 2 weeks of the culture initiation (22).

2.4.2 *Osmoticum*

The osmotic pressure in the culture medium should not differ from that used for the isolation and washing steps. Any osmotic shock will reduce the viability of the protoplasts and reduce the division frequency. The osmotic pressure should be reduced in the weeks following culture of the protoplasts, together with the level of plant growth regulators. In general, the amount of osmotic stabilizer which is added to the medium, in addition to the sucrose or glucose added to supply a carbohydrate source, should be reduced by ⅓ to ¼ for each week following the first week of culture. This may not be optimum for every culture system.

The type of osmoticum used will depend on the particular protoplasts. One of sorbitol, mannitol, glucose or sucrose are generally used. The bulk of the osmotic stabilization can be made up of any of these four. The use of sucrose will in general increase the extent of chloroplast dedifferentiation in the cultures. Recently the use of mellibiose as osmotic stabilizer has been suggested (26).

2.4.3 *Temperature*

Most protoplasts divide best at a temperature of between 22 and 28°C. However, there may be beneficial effects from culture at low temperatures (e.g. 12°C) for a short period after isolation.

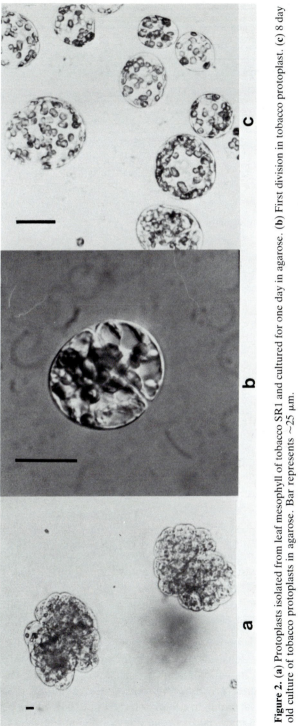

Figure 2. (a) Protoplasts isolated from leaf mesophyll of tobacco SR1 and cultured for one day in agarose. **(b)** First division in tobacco protoplast. **(c)** 8 day old culture of tobacco protoplasts in agarose. Bar represents ~25 μm.

Table 1a. Composition of the culture media: salt and vitamin components.

	Medium					
	T/N^b	LS	K3	H/J^c	W5	F
Macroelements[a]: (mg/ml final concentration)						
KCl					380	368
KNO_3	950	1900	2500	1900		
KH_2PO_4	68	170		170		
NH_4NO_3	720	1650	250	600		
$CaCl_2 \cdot 2H_2O$	220	440	900	600	18 375	18 400
$MgSO_4 \cdot 7H_2O$	185	370	250	300		
$(NH_4)_2SO_4$			134			
NaCl					9000	8000
$Na_2HPO_4 \cdot 2H_2O$						125
Microelements[a]: (mg/l final concentration)						
Na_2EDTA	74.6	74.6	74.6	74.6		
$FeCl_3 \cdot 6H_2O$	27.0	27.0	27.0	27.0		
H_3BO_3	10.0	6.2	3.0	3.0		
KI		0.83	0.75	0.75		
$MnSO_4 \cdot H_2O$	17.25	16.9	10.0	10.0		
$ZnSO_4 \cdot 7H_2O$	10.0	8.6	2.0	2.0		
$CuSO_4 \cdot 5H_2O$	0.025	0.025	0.025	0.025		
$Na_2MoO_4 \cdot 2H_2O$	0.25	0.25	0.25	0.25		
$CoCl_2 \cdot 6H_2O$			0.025	0.025		
$CoSO_4 \cdot 7H_2O$		0.03				
Vitamins and other organics[d]: (mg/l final concentration)						
Biotin	0.05			0.01		
Pyridoxine HCl	0.5		0.1	1.0		
Thiamine HCl	0.5	0.04	1.0	10.0		
Nicotinamide				1.0		
Nicotinic acid	5.0		0.1			
Folic acid	0.5			0.4		
Glycine	2.0			0.1		

2.4.4 Light

The light regime can be important in many cultures. Some cultures require that they be grown in the dark, at least to begin with, and high light intensities (>20 $\mu E/m^2/sec$) will tend to reduce protoplast vigour.

2.4.5 Use of feeder layers and agarose systems

Feeder layers have been and are still being used for culture of cells and protoplasts from low densities as well as for culture of cells arising from co-cultivation. Methods in which the protoplasts do not need to be physically separated from the nurse are those employing X-irradiated protoplasts (27) or auxotrophic protoplasts which are later selected out (28). Filter papers are used to separate the nursed material from the nurse, on agar plates in the methods of

Table 1b. Composition of the culture media: sugars and hormones.

	Medium							
	T/N^b	LS	K3	K3C	K3E	H^c	J^c	W5
Sugars and sugar alcohols: (g/l final concentration)								
Sucrose	10.0	30.0	102.96	102.96	36.0	0.25	0.25	
Glucose						68.40	21.0	9.0
m-Inositol	100.0	100.0	100.0	100.0	100.0			
Xylose			0.25	0.25	0.25	0.25	0.25	
Hormones: mg/l final concentration								
2,4-D		0.05	0.1	0.05	0.05	0.1		
NAA		2.0	1.0	2.0	2.0	1.0	1.0	
BAP		0.1	0.2	0.1	0.1	0.2	0.25	
Kinetin		0.1		0.1	0.1			
Final pH (with KOH or HCl)								
	5.0	5.8	5.8	5.8	5.8	5.8	5.8	6.0

Where the inorganic component is common i.e. media H/J, these have been given together.
[a] Macroelements are usually made up as a 10× concentrated stock solution, and microelements as a 1000× concentrated stock solution.
[b] Nitsch 1972 is this medium with half the concentration of Na_2EDTA, 27.9 g/l $FeSO_4 \cdot 7H_2O$ in place of the $FeCl_3$, 20 g/l sucrose and pH 6.0.
[c] These media contain a complex mixture of defined additions as follows:
Citric, fumaric and malic acid (each 40 mg/l final conc.) and sodium pyruvate (20 mg/l) are prepared as a 100× concentrated stock solution, adjusted to pH 6.5 with NH_4OH, and added. Adenine (0.1 mg/l), and guanine, thymine, uracil, hypoxanthine and cytosine (0.03 mg/l) are prepared as a 1000× concentrated stock solution, adjusted to pH 6.5 with NH_4OH and added. The following amino acids are added using a 10× stock solution (pH 6.5 with NH_4OH) to yield the given final concentrations: glutamine (5.6 mg/l), alanine, glutamic acid (0.6 mg/l), cysteine (0.2 mg/l), asparagine, aspartic acid, histidine, isoleucine, leucine, lysine, methionine, phenylalanine, proline, serine, threonine, tryptophan, tyrosine and valine (0.1 mg/l). The vitamin stock also contains the following vitamins to yield the given final concentrations (mg/l) in the medium: ascorbic acid (2.0), calcium pantothenate and choline chloride (1.0), riboflavin (0.20), *p*-aminobenzoic acid and vitamin B12 (0.02), vitamin A and vitamin D3 (0.01). These media also contain mannitol, sorbitol, cellobiose, fructose, mannose, rhamnose and ribose, each at 0.25 g/l.
[d] Vitamin stock solutions are normally prepared 100× concentrated.

Horsch and Jones (29). The filter paper also probably adsorbs brownish materials produced by the protoplasts and helps gas exchange.

Agarose has been found to be an excellent medium for the growth of plant protoplasts and cells (2,30). This led to the development of the 'bead type' culture system for culture of protoplasts (30) which consists of embedding the protoplasts in a low melting point agarose (e.g. SeaPlaque) and, after a suitable period of time, placing this agarose containing the protoplasts into liquid medium, which is shaken on a rotatory shaker to aerate the medium. The advantages of this system are that, in addition to an improved growth response in many species, one can replace the medium surrounding the growing cells without disturbance. This is particularly valuable when carrying out selection, in that one can maintain the level of a drug in the medium at a high level in order to preclude any growth of partially tolerant wild-type cells due to decay of the drug.

2.5 **Selection**

The selection protocol chosen will rely on the particular selectable marker used in the transformation system. In general, selection should be applied after leaving a suitable interval for the protoplasts to form a cell wall and start dividing. The time chosen is usually in the order of 7–10 days. At this time, cell colonies formed from the individual protoplasts will have reached a size of 8–20 cells. Selection with kanamycin, G418 (Geneticin; Gibco) or hygromycin is straightforward, consisting of adding an appropriate concentration of the drug to the suspending medium, and waiting for the resistant colonies to appear, growing from a background of inhibited, non-transformed cells.

A number of selection protocols have been elaborated. They are of two types. One relies on adding the drug to the medium after 7–10 days, upon dilution of the cell clumps, and the other relies on the 'bead type' culture technique (see above) in which the medium can be refreshed weekly with fresh drug. The latter method has given a cleaner result in our hands, with less chance of background interference by colonies which are not transformed being able to survive selection.

Drugs such as chloramphenicol are more difficult to use for selection. In this case, the drug may prove ineffective in selecting transformed cells when applied over long periods of time, but a short application of a high level can achieve selection (31).

2.6 **Regeneration to plants**

In model systems using tobacco or petunia, plant regeneration is simple, and can be achieved by manipulation of the cytokinin:auxin ratio. However, the process of plant regeneration is more complicated for most other species and is an individual problem which researchers must solve using their own knowledge and the published literature.

3. DIRECT GENE TRANSFER

Direct gene transfer has proved to be a simple and effective technique for the introduction of foreign DNA into plant genomes. Here we will discuss some of the factors important in carrying out direct gene transfer.

3.1 **Choice of vectors and markers**

The subject of vector construction is adequately dealt with in numerous methodology books (e.g. ref. 32) and in this volume. Here we will discuss only the choice of components of a vector for direct gene transfer.

3.1.1 *Vector*

The plasmid which is used to propagate the genes to be used for direct gene transfer does not appear to be important. What is required is that it gives a good yield of plasmid in order to be able to produce large quantities without undue effort. The plasmid pABD1 used in our examples is constructed using pUC8 as the basic replicon.

3.1.2 *Marker*

The selectable marker to be used will depend on the plant species and the markers available. The most effective markers for stable transformation to date are neomycin phosphotransferase [known as NPT or APH(3′)II from transposon 5 (Tn5) (3) and hygromycin resistance (33,34)]. Kanamycin has often been replaced by G418 for selection for NPT II activity in graminaceous monocots (35). The use of a mouse methotrexate resistance gene as a selectable marker in plants has recently been suggested.

Chloramphenicol acetyl transferase (CAT) from Tn9 (36) is not to be recommended for selection (31). There are a few other markers which have been described in the literature, such as a bacterial methotrexate resistance (37) and β-galactosidase (see Chapter 6). However, these have not proved to be generally useful. CAT (38) and the luciferase system which has recently been shown to function in plants (39) are good markers for transient expression experiments as the assays for these gene products are both sensitive and relatively easy to perform.

3.1.3 *Promoter*

The choice of promoter to which the gene is to be attached will depend on the desired effect. For stable transformation, the promoter for the selectable gene should be a constitutive and high level one, allowing easy selection of transformants. The 35S RNA promoter from cauliflower mosaic virus is being increasingly used (31), as this is generally considered to offer a higher expression level than the previously used 19S promoter (3) from the same virus. The nopaline synthase promoter in its 'improved' form is also being used (40).

The promoter for transient expression will be chosen on the basis of what is to be studied. For preliminary experiments to establish the system, the 35S promoter is probably the most useful, due to the high expression achieved. However, most transient expression systems are set up to study promoter function, and thus the promoter, or a component of the promoter system under study will naturally be used.

Promoters for non-selectable genes to be inserted will depend on whether a constitutive expression is required, or whether the researcher wishes to express the gene in a particular region (organ or tissue) of the plant. In this case the promoter will be chosen for specificity for that particular organ.

3.1.4 *Use of autonomously replicating sequences (ars)*

These have been sought for some time in plant systems, as a possible means of producing a vector system for use in a similar way to the plasmids of lower organisms. There have been a number of unpublished reports of researchers finding increased transformation frequencies due to the incorporation of so-called *ars* in their vectors.

A second application is to the transformation of chloroplasts, in that the probable replication origin of chloroplast DNA have been cloned and inserted into a vector which has been used in order to attempt transformation of chloroplasts. So far, there are no clear results from any of these experiments.

3.1.5 *Choice of carrier DNA*

The role of carrier DNA in promoting DNA uptake and/or interegration is not known. The carrier DNA which has most widely been used in direct gene transfer experiments is calf thymus DNA. This arose historically as it had been used for DNA transformation of animal cells. In the case of the latter, complex eukaryotic DNA is normally used as a carrier. However, in plants this is not necessary, and we have successfully used plasmid DNA as carrier in many cases. This is the case for example when carrying out co-transformation experiments (41).

3.2 **PEG-mediated transformation**

PEG was used in some of the original experiments in transfer of Ti plasmid (42), and later, *Escherichia coli* plasmids (3), into plant cells. At that time the efficiency of transfer was low (1×10^{-5} of recoverable microcolonies). The use of PEG has been developed so that it now rivals the efficient electroporation method of transfer of DNA into protoplasts (43). This has been achieved by optimizing the salt concentration and other factors of the protocol. PEG based methods can also be used to carry out transient expression as described below for electroporation.

3.3 **Electroporation**

Electroporation has proved to be an easy and efficient method for introducing foreign genes into plant protoplasts in a way which allows transient expression and integrative transformation to be achieved. The former is important as it offers a quick assay for expression of a gene in plants for use in some promoter studies. The latter allows one to incorporate both selectable and non-selectable genes into plants quickly and efficiently (41,44).

There are two basic methods of electroporation used to stably transform plant cells (38,44). These use either a short or a long pulse of an electric field to induce pore formation in the membrane of protoplasts. The long pulse method uses a low voltage electric field, and the short pulse method a very high voltage field. Typical values for tobacco are: (respectively) 400 V with an exponential decay constant of 10 msec; or 1400 V with a decay constant of 10 μsec. It is not clear which system is the most efficient, but preliminary experiments have suggested that the short pulse system is more efficient for stable transformation and the long pulse for transient expression.

The voltage to be used in each protoplast system depends on the source of the protoplasts and their diameter. A rule of thumb for the short pulse method is that 1400 V was found to be suitable in our laboratory for SR1 tobacco protoplasts, which had an average diameter of 42 μm. For every doubling of diameter, the voltage must be decreased 2-fold, and vice versa. Thus a protoplast of 21 μm would require in the region of 2800 V. The precise parameters must be determined for each individual type of protoplast. In general, we have found that cell suspension derived protoplasts require slightly higher voltages than expected from data derived from mesophyll protoplasts.

The first experiment to be done in approaching a new species or source of protoplasts is to construct a curve showing the overnight survival of treated protoplasts against the voltage used. The treatment giving around 50% survival, as related to the control untreated sample, will be close to that required for the most efficient stable transformation. These relationships do not necessarily hold for the long pulse method.

3.4 Transient expression

Transient expression of a suitable gene construct can probably be carried out using every type of protoplast. There are many accounts in the literature. In principle the method requires an efficient means of delivering DNA into protoplasts and a simple assay for the gene product. Following treatment of the protoplasts with the DNA in suitable uptake conditions, the cultures are incubated for a time, usually in the region of 1–7 days, and the activity of the gene product measured. Alternatively, the presence of the desired protein in the cells can be determined by use of antibody techniques, as is routinely done for tobacco mosaic virus (TMV) and other viruses (45). The advantage of the latter method is that it can be used to give a measure of the percentage of cells taking up and expressing DNA, as opposed to measuring the average production of the desired protein.

3.5 Graminaceous monocot transformation

Transformation of graminaceous monocots with concomitant regeneration of transformed plants has not been reported at the time of writing. It appears that radical changes in the host range of *Agrobacterium tumefaciens* will be necessary for this bacterium to be used as a route.

Direct gene transfer has been used to transform graminaceous monocot protoplasts on a number of occasions, with non-morphogenic cell cultures being produced. To date, only a limited number of graminaceous monocots have been shown to regenerate into organized structures from protoplasts. It is expected that the recent advances in rice protoplast culture (18,19) will lead to the production of transformed rice plants in the near future by utilizing direct gene transfer.

4. TRANSFORMATION VIA CO-CULTIVATION WITH A.TUMEFACIENS

Transformation of protoplast derived cells via co-cultivation (46) is finding less use of late, due to the development of the leaf disc transformation method (see

Chapter 6) and of direct gene transfer to protoplasts. We will therefore only discuss the subject briefly here.

4.1 **Growth of bacteria**

A.tumefaciens can be grown in a number of ways for use in co-cultivation. The commonly used wild-type strains are available from culture collections, as are strains carrying selectable markers for genetic engineering of plants from the laboratories publishing in this area.

A.tumefaciens can be grown on agar plates, or in liquid cultures (see Chapter 6) but they must be grown at 28°C or below. Otherwise, they lose their Ti plasmid which carries the DNA to be transferred to the plant cells. *A.tumefaciens* cultures grow slowly, and some 16 h is normally required to obtain a late log phase culture from a small inoculum. Storage of *A.tumefaciens* can be carried out at low temperature in glycerol in the same way as *E.coli* (32). Agar plates can be stored sealed with parafilm at 4°C for 1–2 months.

Cultures for use in co-cultivation are normally grown overnight from a ½₀ dilution of a liquid culture grown from a single colony. Antibiotics are added to maintain the desired constructs in the cells. They can, however, cause a decrease in the ability of the bacteria to transfer DNA, and are thus often left out of the overnight culture. The bacteria in culture medium can be added directly to the plant cells or can be first washed with 0.9% w/v NaCl solution to remove contaminating bacterial culture medium.

4.2 **Co-cultivation**

Bacteria are added to 3-day-old protoplast cultures of tobacco or petunia, at a concentration which gives approximately 400 bacteria per original plant protoplast. More bacteria can lead to difficulties, caused by overgrowth of the plant cells by bacteria, and less bacteria give a lower transformation efficiency.

4.3 **Removal of bacteria**

Following an incubation period of 32 h, the bacteria must be prevented from overgrowing the plant cells. The incubation period can be longer than 32 h, and incubations of up to 72 h have been used with tobacco protoplasts, but tend to reduce the viability of the plant cells. Shorter times than 32 h will reduce the number of transformants recovered (Shillito, unpublished data). The culture is washed one or two times with culture medium, and resuspended in fresh medium at the original or a lower (3×) density. Antibiotics are then added, to kill the remaining bacteria. A number of antibiotics have been used. The most common are carbenicillin (at 500 mg/l) and cefotaxime (Claforan, Hoechst; 250–500 mg/l). The former has found less use of late due to the use of the β-lactamase gene as a marker for strain construction. However, we have found this to be an effective antibiotic even against bacteria carrying a resistance gene, particularly when used in combination with vancomycin at 250 mg/l. We have had some problems with cefotaxime due to persistence of the bacteria over long periods in the presence of this bacteriostatic. The need to remove the bacteria and maintain selection pressure against them is demonstrated by the study of Pollock *et al.* (47) in which they showed that it was possible to transform callus of *Petunia hybrida* with *A.tumefaciens* added to the callus or with residual bacteria from co-cultivation.

5. UPTAKE OF ISOLATED ORGANELLES

To date there has been no published account of the uptake of functional organelles other than nuclei by protoplasts. This is in spite of much effort in this area by a number of investigators. The methods used to try and achieve this aim include treatment of protoplasts with PEG in the presence of the organelle. In the case of chloroplasts, this has normally been a preparation of mature chloroplasts from leaf material. As little is known about the way in which chloroplasts replicate, it may be that these experiments are being carried out with little chance of success. A much better understanding of the mechanism of replication of organelles is required if one wishes to carry out these experiments.

Transformation by uptake of isolated nuclei has been demonstrated (48). The transformation efficiency in this case is in the range of 1×10^{-5} per recoverable microcolony.

6. GENETIC ANALYSIS OF PLANTS PRODUCED BY TRANSFORMATION

When regeneration of fertile plants from tissue culture is possible then the introduced trait can be followed in its transmission to progeny. With a gene such as kanamycin resistance this is relatively simple (e.g. ref. 1). For a gene which is not functional at the seedling level (e.g. a storage protein gene), it is more laborious, involving screening of large numbers of plants by Southern analysis, or waiting until the plant expresses the gene in the appropriate organ. However, genetic analysis is an important part of gene transfer, in that it will only be possible to use genes in an agricultural context if the genes are stable and stably expressed through many generations. Genetic analysis has been carried out for a limited number of genes introduced by *A.tumefaciens* and via direct gene transfer.

The earliest opportunity to observe segregation at meiosis is by culture of the male gametes via anther culture (49,50). Haploid plantlets developed from microspores can be tested under selective conditions by transfer to kanamycin containing media when they are at the 'seedling' stage. For a single dominant gene one expects approximately 50% of the plants to be resistant due to the segregation of the trait during meiosis. Analysis at the level of the gametes is not often carried out, but it can throw some light on the mechanisms involved when used to analyse plants which are showing disturbed segregation ratios in conventional genetic analysis.

7. PROTOCOLS

7.1 **Preparation of DNA for transformation**

7.1.1 *Preparation of plasmid pABD1 DNA for transformation*

(i)   Take a preparation of pABD1 plasmid DNA in an Eppendorf type centrifuge tube at a concentration of 100–300 μg/ml and cut it with *Sma*I overnight at room temperature (32).

(ii)   Add ¹⁄₁₀ vol of 3 M sodium acetate, followed by 2.5 vols of absolute ethanol. Mix gently, and place at 4°C for 10–30 min to allow the DNA to precipitate. Where very small amounts of DNA at low concentrations (<0.01 mg/ml) are being used, it may be necessary to leave the preparation overnight at −20°C.

(iii)   Centrifuge for 15 min in a microcentrifuge. Remove the supernatant under sterile conditions, and fill the tube with 70% ethanol.

(iv)   Centrifuge again for 2 min, remove the supernatant under sterile conditions, and refill the tube with absolute ethanol. Aliquots can be stored at this stage at −20°C for later use. Centrifuge again (1 min).

(v)   Remove the supernatant, centrifuge quickly, and remove the remaining drops of ethanol. Dry the DNA in the air flow in a sterile flow hood (30–60 min depending on the quantity of DNA).

(vi)   Add sufficient sterile distilled water (autoclaved) to bring the DNA concentration to 0.4 mg/ml. Allow to dissolve for a minimum of 30 min.

7.1.2 *Preparation of carrier calf thymus DNA for transformation*

(i)   Dissolve 20 mg of calf thymus DNA (Sigma D-1501) in 10 ml sterile distilled water *with shaking*.

(ii)   Using a syringe with an 18-gauge needle, shear the DNA by drawing and expelling the solution through the needle. Shear the DNA until it has a size range of 3–15 kb, with most of the DNA being in the 7 kb range. Too large a size of carrier DNA causes breakage of protoplasts during the transformation step. Check the size of the DNA by running it in an agarose gel.

(iii)   Place 500 μg quantities (250 μl) of the DNA solution in 1.5 ml Eppendorf type centrifuge tubes, and add 25 μl of 3 M sodium acetate, followed by 700 μl of absolute ethanol. Mix gently and place at 4°C for 1 h.

(iv)   Carry out steps (iii) to (v) for plasmid DNA (Section 7.1.1).

(v)   Add 250 μl of sterile distilled water and dissolve overnight at 4°C.

Both carrier and plasmid DNA solutions are stable for up to 2 months if kept sterile at 4°C. The two are normally mixed 1:1 before addition to the protoplasts. Any other plasmid DNA can be prepared in a similar fashion, using a restriction enzyme which cuts in an area at least 200 bp away from any area of gene construct required for its desired function in the plant. Other high molecular weight (e.g. tobacco) or plasmid DNA can be prepared as carrier

DNA in the same way as calf thymus DNA. Plasmid DNA does not require shearing.

7.2 Isolation of protoplasts of *Nicotiana tabacum*

Tobacco leaf protoplasts are isolated from sterile shoot cultures of line SR1 (51) grown in a controlled environment chamber (*Table 2*). This provides a steady source of material year round, with only minimal seasonal variation. A reduction in growth rates and a fall in the yield and quality of the protoplasts is, however, to be expected during the winter months.

7.2.1 *Protoplast isolation*

(i)   Remove fully expanded leaves of 4–6 week old shoot cultures (*Table 2*) under sterile conditions and wet them thoroughly on both sides with enzyme solution.

(ii)   Cut the leaves into 1–2 cm squares and float them on enzyme solution (1.2% w/v Cellulase 'Onozuka' R10, 0.4% w/v Macerozyme R10 in K3A medium with 0.4 M sucrose), with the lower epidermis in contact with the solution, in Petri dishes (~1 g leaves in 12 ml enzyme solution in a 9 cm diameter Petri dish).

(iii) Seal the dishes with Parafilm and incubate them overnight at 26°C in the dark.

(iv)   Agitate the digest gently and then let stand for a further 30 min to complete digestion.

(v)   Filter the digest through a 100 µm pore size stainless steel sieve and wash through with ½ vol. of 0.6 M sucrose (0.1% w/v MES, pH 5.6).

(vi)   Distribute the filtrate into capped centrifuge tubes and centrifuge for 10 min in a swing out rotor at approximately 60 g. The protoplasts collect at the upper surface of the medium.

Table 2. Establishment of shoot cultures from seed.

1. Sterilize seed using a 5 min treatment with sodium hypochlorite solution (1.4% w/v sodium hypochlorite containing 0.05% w/v Tween 80). An easy method to carry this out is to place the seeds in a steel mesh sieve (or a fine mesh tea strainer), which can be moved between the solutions used for sterilization and washing.
2. Wash the seeds five times with sterile distilled water, and place individually on T medium (52) solidified with 0.8% w/v Difco Bacto or equivalent agar.
3. Culture at 26°C in 16 h per day light (10–20 µE/m²/sec) in a growth chamber. The use of higher light intensities can lead to a loss of quality in the protoplasts isolated from this material. For culture, any container which remains sterile and allows good light transmission can be used. The following are recommended: GA7 autoclavable containers (Magenta Corp., IL, USA) and PlantCon single use containers (Flow Laboratories, VA, USA) shown in *Figure 1b*, and for larger numbers of plants, autoclavable plastic containers (plant boxes, 195 × 295 mm, Brevet Afiplastex, France).
4. Take stem cuttings of the shoots every 4–6 weeks onto fresh T medium under the same conditions. It may require several subcultures to establish a culture with good expanded (~5 × 3 cm) leaves suitable for protoplast isolation.

(vii)   Remove the medium and pellet from under the protoplasts using a 20 ml plastic syringe with a long needle or cannula (A.R.Howell Ltd, UK). This must be done carefully so as not to disturb the protoplast layer.

(viii) Resuspend the protoplasts in K3A medium containing 0.4 M sucrose. Two tubes from the previous step can be combined at this stage into one.

(ix)   Repeat the centrifugation and resuspension in fresh medium twice so as to wash the protoplasts.

(x)   Before the last wash, take out a sample of 0.1 ml and dilute it with 0.9 ml of the 0.17 M $CaCl_2$ solution given in the alternative step (vi) below. Count the protoplast density in this sample using a haemocytometer and calculate the total yield. This should be in the range of, or better than, 2 million/g freshweight of leaf material used.

(xi)   Resuspend the protoplasts for transformation according to the method to be used.

Alternative to steps (vi)–(ix).

(vi)   Distribute the filtrate into capped centrifuge tubes and cover each with a 1.5 ml overlay of 0.17 M $CaCl_2$ (0.1% w/v MES, pH 5.6). Centrifuge for 10 min in a swing out rotor at 60 g (~600 r.p.m. in most bench top centrifuges).

(vii)   Collect the protoplasts which concentrate at the interface.

(viii) Combine the protoplasts from each two tubes into one, fill the tubes with $CaCl_2$ solution.

(ix)   Wash them twice with the same solution by sedimenting at 60 g for 5 min and withdrawing the supernatant each time. When using method (B) for PEG-mediated transformation (Section 7.3.2), the washes should be carried out using W5 salt solution (*Table 1*).

7.3 Transformation of protoplasts

7.3.1 *Heat shock*

It has been found advantageous, particularly for the electroporation methods of direct gene transfer, to give the protoplasts a heat shock around the time of transformation (*Table 3*). The most convenient time to carry this out is before the distribution of the protoplasts into aliquots.

7.3.2 *Transformation using PEG*

Method A. This is basically the protocol used for electroporation without the electroporation step.

(i)   Resuspend the protoplasts in 0.4 M mannitol solution containing 6 mM $MgCl_2$ (0.1% w/v MES, pH 5.6) at 1.6×10^6 per ml.

(ii)   Carry out the heat shock treatment (*Table 3*).

(iii)   Place 0.66 ml aliquots of protoplast suspension in 10 ml centrifuge tubes.

(iv)   To each tube add 10 µg of linearized plasmid DNA and 50 µg of sheared calf thymus or other suitable carrier DNA.

Table 3. Heat shock.

1. Place the tube containing the protoplasts (resuspended in the appropriate solution) for 5 min into a water bath held at 45°C.
2. Agitate the suspension intermittently to ensure that the protoplasts are evenly treated.
3. On removal from the bath, place the tube in ice water until the temperature returns to ambient.
4. All the following steps are then carried out at room temperature.

(v)   Mix gently, and then add 0.33 ml of PEG solution [40% w/v PEG (mol. wt 6000; Fluka or Merck) in 0.4 M mannitol buffered with 0.1% w/v MES to pH 5.6].

(vi) Mix gently, and incubate for 10 min, shaking gently once after 5 min.

(vii) At the end of this time, place 0.33 ml aliquots of this mixture into 6 cm Petri dishes and add medium as described in *Table 4*.

Method B. This is a modification of the original method for transformation of protoplasts of *N.tabacum* (3), which was in turn a modification of previous methods (42). We have added a heat shock and changed the order of addition of DNA and PEG.

(i)   After the last wash (Section 7.2.1), adjust the protoplast concentration to 2×10^6 per ml in K3A medium and dispense aliquots of 1 ml into 10 ml sterile plastic centrifuge tubes.

(ii)   Administer a heat shock (*Table 3*).

(iii) Add DNA solution to the aliquots (10 μg of pABD1 plus 50 μg of calf thymus DNA in 50 μl sterile distilled water). Mix gently.

(iv)   Add 0.5 ml of PEG solution [40% w/v PEG 6000 in F medium (*Table 1*)] with shaking. Incubate for 30 min at room temperature with occasional gentle mixing.

(v)   Add five aliquots of 2 ml of F medium (*Table 1*) at intervals of 5 min. We have noted that the pH of F medium drops to 4.3–4.6 after autoclaving. Since this is likely to be harmful to many protoplasts, we recommend adjustment of the pH after autoclaving to 5.8 with sterile additions of KOH.

(vi)   Sediment the protoplasts by centrifugation for 5 min, resuspend them in 2 ml of K3A culture medium and transfer in 0.33 ml aliquots to 6 cm Petri dishes. Add medium and culture as described in *Table 4*.

This protocol gives transformation efficiencies in the range of 10^{-4}–10^{-3}.

Method C. This method was only recently developed (43). It is comparable to electroporation in that it gives transformation frequencies with tobacco SR1 in the range of 1×10^{-2} per developing microcallus. It has so far been used for leaf mesophyll protoplasts from SR1 tobacco and *N.plumbaginifolia*, the latter being transformable only at 1×10^{-3}.

The protoplasts should be washed with W5 solution [see Section 7.2.1 alternative step (ix)] and must be in W5 or a similar solution for at least 30 min prior to attempting transformation. The protoplasts can be stored in the last wash of W5 solution for up to 8 h at 8°C before use without loss of transformation competence. The exact concentration of $MgCl_2$ used in the mannitol solution is critical, and must be optimized (range 5–50 mM) for each protoplast type. A heat shock may not increase the transformation efficiency of this method.

(i)   Prepare the PEG solution in advance: dissolve 40% w/v PEG 4000 (Merck) in 0.4 M mannitol, 0.1 M $Ca(NO_3)_2$, pH 8.0 with KOH. The pH of this solution takes 3–4 h to stabilize. If excessive bursting is experienced with this method, the substitution of PEG with a mol. wt of 6000 may reduce it.

(ii)   Resuspend the protoplasts following washing with W5 solution at 1.6×10^6 per ml in 0.5 M mannitol containing 15 mM $MgCl_2$ and 0.1% w/v MES (pH 5.6 with KOH). It is particularly important that as little as possible of the W5 solution is carried over with the pellet.

(iii) After resuspending the protoplasts in the mannitol solution, distribute the protoplasts into 0.3 ml aliquots.

(iv)   Add 30 μl of DNA solution containing 4 μg of plasmid DNA and 10 μg of carrier DNA. Mix gently, and then add 0.3 ml of PEG solution. Incubate at room temperature with occasional shaking.

(v)   Slowly add 10 ml of W5 solution (1 ml, 3 ml, 6 ml at 3 min intervals). For SR1 and other robust protopasts, one may proceed directly to plating of the protoplasts as described in *Table 4* without carrying out steps (v) and (vi).

(vi)   Centrifuge at 60 *g* for 5 min. Resuspend the pellet in 0.3 ml of K3A medium, and proceed as described in *Table 4*.

7.3.3 *Transformation using electroporation*

Electroporation can be carried out using a number of different types of apparatus which have been described in the literature. Here we describe the method using high voltage pulses (~1.5 kV/cm) applied for a short time (exponential decay constant 10 μsec) to introduce DNA. This method has been in use for a period of 2.5 years and over 200 experiments in our hands, and has proved reliable and easy to adapt to different protoplast types from a number of species. Heating of the medium due to the pulse is negligible (<0.5°C) so that the whole procedure can be carried out at room temperature. The resistance across the chamber must be controlled so as to obtain repeatable pulse lengths. The resistance can be measured using any good alternating current multimeter. A typical instrument which we have used in our experiments is the AVO B183 LCR meter (Thorn–EMI Ltd, UK) which operates at 1 kHz when measuring resistances in this range.

Protoplasts are treated with high voltage electrical pulses in the chamber of a 'DIA-LOG' Elektroporator (DIA-LOG GmbH, FRG). This chamber is cylindrical in form with a distance of 1 cm between parallel steel electrodes and has a pulsed volume of 0.32 ml (53). A chamber with a volume of 1 ml is also available. For this 1 ml chamber the volumes given below should be tripled, and the resistance value should be in the range of 0.33–0.36 kΩ.

(i)   Prepare a PEG solution in advance: 24% w/v PEG 6000 (Merck) in 0.4 M mannitol. Prepare this solution with sufficient $MgCl_2$ added (~30 mM) to bring the resistance when measured in the chamber of the electroporator to between 1.2 and 1.3 kΩ. Filter-sterilize the solution. This can be speeded up by warming the solution to 50°C.

(ii)   Transfer an aliquot of 0.34 ml of the protoplast suspension to the chamber of the electroporator, and measure the resistance. This should be in the region of 1–4 kΩ. If the value is lower than 1.5 kΩ, then too much wash solution is being carried over into the electroporation solution with the protoplast pellet.

(iii)   Add the appropriate amount of $MgCl_2$ solution to the protoplast suspension to adjust the resistance to a value of 1–1.1 kΩ.

(iv)   Carry out the heat shock (*Table 3*).

(v)   Dispense aliquots of 0.25 ml of protoplast suspension into 5 ml capped polycarbonate tubes.

(vi)   Add 20 μl of DNA solution containing 4 μg of plasmid DNA and 10 μg of carrier DNA. Mix gently, and then add 0.125 ml of PEG solution. Shake gently to mix the solutions.

(vii)   Ten minutes after addition of the DNA and PEG, transfer samples to the chamber of the electroporator and pulse three times at 10 sec intervals with pulses of an initial field strength of 1.4 kV/cm.

(viii) Return each sample to a 6 cm diameter Petri dish and wait for 10 min.

(ix)   Add culture medium as described in *Table 4*.

7.4 Protoplast culture and selection

Plating and culture of protoplasts is described in *Table 4*.

7.4.1 *Selection at the microcolony level*

Selection is carried out in the bead type cultures (*Table 4*). For selection of kanamycin resistant transformants, add 50 μg/ml kanamycin sulphate to all the media used in bead cultures. For hygromycin selection, use 20 μg/ml hygro-mycin sulphate.

(i)   Replace half of the medium surrounding the agarose beads (*Table 4*) with fresh medium after 1 week in the bead type culture. Use a 1:1 mixture of K3C and J200 media. To this medium add the appropriate amount of filter-sterilized drug solution for selection.

(ii)   After a further week, again replace half of the medium, this time with a 1:1 mixture of K3E and J200, and with the appropriate drug addition.

(iii)   Repeat step (ii) 1 week later. This progression of media ensures that the osmotic pressure and hormone requirements of the developing colonies are satisfied.

(iv)   The first resistant colonies should be seen a total of 3–5 weeks after the initiation of the experiment. They show as dense white, small-celled colonies in a background of dying colonies with expanded cells.

Table 4. Plating and culture of protoplasts.

Prepare agarose medium in the following way:

1. Autoclave 0.7 g of SeaPlaque agarose (FMC Corp., MN, USA) without medium, in a flask containing a magnetic stirrer bar.
2. Remove from the autoclave before the temperature drops below 80°C, and shake the flask vigorously to break up the agarose. This can be stored for 4–8 weeks at room temperature before use.
3. When required, add 50 ml of K3A medium, and heat the flask to 70–80°C until the agarose is dissolved.
4. Add 50 ml of H460 medium, and store in a water bath at 45°C for 1–2 h. Any medium left over can be allowed to solidify, and re-heated once. Repeated reheating will reduce the gelling ability of the agarose, causing break-up of the beads during the subsequent culture phase.

Plating of protoplasts
1. To each 6 cm diameter Petri dish containing protoplasts, add 3 ml of a 1:1 mixture of K3A medium and H460 medium containing agarose, prepared as described above.
2. Allow the medium to solidify. Solidification is best carried out by placing the dishes containing the protoplasts on a metal plate cooled to 16°C. This ensures rapid solidification. Any movement of the dish during gelling of the agarose should be avoided as it leads to shearing stresses on the protoplasts, and causes reduced viability.

Culture of protoplasts
1. Seal the dishes with Parafilm, and incubate overnight at 26°C in the dark.
2. Move the dishes to low light (5–10 μE/m^2/sec 'Sylvania' cool white fluorescent) for a further 6 days.
3. Cut the agarose containing the cell clusters into quadrants and place all the quadrants from one dish into 30 ml of the medium used for plating (without agarose) in a tall 9 cm diameter Petri dish (90 × 25 mm) or preferably a Semadeni container (101 × 54 mm polystyrol vessels with lid; Semandeni AG, Switzerland).
4. Place the cultures on a gyrotatory shaking platform (60 r.p.m.; 2.5 cm throw) in the dark at 26°C. This constitutes the 'bead type' culture system (30). Cultured protoplasts divide and form colonies in the agarose (*Figure 2*).

(v) When they attain 1–2 mm in diameter, remove the colonies from the bead culture, and place them on Linsmaier and Skoog (LS; *Table 1*) medium solidified with 0.6% w/v SeaPlaque agarose (0.1 mg/l 2,4-D, 1 mg/l NAA, 0.1 mg/l BAP, 0.1 mg/l kinetin), containing half of the drug concentration used for selection.

(vi) The resulting calli can be subsequently subcultured indefinitely on the same medium containing the full drug concentration and solidified with cleaned agar (30).

7.4.2 *Selection at the colony level*

Where transformation frequencies are in the 1% range, it is possible to select transformants at the colony level, either by screening for a drug resistance, or for an enzyme activity.

(i) At the point where sectors of agarose-containing microcolonies are placed in liquid medium (*Table 4*, culture step 3), place only half a quadrant in each 30 ml culture.

(ii)   Carry out the bead culture as described in Section 7.4.1, but omitting the selecting drug.

(iii) After 5 weeks, the beads will have broken up, and most of the colonies will be floating free in the medium.

(iv)   Take these colonies individually (using forceps), and place them on LS medium (*Table 1*) solidified with agarose as in Section 7.4.1. For drug resistance screening, the drug can be incorporated into this medium (e.g. kanamycin sulphate at 50 μg/ml), and the growing colonies scored and cultured further. For an enzyme screening, the calli need to be subcultured until they reach a size suitable to be assayed. This depends on the particular enzyme (e.g. nopaline synthase) which is to be assayed.

7.4.3 *Regeneration of plants*

Regeneration (*Table 5*) is more difficult to achieve in the presence of the selecting drug. By the time selection is attempted, the material should consist only of the clonal transformed cells, so that there will be no danger of regenerating non-transformed shoots. Therefore regeneration may be initiated in the absence of selection. Where the experimenter believes that it is particularly necessary because of possible instability of the gene, selection should be continued throughout. The shoots regenerated can subsequently be placed on and maintained on medium with selective drug.

7.5 **Genetic analysis of plants produced by transformation**

The inheritance of a functional kanamycin resistance gene can be followed by its expression in the progeny of a transformed plant. This is only the case if the gene is not turned off or lost during development or the reproduction process. In the case of the NPT II gene introduced from plasmid pABD1 into *N.tabacum* via direct gene transfer, and a number of other such genes with constitutive promoters, it has been found that the coded function segregates in most cases as a dominant Mendelian locus. Protocols are given for the analysis of kanamycin resistance in tobacco seedlings (Section 7.5.1) and pollen-derived haploid plants (*Table 6*). Similar protocols can be used for the analysis of other selectable markers such as hygromycin.

7.5.1 *Genetic analysis of a functional kanamycin resistance gene in tobacco SR1*

(i)   Grow plants of *N.tabacum* in an insect-free greenhouse to flowering.

(ii)   Self the plants and cross them in both directions with wild-type SR1 plants. The reciprocal crossing test will identify non-Mendelian (or uniparental) inheritance. Uniparental inheritance is to be expected in cases of transformation of chloroplasts or mitochondria. Selfing can usually be carried out by bagging part of the flower head. However, it may be necessary to do this by hand due to heterostyly often seen in regenerated plants. Crossing with the wild-type is achieved by pollinating buds just before opening with the desired pollen from a mature flower: slit

Table 5. Regeneration of plants.

1. Initiate shoot regeneration by placing pieces of callus of ~1 cm in diameter on LS medium (*Table 1*) containing 0.2 mg/l BAP as the sole phytohormone, and solidified with cleaned agar.
2. Culture the calli in the dark for 1 week (26°C), followed by light (10–20 μE m^{-2} sec^{-1} cool white fluorescent) until shoot primordia are seen (2–4 weeks).
3. Transfer the calli to the same LS medium without hormones, in order to allow the shoots to grow out.
4. Cut off the resulting shoots when 2–4 cm in length and culture them on T medium in the same conditions as used for the original shoot cultures (*Table 2*).
5. The resulting shoots can be subcultured indefinitely as shoot cultures.
6. When roots have formed on a shoot, plantlets can be transferred to soil: wash off the agar from a well rooted 3 week-old shoot, and pot this up in a 3–4 inch pot in potting compost.
7. Cover the plant with an upturned beaker, or place in a humid atmosphere for about 1 week for hardening off.
8. Gradually remove the cover, or reduce the humidity, and grow the plant in the greenhouse, potting up to a larger pot as necessary.

Table 6. Analysis of pollen-derived plantlets of tobacco SR1.

1. Harvest flower buds when the pollen is just before the stage of the first pollen mitosis. This corresponds in SR1 to a flower bud length of 12–14 mm. Place the buds in the cold at 4°C for 2 days.
2. Remove the calyx carefully and sterilize the buds by treatment with sodium hypochlorite (1.4% w/v containing 0.05% v/v Tween^R 80) for 5 min, followed by five washes with sterile distilled water.
3. Remove the anthers without damaging them, and place them on Nitsch (49) medium (*Table 1*), five per 6 cm Petri dish. Seal the dishes with Parafilm and incubate at 25°C in dim light (10 μE m^{-2} sec^{-1}).
4. When the young plants emerge from the anthers (3–5 weeks), gently tease them apart and place them on T medium (*Table 1*) solidified with agar and containing 300 μg/ml kanamycin sulphate.
5. Score the number of green (resistant) and bleached (sensitive) plantlets.
6. Compare the counts to the expected segregation ratios for one or more independent loci for the resistance character using the chi-square test.

the petals open, remove the stamens and apply fresh pollen to the pistil. Reclose the flower. If the procedure is carried out carefully, it is not normally necessary to bag the flowers, as fertilization occurs before the pistil emerges. Label each fertilization carefully (*Figure 3a*).

(iii) As the seed capsule is often sticky and causes problems of fungal infection at later stages, it is recommended that the seeds be separated from the capsule as they are collected. Collect the seeds from individual capsules into seed envelopes, and store them in a dry place at room temperature.

(iv) Sow non-sterilized seed on half strength NN69 medium containing 300 μg/ml kanamycin sulphate and solidified with 0.8% cleaned agar (1). Sterilization is not normally necessary and can cause variability in response of the seedlings.

(v) The seeds should germinate at a high frequency. The proportion of non-germinating seeds should be determined, as non-germination can affect

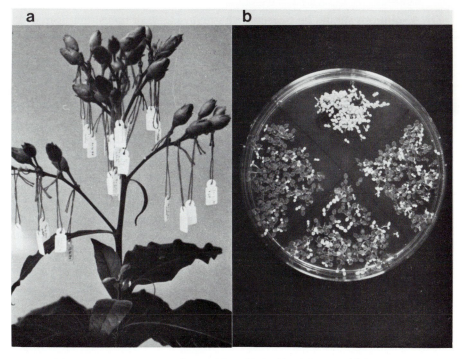

Figure 3. (a) Flowering plant of *N.tabacum* SR1 with fertilized flowers. (b) Genetic analysis of transformed tobacco plants. An agar plate containing 300 µg/ml kanamycin sulphate is shown one week after seeding. The quadrants were seeded with (from the top clockwise) wild-type seeds, seeds from the two reciprocal crosses of the transformant with the wild-type, and seeds obtained from selfing of the transformant. The segregation ratios are (resistant:sensitive): 0:1, 1:1, 1:1 and 3:1, respectively.

the segregation ratios. After 1 week, score the seedlings for resistance to kanamycin. Green seedlings are resistant, white sensitive (*Figure 3b*).

(vi) Compare the counts to the expected segregation ratios for one or more independent loci for the resistance character using the chi-square test.

Normally, a high proportion of the regenerated plants are fertile and 86% pass the introduced genes to their progeny in a dominant Mendelian fashion with one or more loci, but a small proportion of the plants are disturbed in their fertility (Shillito and Potrykus, in preparation).

7.6 CETAB small scale DNA isolation method

This method (54) relies on the fact that CETAB will hold DNA in solution at high salt concentrations, but on lowering the salt concentration, precipitates with the DNA. DNA prepared in this way can be used for Southern analysis and will be cut by most restriction enzymes if restriction is carried out at low DNA concentration (≤ 0.1 µg/µl). The method avoids having to run $CsCl_2$ gradients on small amounts of DNA and allows one to prepare DNA from up to 50 samples simultaneously.

DNA solutions should be treated with great care at all times: that is in order to prevent shearing of the high molecular weight DNA they should not be shaken or pipetted vigorously. It is necessary to maintain the isolated DNA in as large fragments as possible in order to obtain sharp bands on Southern analyses.

(i)   Collect 500–800 mg of leaf or callus material in a 2.5 ml Eppendorf tube and cover with aluminium foil.

(ii)   Place in a freeze dryer overnight or until completely dry.

(iii)   Grind the dry material to powder in the Eppendorf tube with a steel (or glass) rod.

(iv)   Add 1 ml of CETAB extraction buffer (*Table 7*).

(v)   Heat the tube for 15–30 min at 60°C in a water bath.

(vi)   Cool and then add 1 ml of chloroform/isoamyl alcohol (24:1).

(vii)   Mix gently, and centrifuge for 30 sec in a microcentrifuge.

(viii)   Transfer the top phase to a fresh 1.5 ml vol. tube, measuring the volume at the same time.

(ix)   Add $\frac{1}{10}$ vol. of 10% CETAB (*Table 7*). It is important that this volume ratio is precise.

Table 7. Solutions for CETAB DNA isolation[a].

1 × CETAB extraction buffer	
For 100 ml buffer:	1 g CETAB (1%)
	5 ml 1 M Tris–HCl, pH 8.0 (50 mM)
	2 ml 0.5 M EDTA (10 mM)
	14 ml 5 M NaCl (0.7 M)
	0.5 g PVP mol. wt 360 000 (0.5%)
10% CETAB	
For 100 ml buffer:	10 g CETAB (10%)
	14 ml 5 M NaCl (0.7 M)
CETAB precipitation buffer	
For 100 ml buffer:	1 g CETAB (1%)
	5 ml 1 M Tris–HCl, pH 8.0 (50 mM)
	2 ml 0.5 M EDTA (10 mM)
High salt TE	
For 100 ml buffer:	1 ml 1 M Tris–HCl, pH 8.0 (10 mM)
	0.2 ml 0.5 M EDTA (1 mM)
	20 ml 5 M NaCl (1 M)
$\frac{1}{10}$ TE	
For 100 ml buffer:	0.1 ml 1 M Tris–HCl, pH 8.0 (1 mM)
	0.02 ml 0.5 M EDTA (0.1 mM)

[a]Final concentrations are given in brackets.

(x)   Repeat the chloroform extraction as in steps (vii) and (viii).

(xi)   Add 1 vol. of CETAB precipitation buffer (*Table 7*).

(xii) Mix gently and centrifuge for 1 min in a microcentrifuge. Remove and discard the supernatant.

(xiii) Dissolve the pellet in 0.4 ml of high salt TE (*Table 7*) at 65°C for 1 h. Cool to room temperature.

(xiv) Re-precipitate the DNA by adding 2 vols of ice-cold absolute ethanol.

(xv)   Centrifuge for 5 min, remove the supernatant and wash the precipitate with 70% ethanol and 100% ethanol.

(xvi) Dry the DNA in the airstream from a sterile flow bench.

(xvii) Dissolve the DNA overnight in 250 μl of ¹/₁₀ TE.

(xviii) Extract the solution with phenol/chloroform/isoamyl alcohol (25:24:1), centrifuge for 30 sec, and remove the water phase to a fresh tube.

(xix) Extract with chloroform/isoamyl alcohol (24:1), spin for 30 sec, and remove the water phase to a fresh tube.

(xx)   Add ¹/₁₀ vol. 3 M sodium acetate followed by 2 vols of absolute ethanol to precipitate the DNA.

(xxi) Centrifuge and wash as before [steps (xv) and (xvi)], and dissolve the pellet in 20 μl ¹/₁₀ TE overnight.

(xxii) Run the DNA in an agarose gel to determine the quality and amount of DNA.

The amount of DNA can also be determined using spectrophotometric means or via other techniques such as the diphenylamine reaction (55).

8. ACKNOWLEDGEMENTS

The authors thank all who contributed to the methods described in this publication, particularly I.Negrutui for allowing use of data before publication. Thanks are also due to C.T.Harms for critical reading of the manuscript.

9. REFERENCES

1. Potrykus,I., Paszkowski,J., Saul,M., Petruska,J. and Shillito,R.D. (1985) *Mol. Gen. Genet.*, **199**, 169.
2. Lörz,H., Baker,B. and Schell,J. (1985) *Mol. Gen. Genet.*, **199**, 178.
3. Paszkowski,J., Shillito,R.D., Saul,M., Mandak,V., Hohn,T., Hohn,B. and Potrykus,I. (1985) *EMBO J.*, **3**, 2717.
4. Hain,R., Stable,P., Czernilofsky,A.P., Steinbiss,H.-H., Herrera-Estrella,L. and Schell,J. (1985) *Mol. Gen. Genet.*, **199**, 161.
5. Hain,R., Steinbiss,H.-H. and Schell,J. (1984) *Plant Cell Rep.*, **3**, 60.
6. Deshayes,A., Herrera-Estrella,L. and Caboche,M. (1985) *EMBO J.*, **4**, 2731.
7. Reich,T.J., Iyer,V.N. and Miki,B.L. (1986) *BioTechnology*, **4**, 1001.
8. Crossway,A., Oakes,J., Irvine,J., Ward,B., Knauf,V. and Shewmaker,C. (1986) *Mol. Gen. Genet.*, **202**, 179.
9. Eriksson,T. (1985) In *Plant Protoplasts*. Fowke,L. and Constabel,F. (eds), CRC Press Inc., FL, p. 1.
10. Binding,H. (1975) *Physiol. Plant.*, **35**, 225.
11. Nagy,J.I. and Maliga,P. (1976) *Z. Pflanzenphysiol.*, **78**, 453.
12. Xhu,Z.-H., Davey,M.R. and Cocking,E.C. (1981) *Z. Pflanzenphysiol.*, **104**, 289.
13. Davey,M.R. (1983) In *Protoplasts 1983: Lecture Proceedings*. Potrykus,I. *et al.* (eds), Birkhauser, Basel, p. 19.

14. Potrykus,I. (1979) *Proceedings 5th International Protoplast Symposium*, Szeged, Pergamon Press, London, p. 243.
15. Dale,P.J. (1983) In *Protoplasts 1983: Lecture Proceedings*. Potrykus,I. *et al.* (eds) Birkhauser, Basel, p. 31.
16. Vasil,V. and Vasil,I.K. (1980) *Theor. Appl. Genet.*, **56**, 97.
17. Imbrie-Milligan,C.W. and Hodges,T.K. (1986) *Planta*, **168**, 395.
18. Abdullah,R., Cocking,E.C. and Thompson,J.A. (1986) *BioTechnology*, **4**, 1087.
19. Toriyama,K., Hinata,K. and Sasaki,T. (1986) *Theor. Appl. Genet.*, **73**, 16.
20. Nagata,T. and Takebe,I. (1970) *Planta*, **92**, 301.
21. Zieg,R.G. and Outka,D.E. (1980) *Plant Sci. Lett.*, **18**, 105.
22. Wernicke,W., Lörz,H. and Thomas,E. (1979) *Plant Sci. Lett.*, **15**, 239.
23. Lu,C.-Y., Vasil,V. and Vasil,I.K. (1981) *Z. Pflanzenphysiol.*, **104**, 311.
24. Kao,K.N. and Michayluk,M.R. (1975) *Planta*, **126**, 105.
25. Binding,H., Nehls,R. and Jorgensen,J. (1982) In *Plant Tissue Culture*. Fujiwara,A. (ed.) Japanese Association of Plant and Tissue Culture.
26. Dracup,M., Gibbs,J. and Greenway,H. (1986) *Abstracts of 6th International Conference, International Association Plant Tissue Culture*. August 1986, Minneapolis, USA, p. 8.
27. Raveh,D., Huberman,E. and Galun,E. (1973) *In Vitro*, **9**, 216.
28. Menczel,L., Lazar,G. and Maliga,P. (1978) *Planta*, **143**, 29.
29. Horsch,R.B. and Jones,G.E. (1980) *In Vitro*, **16**, 103.
30. Shillito,R.D., Paszkowski,J. and Potrykus,I. (1983) *Plant Cell Rep.*, **2**, 244.
31. Pietrzak,M., Shillito,R.D., Hohn,T. and Potrykus,I. (1987) *Nucleic Acids Res.*, **14**, 5857.
32. Maniatis,T., Fritsch,E.F. and Sambrook,J. (1982) *Molecular Cloning: A Laboratory Manual*, Cold Spring Harbor Laboratory, New York.
33. Gritz,L. and Davies,J. (1983) *Gene*, **25**, 179.
34. Waldron,C., Malcolm,S.K., Murphy,E.B. and Roberts,J.L. (1985) *Plant Mol. Biol. Rep.*, **3**, 169.
35. Potrykus,I., Saul,M., Petruska,J., Paszkowski,J. and Shillito,R.D. (1985) *Mol. Gen. Genet.*, **199**, 183.
36. Alton,N.K. and Vapnek,D. (1979) *Nature*, **282**, 864.
37. Herrera-Estrella,L., Depicker,A., Van Montague,M. and Schell,J. (1983) *Nature*, **303**, 209.
38. Fromm,M., Taylor,L.P. and Walbot,V. (1985) *Proc. Natl. Acad. Sci. USA*, **82**, 5824.
39. Ow,D.W., Wood,K.V., DeLuca,M., De Wet,J.R., Helsinki,D.R. and Howell,S.H. (1986) *Science*, **234**, 856.
40. An,G., Watson,B.D., Stachel,S., Gordon,M.P. and Nester,E.W. (1984) *EMBO J.*, **4**, 277.
41. Schocher,R.J., Shillito,R.D., Saul,M.W., Paszkowski,J. and Potrykus,I. (1986) *BioTechnology*, **4**, 1093.
42. Krens,F.A., Molendÿk,L., Wullems,G.J. and Schilperoort,R.A. (1982) *Nature*, **296**, 72.
43. Negrutiu,I., Shillito,R.D., Potrykus,I., Biasini,G. and Sala,F. (1987) *Plant Mol. Biol.*, **8**, 363.
44. Shillito,R.D., Saul,M.W., Paszkowski,J., Müller,M. and Potrykus,I. (1985) *BioTechnology*, **3**, 1099.
45. Otsuki,Y. and Tkebe,I. (1969) *Virology*, **38**, 497.
46. Marton,L., Wullems,G.J., Molendÿk,L. and Schilperoort,R.A. (1979) *Nature*, **277**, 129.
47. Pollock,K., Barfield,D.G., Robinson,S.J. and Shields,R. (1985) *Plant Cell Rep.*, **4**, 202.
48. Saxena,P.K., Mii,M., Crosby,W.L., Fowke,L.C. and King,J. (1986) *Planta*, **168**, 29.
49. Nitsch,J.P. (1972) *Z. Pflanzenzüchtung*, **67**, 3.
50. Heberle-Bors,E. (1986) *Theor. Appl. Genet.*, **71**, 361.
51. Maliga,P., Sz.Breznovitz,A. and Marton,L. (1973) *Nature New Biol.*, **244**, 29.
52. Nitsch,J.P. and Nitsch,C. (1969) *Science*, **163**, 85.
53. Neumann,E., Schaeffer-Ridder,M., Wang,Y. and Hofschneider,P.H. (1982) *EMBO J.*, **1**, 841.
54. Murray,H.G. and Thompson,W.F. (1980) *Nucl. Acids Res.*, **8**, 4321.
55. Richards,G.M. (1974) *Anal. Biochem.*, **57**, 369.

CHAPTER 8

Transposable elements and gene-tagging

NANCY S.SHEPHERD

1. INTRODUCTION

1.1 What is gene-tagging?

Transposable elements are mobile segments of DNA which inactivate or alter gene expression by insertion into a genetic locus. Transposon-induced mutations have been described in various plant species and a few plant transposable elements have been cloned and described at the molecular level (for review see ref. 1). Gene-tagging, a technique originally described for work with *Drosophila* (2), is a method by which a transposon-induced mutant allele of a desired gene is physically isolated through the use of transposon-specific sequences as a hybridization probe. A spectacular advantage of this method over more conventional ones is that one may clone a gene without knowledge of the gene product. Thus the method is applicable to the majority of known plant genetic defects where a mutant plant phenotype is recognizable, but the gene product (and its function) is not known. The method involves genetic techniques both to move a transposon into a 'target' gene, and to demonstrate that the resulting mutation is due to a particular element insertion. The method also involves molecular techniques to identify and isolate the target gene. In this chapter I will outline the basic steps involved in gene-tagging, discuss some of the successful gene-tagging experiments and present items to consider in designing new experiments. Although the information presented will be primarily from the plant *Zea mays* (maize or corn), it is hoped that it will lend guidance to researchers interested in extending the methodology to other plant species as well. It is understood that the cloning of a DNA fragment responsible for a specific mutant plant phenotype will only be a first step in the molecular analysis of the role of that gene in the plant.

An alternate method of cloning a gene using gene-tagging methodology is to introduce 'foreign' DNA into the plant (a transgenic plant), identify mutations due to this DNA and then to re-clone the 'foreign' DNA along with the flanking chromosomal regions. This latter method will be discussed briefly in Section 6.

1.2 Successful gene-tagging experiments

The first successful gene-tagging experiment in plants, reported in 1984, was the cloning of the *bz* locus of maize using a transposon called *Ac* (3). In the next 2 years eight different genetic loci of maize and one gene of *Antirrhinum majus* (*pal-rec*) had been cloned through the use of a variety of transposable elements (*Table 1*). This may seem remarkable when it is noted that of 21 different loci of maize for which mutable alleles have been reported in the literature (see Table 3 of ref. 1), eight have been cloned via transposons, while only three have been cloned using conventional cDNA methods. However, it must be realized that for the most part the creation, identification and analysis of the transposon insertions present in the mutable alleles listed in *Table 1*, required many years of genetic research.

In designing new experiments where it is necessary to create and identify new mutants of specific target genes, much can be learned from the experience of these geneticists.

(i)   It is much easier to work with target genes where the mutant phenotype can be scored as a kernel characteristic (all maize genes listed in *Table 1* are such).

(ii)   All successful gene-tagging experiments listed in *Table 1* had 'target' genes where instability of gene expression could be easily recognized in somatic tissue (see Section 1.3).

Table 1. Successful gene-tagging experiments.

Mutable allele	Element[a]	Probe[b]	Method[c]		Reference
			1 2 3 4 5 6 7		
bz-m2	*Ac*	*Ac*	x x x x x		3
bz-m2	*Ac*	*Ds5933*	x x x x		4
a1-m(papu)	*En*	*Spm-I8*			
and *a1-Mum2*	*Mu*	*Mu1*	x x		5
pal[rec]	*Tam3*	*Tam3*	x x x x		6
c-m668655	*En*	*En1*	x x		7
c1-m5	*Spm*	*Spm-I8*	x x x x x x x		8
c2-m1	*Spm*	*En1*	x x x x		9
bz2-m	*Ds*	*Ds5933*	x x x x x x		10
P-vv	*Mp(Ac)*	*Ac*	x x x		11[d]
r-nj:m1	*Ac*	*Ac*	x x x x x		12
O2-m	*Spm*	*Spm-I8*	x x x x		70

[a] Element thought to be present within the mutable allele.
[b] Element from which a radioactive DNA probe was made for the initial screen of the genomic library.
[c] A summary of the methods used for identifying the correct clone (see text for description).
1. Use of a progenitor or wild-type gene (Section 5.1).
2. Use of a revertant allele (Section 5.1).
3. Use of a deletion strain (Section 5.1).
4. Use of other mutable alleles of the target gene (Section 5.1).
5. Correlation of mutant phenotype with a band in a Southern experiment in a number of individuals of a segregating population (Sections 4.2.1 and 5.2).
6. Use of DNA modification-sensitive restriction enzymes (Section 4.2.1).
7. Correlation of a transcript or of a DNA sequence homology with that expected (Section 5.4).
[d] Combined results of Peterson,T., Schwarz,D., Lechelt,C., Laird,A. and Starlinger,P. as published in ref. 11.

(iii)   Previously characterized alterations of the locus (i.e. mutants) are extremely valuable in both the creation of a transposon-induced mutant allele and in the assignment of a particular DNA clone to a specific genetic locus (see Method column of *Table 1*).

The ability to use endogenous plant transposons to clone various genetic loci has therefore been dependent upon an underlying foundation of genetic research.

1.3 Description of basic method

There are four major steps involved in the cloning of a gene via a transposon from any plant species.

(i)   A molecular probe (a clone) for an endogenous, active transposon must be available (Section 2).

(ii)   This transposon DNA must be homologous to DNA sequences present within a transposon-induced mutation of the desired 'target' gene (Section 3).

(iii)   A genomic library is constructed from the mutant plant material and specific clones homologous to the transposon DNA are isolated (Section 4).

(iv)   Since most transposons isolated to date are repetitive in the plant genome, several methods are employed to identify which clone also contains DNA from the mutant target gene (Section 5).

A hypothetical illustration of the gene-tagging method is given in *Figure 1*. The fully coloured maize kernel on the left is a phenotype which results when genes necessary for anthocyanin pigmentation are expressed. One of the genes required is that known as the *A* or *A1* gene. To produce a transposon-induced mutation of the target *A* gene, one would use plants carrying the wild-type *A* allele and an active transposon, in this case one called *Mrh* (14). Using the appropriate genetic cross (explained in Section 3.2.1) one would screen for mutants of the target *A* gene. Since the phenotype of the *A* gene can be scored on the cellular level, two classes of mutant kernel phenotypes could be distinguished—'unstable' and 'stable'. The spotted kernel shown in the middle photo is an example of an 'unstable' mutant phenotype, for only some somatic cells are inhibited in anthocyanin production, while other cells do indeed produce the pigment. Such somatic instability of gene expression from mutant to wild-type (or near wild-type) expression is suggestive, but does not prove that the particular mutation is transposon-induced (1). The inhibition or alteration of gene expression may be due to the presence of the transposon within the gene, while the release of this inhibition may be due to the excision of the transposon during somatic tissue growth.

The second phenotypic class is shown in the right-hand photo. Here all cells of the kernel aleurone seem uniformly colourless and the kernel exhibits a 'stable' mutant phenotype. For many 'target' genes only this class of mutant phenotype will be recognized, for unlike the *A* gene, the mutant phenotype may not be distinguishable on the cellular level. Although transposon insertion can result in

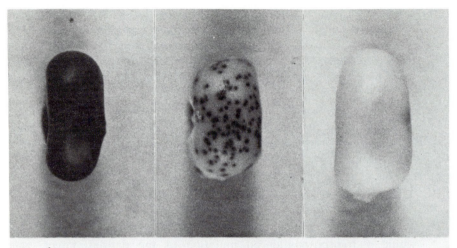

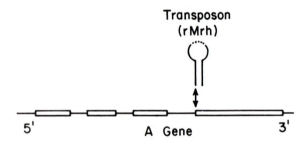

Figure 1. Example of transposon-induced instability of the *A* gene of maize. Structure of the *A* gene (bottom) is from ref. 13 with exons indicated by boxes. The fully pigmented kernel (top left) from line C is homozygous for the wild-type *A* allele as well as other genes necessary for anthocyanin pigmentation of the kernel aleurone (13). The spotted (top centre) and colourless (top right) kernels are homozygous for the *a-mrh* allele (14) and contain a non-autonomous transposable element, *rMrh*, within the gene as indicated below (Shepherd, unpublished results). The difference in phenotype of these two kernels is due to the presence or absence of the autonomous *Mrh* element at a different location in the genome (14).

a phenotype that is constant during somatic tissue growth, many other alterations of the gene (e.g. point mutation, deletion, gene rearrangement) would also be included in this class of mutants. An inability to detect somatic instability of the desired target gene is a disadvantage, since there is no quick way to identify a transposon-induced mutant among several that may appear in the screened progeny.

Assuming that a transposon-induced mutation has occurred in the example given above, then the next question is which transposon is responsible and is there a molecular probe available? Although one may be aware of only one active transposon in the parents (in this case *Mrh*), the mutation may not be due to insertion of a *Mrh* homologous sequence. Maize genetics can be useful in determining which transposon has indeed caused the mutation. This will be discussed more fully in connection with *Figure 1* in Section 1.4.

190

Finally, if the DNA of the *Mrh* transposon is available as a hybridization probe [requirement (i)], and if the mutable allele of *A* is indeed transposon-induced *and* that transposon is homologous to *Mrh* [requirement (ii)], then one can construct a genomic library of the mutant plant material and select clones that hybridize to the *Mrh* probe [requirement (iii)]. Perhaps one of the clones will also be identified as carrying part of the desired target gene [requirement (iv)].

At the moment gene-tagging using transposons in plants is an adventure due to many uncertainties. Research concerning plant transposons is in its infancy. Major questions such as which portions of an element are indispensable for its excision and transposition capability, what determines the target site, how other genes affect the frequency of transposition, and how the elements are regulated, are of extreme importance in making the gene-tagging process more efficient. It is thus necessary to review our present understanding of transposons in plants.

1.4 **Plant transposons—a brief review**

A review which lists mutable alleles of many plant species, gives a detailed discussion of how to recognize a mutation due to a transposable element, and describes the structure of several of the elements at a molecular level was published in 1986 (1). Only a few basic points relevant to gene-tagging will be repeated here.

The schematic drawing of the transposon in *Figure 1* shows a stem-loop structure for the element. Most of the cloned elements listed in *Table 2* have an inverted repeat at or very near the termini of the element. In general, this region is present in a higher copy number in the genome than is a central portion of the element (represented by a dotted line in *Figure 1*). This difference in the repetitive nature of different portions of an element will be relevant in choosing a molecular probe for the transposon (Section 4).

In maize some elements are capable of self-encoded excision and transposition. Such elements are termed 'autonomous' and are represented by the 'complete' transposon structure. Recent molecular studies have shown that other elements are homologous to the ends of the 'complete' transposon, yet may have internal DNA deletions (and/or insertions) (15–18). These 'non-autonomous' elements excise and transpose only when the complete, 'autonomous' element is present in the genome. This has led to a better understanding of what had been described earlier from genetic studies as the two-element transposon system in maize.

For example, the autonomous maize element *Ac(Mp)* is capable of catalysing its own excision and transposition, while elements which resemble *Ac* in the DNA sequence of their termini, the *Ds* elements (*Ds, Ds1* and *Ds2*), are non-autonomous and require an active *Ac* element to be present in the genome before they move their position. Thus *Ac* and *Ds* comprise a two-element system in maize. Similarly, the *En* and *Spm* elements are functionally autonomous and identical elements [thus the notation *En(Spm)* is used in this chapter]. The non-autonomous element of this two-element system is the *I* (or

Table 2. Cloned plant transposons and other insertion elements[a].

Zea mays two-element system:	Autonomous	Non-autonomous
	Ac(Mp)	**Ds**, *Ds1*, *Ds2*
	En(*Spm*)	**I**(*rSpm*)
		rDt
		rMrh
others:	**Mu1**(*Mu1.4*, *Mu1.7*, *Mu3*, *Mu4*, *Mu5*, *Cin1*, *Cin2*, *Cin3*, *Cin4* *Bs1*, *TZ86*, *MISD-1*, *Wx-RFLP*	
Antirrhinum majus	*Tam1*, *Tam2*, **Tam3**	
Glycine max	*Tgm1*	
Lycopersicon esculentum	*CR1*	
Petroselinum crispum	*Tpc1*	
Pisum sativum	*Pis1*	

[a] Elements shown in bold type have been successfully used as a hybridization probe in a gene-tagging experiment (see *Table 1*). All independently cloned isolates of elements such as *Ac*, *Ds*, *En*, and *I* are not listed although there may be sequence differences. References for the cloning of the above elements are either listed in Table V of ref. 1 or are as follows: *Ds2* (17), *En*(*Spm*) (18), *rDt* (Sorrentino, O'Reilly, Schwarz-Sommer, Saedler and Shepherd in ref. 19), *rMrh* (Shepherd, unpublished), *Mu1.4* and *Mu1.7* (20), *Mu3* (21), *Mu4* and *Mu5* (71), *Cin2* and *Cin3* (22), *Cin4* (13), *Bs1* (23), *MISD-1* (24), *Wx-RFLP* (72), *Tam3* (25), *CR1* (73), *Tpc1* (74), and *Pis1* (75).

rSpm element). More recent work with the *Tam* elements from *A.majus* has also suggested a two-element system of transposition; excision of the *Tam2* element requires the presence of another element (perhaps *Tam1*) in the genome (26).

Many different two-element systems have been defined in maize, with each consisting of an autonomous and a non-autonomous element (Table IV of ref. 1). The two-element systems exhibit a specificity which can be extremely useful in a gene-tagging experiment. For example, the less extensively studied *Mrh/rMrh* two-element system shown in the photographs of *Figure 1* is analogous to the *Ac/Ds* and *En/I* system. The non-autonomous *rMrh* element is stable in its chromosomal location in the *A* gene even if *Ac* or *En*(*Spm*) is present elsewhere in the genome. Only the presence of *Mrh* will induce excision or transposition of *rMrh* (spotted versus colourless kernel phenotype). Therefore if a newly found mutation in a desired target gene (exhibiting a stable or unstable phenotype) is due to the autonomous element *Mrh*, then the appropriate genetic crosses to a strain carrying a tester allele (for example a cross to a plant homozygous for *a-mrh* and lacking *Mrh*) could reveal the presence of the autonomous *Mrh* element linked to the mutant target locus. Similarly, if the mutated target gene contains a non-autonomous element, the identity of that element can be genetically determined by crossing the mutant with tester plants, each known to contain a particular autonomous element. The genetic crosses which should be performed to establish the relationship of a newly found mutable allele to an established two-element system, are given in Figure 6 of ref. 27. A specific example of how to work with the *Ac/Ds* two-element system in determining if a mutable allele contains a *Ds* element is found in ref. 28. However, note that to identify a non-autonomous element that is present at the mutated target gene, one must be able to recognize somatic instability in the phenotype of the target gene. Not all target genes meet this requirement (e.g. how would somatic instability of a gene causing a dwarf phenotype be recognized?).

The maize genome also contains other transposons and insertion sequences which have not been definitively associated with a two-element system. Although the *Mu* elements have been shown to transpose (for review, see 29) and may be associated with the *Cy* two-element system (P.S.Schnable and P.A.Peterson in ref. 11), the control of this transposition is not well understood (29). Still other elements of maize or the *Tgm1* element of soybean, have been found as inserts causing a stable mutation of a known gene, but the ability of the identified element to undergo further transpositions and thus its usefulness in gene-tagging is uncertain. These elements are included in *Table 2* as a reminder that when trying to extend gene-tagging methodology to other plant species, the finding of transposon-like structures may be easy but the ability to use those elements for gene-tagging depends upon their mobility and repetitiveness in the genome.

2. MOLECULAR PROBE FOR THE TRANSPOSON

Transposons have been cloned from an extremely small number of plant species (*Table 2*). To expand gene-tagging techniques to other plants, an endogenous transposon must be cloned (an exception is discussed in Section 6). There are several methods of approaching this problem, the most straightforward being to clone a gene that is known to be mutable (expressing an 'unstable' phenotype). The first transposons to be cloned in maize were found by first obtaining the wild-type gene through the identification of a cDNA and then a genomic clone, and then cloning a known mutable allele of that gene (30, and Table V of ref. 1).

The three transposons from *Antirrhinum*, *Tam1*, *Tam2* and *Tam3*, were all cloned from mutants of the *Nivea* gene (25,31,32). It is interesting to note that the wild-type *Nivea* gene was itself cloned using a heterologous probe from parsley and a known mutable allele of the *Nivea* locus helped in the identification of the gene (33). Several of the genes now cloned in maize may be especially attractive for use as a heterologous probe. A gene homologous to the maize *Waxy* gene has been isolated from the rice genome (R.Okagaki and S.Wessler, unpublished). Since there are reports in the literature of unstable mutants of rice, wheat, barley and pearl millet, with a phenotype similar to that expected for a 'Waxy' mutation, the use of this gene as a heterologous probe may prove useful in isolating transposons in these plant species. Similarly, the *A* gene of maize and the *Pal* gene of *Antirrhinum* are homologous (13,34), and thus these genes, as well as several other genes involved in the anthocyanin pathway, may be useful as probes in a heterologous plant system.

Once an active transposon has been cloned, its usefulness in gene-tagging depends upon being able to recognize its insertion into the target gene. This is necessary, for although one may have tried to induce a mutation using a specific transposon, the resulting mutation may be due to another (e.g. refs 28,35). If the cloned transposon is homologous to only a few sequences in the genome, then a correlation between the genomic location of the transposon DNA and the newly induced mutant phenotype may be fairly easily established using standard DNA hybridization techniques on segregating progeny (see Section 5.2).

However, since many cloned transposons are not unique in the genome but rather hybridize to a number of genomic sequences, the molecular approach is difficult. In this case, it is useful to employ genetics to determine which element has indeed caused the mutation. In maize this may be accomplished using what is known concerning the two-element transposon system, as discussed in Section 1.4.

3. TRANSPOSON-INDUCED MUTATION

3.1 Selection of the target gene

There are many good reasons why one would like to clone a particular genetic locus. However, some loci are easier to use as target genes for transposon-mutagenesis. As will be discussed in Section 3.2.3, there is a great advantage in choosing a target gene that can be recognized at the seed or seedling stage. If this is not the case, then it is useful if a screen can be applied such that only the mutant progeny will survive or be recognized (e.g. see Section 3.2.2).

Second, it is no small point to say that being able to observe an unstable somatic phenotype is helpful (e.g. spots on a kernel) since this suggests a transposon-induced mutation rather than a point mutation, gene rearrangement, deletion, etc. One must be concerned about this, because for a particular gene the frequency of obtaining transposon-induced mutations may be below the 'spontaneous' mutation rate. Many putative mutants may be identified but only a small percentage of those may be transposon-induced. The *Rp1* locus of maize seems to be an example of this phenomenon, for the various dominant alleles of this locus mutate in the absence of a known active transposon at relatively high frequencies ranging from 8×10^{-3} to 1.6×10^{-4} (A.J.Pryor, unpublished results).

This phenomenon may not be unusual, for R.A.Brink and R.B.Ashman (as referenced in 36) designed an experiment to see whether the frequency of new mutations arising at a specific locus in maize (such as the *Wx* gene) was increased by having the element *Ac(Mp)* present in the genome. It was found that the frequency of mutation at *Wx* was not increased in those families containing *Ac(Mp)* as compared to the appropriate control population lacking the active element. Both stable and unstable *wx* mutants were recovered from the experiment. It is interesting to note that two unstable *wx-mutable* alleles (from the *Ac*-plus population) were shown to be autonomously controlled by *Ac(Mp)*. Therefore, the ability to recognize an unstable phenotype can be extremely helpful in choosing which of the putative mutants may be transposon-induced.

The ability to recognize an unstable phenotype of the target gene is also important because many of the transposons have been shown to leave the mutant gene entirely via imprecise excision such that a stable mutant phenotype results (1,34). It is of course helpful to be able to recognize this phenomenon (e.g. from a spotted kernel to a colourless kernel phenotype), for the transposon must be present at the locus when one begins the gene cloning experiment.

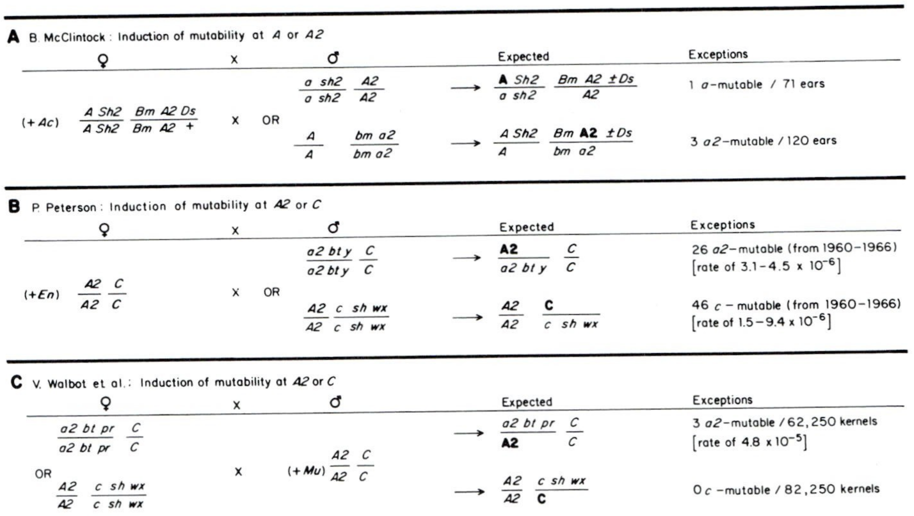

Figure 2. Induction of mutable alleles using *Ac*, *En* and mutator. In the crosses shown the maize genes *A*, *A2*, and *C* are the 'target' for transposon mutagenesis. These genes are required for anthocyanin pigmentation of the kernel aleurone. Loci at which mutable alleles were detected are shown in bold type. All other genes shown were used as flanking markers. See text for details. References for the crosses are B.McClintock (28), P.Peterson (27) and V.Walbot *et al.* (37).

If one cannot detect somatic reversion events where the phenotype of certain patches of cells revert to wild-type (suggesting excision of the transposon from the target gene) then one might try examining progeny of the new mutant for a high frequency of germinal reversion events, as an indication that the new mutant was indeed transposon-induced. However, as discussed in ref. 1, transposon-induced mutations do not always display a high frequency of germinal reversion events.

3.2 Induction of mutations

3.2.1 *Transposon-tagging a specific target gene*

Regardless of the transposon system used, the method of creating the mutable alleles is remarkably similar as outlined in the examples of *Figure 2* for the *Ac* (panel A), the *En* (panel B) and the *Mu* (panel C) transposons. The method involves a cross between two plants: one plant carries the active transposon and is homozygous wild-type for the target gene, while the other plant carries a homozygous stable recessive allele of the target gene. The resulting progeny from the cross will be heterozygous for the target gene (wild-type/stable mutant allele) and thus a wild-type phenotype is expected. If the sample size of the experiment is large enough, a few exceptional progeny not exhibiting a wild-type phenotype may be found, suggesting that something has occurred to alter the wild-type allele of the heterozygote. This phenotype may reflect the insertion of the active transposon into the wild-type allele of the target gene, such that the exceptional progeny is a heterozygote of mutated wild-type allele/stable mutant allele.

In the example for the use of the *Ac* transposon (panel A, *Figure 2*), the maize plant used as female in the cross carried an active *Ac* element and was homozygous for both the dominant wild-type *A* and *A2* genes (both genes are necessary for synthesis of purple anthocyanin pigmentation). To detect transposon insertions into the *A* gene, the male pollen donor was homozygous recessive for the target *A* gene (*a/a*), and to detect transposon insertions into the *A2* target gene the male pollen donor was homozygous recessive (*a2/a2*). The expected progeny were *A/a* and *A2/a2*, respectively, and exhibited a purple pigmentation of the kernel. Out of 71 ears showing only purple kernels due to the *A/a* genotype, one exceptional kernel was found, and it exhibited small areas of purple anthocyanin pigmentation on an otherwise colourless kernel. B.McClintock characterized the mutation by appropriate genetic crosses to show that the new allele, designated *a1-m4*, resulted from the presence of a *Ds* element within the *A* locus (28). Note that in this case, the *Ds* component of the two-element *Ac/Ds* system was the cause of the mutant phenotype, rather than the autonomous *Ac* element itself being inserted. This experiment exhibits several points which should be made concerning the production of new mutants.

(i)   The phenotype of the target gene could be recognized on the kernel, rather than requiring the planting of each kernel from the 71 ears of corn.

(ii)   The phenotype was such that instability of gene expression could be observed, thereby suggesting a transposon-insertion into the target gene.

(iii) The genetics of the two-element *Ac*/*Ds* system allowed identification of the element at the target gene.

(iv)   The element present in the target gene was a *Ds* element and not an autonomous *Ac* element.

Of the three new *a2*-mutable alleles found in the same experiment, only one contained an element from the *Ac*/*Ds* two-element family, and it also contained a *Ds* insertion rather than an active *Ac* element in the target gene (28). Similarly, of four mutable alleles of the *O2* gene found by Motto *et al.* (35), only one mutable, *o2-m2*, contained an element of the *Ac*/*Ds* system and it also was a non-autonomous *Ds* element. The major point of this is to emphasize that although a particular transposon such as the *Ac* element may be active in the genome, the resulting mutations may be due to the more repetitive *Ds* element or even an unknown factor, thus complicating the cloning procedure.

The induction of mutants using the *En*(*Spm*) system (panel B, *Figure 2*) is similar to that shown in panel A for the *Ac* system. Again the female parent contains an active transposon (*En*) and is homozygous wild-type for the target gene, either *A2* or *C*. The male parent is homozygous recessive for the target gene. Although in this experiment most of the newly created mutant alleles seemed to be autonomously controlled (i.e. are thought to harbour an active *En* element) (27), the use of an active *En*(*Spm*) element in the cross does not always produce such an element at the target gene. For example, when an active *Spm* element was used in a cross to create mutants at the *O2* gene, of the three confirmed mutants only one mutant contained an active *Spm* element at the target gene, another contained a non-autonomous *rSpm* element, and the third mutant contained an unidentified 3 kb insertion in the *O2* locus (70).

An example of the use of the maize 'Mutator' system in inducing new mutations is given in panel C of *Figure 2*. This particular experiment has been included to show that the cross may be performed in the opposite direction from that given in panels A and B. However, it is important to notice that any accidental self-pollination of the female plant will produce a stable recessive phenotype for the target locus (e.g. *a2*/*a2* or *c*/*c*). The authors eliminated these possible 'false positives' by disregarding all exceptional progeny displaying a stable recessive phenotype and continued to work only with those showing an unstable phenotype. Note that restriction fragment length polymorphisms (RFLPs) may also be used to eliminate mutants arising from accidental self-pollination rather than crossing.

In the examples given in *Figure 2*, a plant was already available which contained a stable recessive allele of the target gene and this plant was used either as a male or female parent in the cross. The exceptional progeny was actually a heterozygote between the newly induced mutant allele and the known stable recessive allele of the gene. The use of flanking phenotypic (or RFLP) markers can help to ensure that the two mutant alleles can be separated in future generations, even if the phenotypes of the two alleles are similar (e.g. both stable).

Finally, it should be obvious that any new mutant (stable or unstable phenotype) should be checked for heritability. Motto *et al.* (35) found that only four out of 198 kernels showing somatic instability at the *O2* locus were heritable. Similarly, Table 3 of ref. 27 substantiates the observation that often hundreds of putative mutants must be followed to obtain only a few confirmed mutable alleles of a given target locus.

3.2.2 *Transposon-tagging dominant genes*

The examples given to this point all involve a target gene where the wild-type gene is dominant to a known, recessive mutant allele. The same type of cross can be used when the wild-type allele is recessive and the mutant phenotype is dominant. In this case, transposon-mutagenesis of the dominant mutant allele will allow detection of the recessive phenotype.

To mutate the dominant *Rp1* locus of maize, a gene which confers resistance to the rust *Puccinia sorghi*, a variation of the above cross has been used by A.J.Pryor of the CSIRO (19). The standard method as outlined in Section 3.2.1 would be to cross a rust resistant plant (homozygous dominant *Rp-g/Rp-g*) carrying an active transposon, with a plant that is homozygous for the recessive allele (*rp1/rp1*), and to screen for a rust sensitive plant among the *Rp-g/rp1* heterozygotes (i.e. the exceptional plant might be *rp-g::transposon/rp1*). However, when performed in this manner, any mutation in the plant (not just mutations at *Rp-g*) which changes the plant to a sensitive phenotype would be taken as possible candidates. Furthermore, even if the alteration was in the *Rp-g* allele, it might later be difficult to identify the difference between the newly created mutant allele (*rp-g::transposon*) and the original recessive allele (*rp1*). Therefore the variation that was employed by Dr Pryor was to use the fact that some dominant alleles of the *Rp1* locus are resistant to certain races of the rust, while sensitive to other races. The cross that was performed was to have the female parent carry the active transposon and be homozygous for one dominant allele (e.g. *Rp-g/Rp-g*) while the other parent was homozygous for another dominant allele (e.g. *Rp-m/Rp-m*). The F1 progeny of the cross are expected to be *Rp-g/Rp-m*, unless the transposon inactivates the *Rp-g* allele to give a few exceptional progeny *rp-g::transposon/Rp-m*. These F1 progeny are then screened with a rust race that can cause disease even if the *Rp-m* allele is present, but cannot cause disease if the *Rp-g* allele is present. Only the F1 progeny carrying the mutated *rp-g::transposon* allele would be sensitive to the disease and would be self-pollinated. When these plants (presumably *rp-g::transposon/Rp-m*) are self-pollinated, the progeny are then screened with another rust race that now recognizes, and is resistant to, the *Rp-m* allele. Only the homozygote (*rp-g::transposon/rp-g::transposon*) would show disease symptoms. The fact that the other progeny are still able to show resistance suggests that the alteration was at the *Rp-g* locus rather than in another gene necessary for disease resistance.

3.2.3 *Logistics*

Before examining an alternate method to that shown in *Figure 2* it is very important to stop and discuss the logistics involved. If one would like to mutate a specific locus and the mutation frequency is expected to be one mutant in 100 000 gametes (1×10^{-5}) then, if the above crossing scheme is followed, approximately 350–500 plants must be pollinated to observe one mutation event on the resulting corn ears (assuming 200–300 F1 progeny per ear). If the mutant phenotype is not recognizable in this F1 seed, but is rather scored by a seedling phenotype, then each of the 100 000 F1 seed must be grown in a seedling screen. If the mutant phenotype is not recognizable until the mature plant develops, then 100 000 F1 seeds must be planted in the field. For corn that is approximately 5 acres to look for one mutant plant! To have a high degree of confidence of recovery of such a mutant, one may wish to increase this number approximately 5-fold, thus the amount of work involved is tremendous.

3.2.4 *A non-targeted approach*

In the previous method, the cross was designed to identify a mutation event in one specific gene. This 'targeted' approach may result in only one or a few homozygous mutant individuals and is therefore not recommended if the desired gene is likely to be embryonic-lethal when homozygous in the mutant form. An alternate, 'non-targeted' method resulting in a population of individuals, heterozygous as well as homozygous, for a newly induced recessive mutation is described below. In this method one is not really selecting a target gene but rather analysing progeny for interesting mutant phenotypes which may be transposon-induced, and later trying to correlate the molecular movement of the known transposon with the mutant phenotype. This approach was used success-fully to clone the maize *vp-mum1* allele using the transposon *Mu1* (McCarty *et al.*, in ref. 76).

   In principle, any of the known transposon systems could be used for the non-targeted approach. In brief, plants containing an active transposon system are self-pollinated or out-crossed to produce the F1. To reveal recessive mutations this material must again be self-pollinated to form the F2 population. The basic idea is outlined for the 'Mutator' system in *Figure 3* and in ref. 77. Here, the initial cross involves a pollen donor containing an active transposon system called 'Mutator', which has been correlated with a DNA insertion element called '*Mu1*' (38,30). Plants containing this active element contain approximately 20–40 copies of *Mu1* hybridizing sequences (29). The female parent in the cross does not contain many copies of the element and is considered to be the 'non-Mutator' line (39). If a transposition event occurs to inactivate any *Gene*, then the F1 progeny would be heterozygous for the event and the effect would not be seen unless the mutational event was dominant. Although the frequency of obtaining a trans-poson-induced dominant mutation is probably lower than that of obtaining a recessive mutation in the same gene, transposons can be responsible for a dominant phenotype. For example, recent molecular evidence has shown that the *C-I* dominant mutant of maize contains a DNA insert within the gene (8) and a

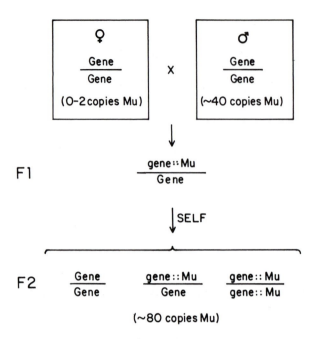

Figure 3. Induction of mutants using Robertson's 'Mutator'—an example of a random, non-targeted approach. The genetic cross indicated is that used by D.Robertson to score for 'Mutator' activity (38). In this example maize plants thought to contain Mutator activity are crossed as male to a non-Mutator line to produce the F1. The copy number of the *Mu1* transposon can vary among plants; however, in general non-Mutator plants contain 0–2 copies of *Mu1* while 'Mutator' plants contain 20–40 copies of the element (e.g. see *Figure 4*, or ref. 29 for review). Transposition of a *Mu1* element into a wild-type *Gene* may result in some F1 progeny heterozygous for a newly mutated locus (*gene::Mu/Gene*). Selfing each F1 plant to produce the F2 results in a doubling of the *Mu1* copy number per plant (29) and allows detection of F2 families segregating for a recessive mutant phenotype.

Tam3-induced mutant of the *Nivea* locus of *Antirrhinum* has produced a derivative allele with a semi-dominant phenotype (78). To score the above material for a recessive mutation, each of the original F1 seeds must be grown to maturity and self-pollinated. If the mutant phenotype is a kernel trait only these F2 ears must be analysed. Here again, the logistics of the experiment are vital, for if the mutation frequency is one gamete in 100 000, it means that 100 000 plants (5 acres of corn) must be self-pollinated to produce 100 000 F2 ears to score for the presence of a particular recessive mutant allele. Furthermore, if the mutant phenotype is a recessive seedling or mature plant trait, then growing several kernels from each of the 100 000 F2 families may be necessary! However, if one is interested in a phenotype which could in principle arise by mutation of, for example, one in any of ten genes in a pathway, then an interesting transposon-tagged mutant may appear by using $\frac{1}{10}$ as many plants. This would then make the approach more feasible.

There are several important points to remember in this approach.

(i)   Both parents must be self-pollinated to ensure that the 'newly' found mutation was not already present in a parent (perhaps as a heterozygous locus).

(ii)   Although as drawn, one might expect the F2 progeny to segregate in a ratio of three wild-type for every recessive mutant, this may not occur if the transposition event did not occur until after the F1 fertilization event (D.Robertson, in ref. 11).

(iii)   The 'Mutator' lines exhibit chromosomal breakage events in certain regions (40) and therefore a DNA insertion may not be responsible for each of the 'newly' found mutants (a point to remember for other systems as well).

(iv)   At the moment, maize genetics does not allow identification of the *Mu1* element as it does for other transposon systems (see Section 1.4).

(v)   As with other transposon systems, although *Mu1* may have been active in the genome, the resulting mutation may be due to a sequence that does not cross-hybridize with a *Mu1* probe.

(vi)   The *Mu1* element copy number may double in the self-fertilized progeny (29) such that correlation of a particular *Mu1* element with the resulting mutant phenotype may be a formidable task. Some possible methods of approaching this will be mentioned in later sections.

In summary, if one is willing to give up the idea of targeting only one particular locus and instead concentrate on learning to identify mutant phenotypes (i.e. possible transposon-induced mutations) and correlating that with molecular evidence, then the alternate 'non-targeted' approach may prove valuable.

3.3 Frequency of obtaining transposon-induced mutations

A few of the reported frequencies for inducing mutations using the *En(Spm)*, *Ac*, and *Mu* systems are given in *Table 3*. It is difficult to compare frequencies from these different experiments because many factors (both environmental and genetical) are known to effect the rate of transposition. In general, however, the rates vary from values of 10^{-6} to 10^{-4} with no particular transposon system giving a consistently higher mutation rate. Note that this range of mutation frequency is consistent with that found by L.J.Stadler in studying the 'spontaneous' mutation frequency of eight different maize genes (43). He observed that the spontaneous mutation frequency is dependent upon the gene studied and varied from a frequency of zero spontaneous mutants per million gametes for the *Wx* gene to a frequency of 492 per million gametes for the *R* gene (i.e. a frequency of 5×10^{-4}).

Since an increase in mutation frequency by several-fold can be extremely important when screening for a particular mutant plant phenotype, three major factors which influence the rate of transposition will be discussed below.

3.3.1 *Importance of chromosomal position*

The highest frequency shown in *Table 3* is that given for the production of two mutants at the bronze locus (41). In this experiment, the target gene (*Bz*) was

Table 3. Mutation frequency.

Element[a]	Target gene	Number of confirmed mutants	Number of gametes	Rate[b]	Reference
En	*A2*	1	274 848	3.6×10^{-6}	27
En	*C*	4	367 584	1.1×10^{-5}	27
En	*C2*	–	–	2.5×10^{-5}	11[c]
Spm	*O2*	3	–	3.7×10^{-6}	70
Spm	*Bz*	2	4582	4×10^{-4}	41
Ac	*R*	4	78 000	5×10^{-5}	12
Ac	*O2*	4	200 000	2×10^{-5}	35
Mutator	*A2*	3	62 250	4.8×10^{-5}	37
Mutator	*Bz2*	1	60 000	1.7×10^{-5}	37
Mutator	*C2*	1	89 500	1.1×10^{-5}	37
Mutator	*Bz*	2	102 750	2.0×10^{-5}	37
Mutator	*R*	6	72 000	8.4×10^{-5}	37
Mutator	*Bperu*	1	178 000	6×10^{-6}	37
Mutator	*Bperu*	3	275 000	1.1×10^{-5}	37
Mutator	*Y*	–	–	2.3×10^{-4}	42
Mutator	*Wx*	–	–	2.9×10^{-5}	42
Mutator	*Y*	–	–	2.2×10^{-4}	42
Mutator	*Wx*	–	–	5.6×10^{-5}	42

[a] Element listed is that which was known to be active in one of the parental lines, but this element was not necessarily found at the target gene (see text).
[b] Number of confirmed mutants/number of gametes.
[c] Results of P.A.Peterson in ref. 11.

located between two active elements (*Spm* and *rSpm*) present on the same chromosome. Of the two mutants obtained at the bronze locus, one contained *Spm* at or near the target gene and the other was shown to contain an *rSpm* element. The reason for the exceptionally high mutation frequency for this particular experiment is not well understood; however, one may speculate that it involves having the target gene physically present on the same chromosome as two elements which were capable of movement. In fact, Nowick and Peterson have documented that the *En(Spm)* element preferentially integrates between 6 and 16 map units away from the parental site of the transposon (44).

This tendency for the transposon to preferentially insert on the same chromosome as the original transposon position was first observed for the *Ac* element of maize. Van Schaik and Brink (45) analysed the distribution sites to which *Ac(Mp)* transposed when the element was removed from the *P* locus of maize. They found that although the element transposed to many sites within the genome, in about two-thirds of the cases the transposed element was linked to *P*. Furthermore, the closer a site was to the original transposon position at *P*, the more likely it was to receive the transposed element. This preference for the same chromosome was also observed by Greenblatt (46). However, his studies showed that for a distance of four map units, the transposition of *Ac(Mp)* from the *P* locus was unidirectional, a point he attributes to the direction of chromosome replication. The *En(Spm)* element did not show a tendency to move preferentially in the proximal or distal direction (44).

Greenblatt used the above observation when designing an experiment to induce *Ac* insertions into the *R-nj* locus of maize. A reciprocal translocation was used to move the *Ac* element, present at the *P* locus on chromosome 1, into close proximity with the *R-nj* locus of chromosome 10. The result was four confirmed mutable alleles (one being a confirmed *Ac* insertion into *R-nj*) out of 78 000 gametes (12). A control experiment trying to induce *Ac* insertions into *R-nj* without first bringing them together on the same chromosome was not performed, thus one can only speculate concerning the benefit of moving an element close to the target site in this particular experiment. It is also difficult to compare this mutation frequency (5×10^{-5}) with other experiments in *Table 3*, for in many experiments the location of the active element(s) was not determined.

There is some indication that certain genes may be more accessible to transposon insertion than others (perhaps due to chromatin structure?). For example, in an experiment spanning a 6 year period, P.Peterson induced mutations in the *A2* and *C* genes of maize using the *En* system (27). Although several different transposon stocks were used to induce the mutations, there was a fairly consistent pattern of producing the *c-mutables* at a higher frequency than *a2-mutables* in parallel experiments (see e.g. the first two rates of *Table 3*). A preference for one target site as compared to another can also be observed in work with the *Mutator* system. For example, the rates given in *Table 3* for induction of mutations in the *Y* gene were consistently higher than the rates given for the *Wx* gene (42). Although genes can also differ in their 'spontaneous' mutation rate, these values for the *Y* and *Wx* genes have been corrected for a control 'spontaneous' mutation rate (although not all mutants were checked for heritability).

Finally, some transposons may exhibit a certain level of target site specificity. For example, only the *A* gene of maize has been found to give a mutable allele which responds to the *Dt* element (for review, see 47). The reason for this is unknown.

3.3.2 *Importance of direction of cross*

New mutations involving transposable elements are usually induced by crossing two plants, one of which contains the active transposon system (Section 3.2). The frequency with which new mutants appear can differ when the cross is performed in opposite directions; that is, in one case the transposon-containing plant is used as a male and in the other case as the female parent. An excellent example of the resulting difference in mutation frequency comes from the experiment cited at the beginning of Section 3.3.1. In this experiment two unstable mutants of the *Bz* gene were found out of 4582 kernels screened when the *Spm(En)*-containing stock was used as the male parent (41). When the cross was performed in the opposite direction with the same stock used as the female parent, no heritable *bz-mutable* alleles were recovered out of 122 200 kernels screened (41). Similarly, the transposition rate of *Ac(Mp)* from a mutable allele of the *R* locus was found to be more numerous when the element was present in the pollen parent (48).

The frequency of inducing mutations using the Mutator system also differs depending upon whether the active *Mu* elements are present in the male or female parent (42). An example of this is shown in the last four lines of *Table 3*. The upper two lines represent use of a Mutator stock as the female parent in inducing mutations at the *Y* or *Wx* gene of maize, whereas the Mutator stock was used as the male parent for the experiment represented in the last two lines. There is a higher mutation rate for the *Wx* locus (but not for *Y*) when the *Mu* elements are present in the male parent (for this particular set of data all mutations used in the calculation would not necessarily be independent events). However, a discordance between the embryo genotype and the phenotype of the kernel endosperm is observed in 30–40% of the cases when the 'Mutator' line is used as the male parent (D.Robertson, in ref. 19) thus, this may account for some of the observed differences.

3.3.3 *Importance of the inducing allele or line*

The frequency of transposition by some elements is affected by the number of copies of the active element in the genome (for review, see ref. 1). For example, both the germinal and the somatic frequency of transposition (as scored by a phenotypic reversion of a mutable allele) are decreased as the copy number of the *Ac* element in the genome is increased. Therefore, in designing an experiment using *Ac* as the active transposon, only one active copy of *Ac* should be present. In contrast, the transposition frequency of a given *En(Spm)* element is not altered by increasing the number of active elements in the genome. When P.A.Peterson was propagating his original *En(Spm)*-containing stocks, to use in the very successful experiment of producing *En*-induced mutants of the *A2* and *C* genes of maize, he crossed the *En*-containing stocks to B-chromosome and K10-chromosome bearing stocks and then selfed to increase the *En* content (27). The number of active *En* elements present in kernels from the same ear where the mutable kernels arose varied from one to three copies of the element. The use of several copies of *En* in the parental material probably increases the chance of obtaining a mutation, but may cause some difficulty in identifying the desired target gene since more than one active element may be present in the plant material (see Section 4.2.1).

With respect to the above mention of the K10 chromosome stock, it is interesting to note that transposition of the *Ac* element from a mutable allele of the *R* gene on chromosome 10 was found to be enhanced when K10, a terminal heterochromatic chromosomal segment, was also present on chromosome 10 (48).

Mutable alleles from many plant species (see Table 1 of ref. 1) are affected by 'modifiers'—other genetic loci of unknown function which alter the timing or frequency of reversion to a wild-type phenotype (i.e. excision of the element). It is reasonable then to assume that certain genetic backgrounds will produce a higher mutation rate than others. This was in fact utilized by I.Greenblatt in inducing mutations using the *Ac* element, for a hybrid between inbred lines (4Co63 and W23) was specifically chosen as the male parent because *Ac* moves at a higher rate in these backgrounds than in any other yet measured (12).

4. CLONING

Assuming that requirements (i) and (ii) of Section 1.3 are met (a transposon has been cloned and one has initially characterized a possible mutation due to that transposon), then one must use the transposon as a hybridization probe to detect pieces of plant DNA harbouring the transposon. This initially involves selecting an appropriate probe(s) for the transposon and probing genomic DNA of the mutant plant in 'Southern' experiments (Section 4.1). At this point experiments may be performed to perhaps identify whether a particular DNA fragment hybridizing to the transposon probe(s) may be more likely to contain the 'target' gene (Section 4.2). And finally, some decisions must be made concerning the actual cloning of transposon-hybridizing sequences (Section 4.3). Detailed methods concerning experiments described below are in references listed in Table 1 or ref. 49.

4.1 Choice of probe

All transposons cloned to date (*Table 2*) show several hybridizing bands when the entire element is used as a radioactive probe in a standard Southern hybridization experiment on total genomic DNA from the plant species from which it was cloned. To facilitate target gene identification, researchers have sought to identify restriction fragments of the transposon which seem less repetitive in the genome, but yet are still capable of hybridizing to the element within the 'target' gene. This is extremely helpful, for the fewer the number of candidate clones, the less work involved in identifying which clone may contain the target gene. However, use of only a portion of the transposon as an initial probe makes a major assumption—that the element inserted in the 'target' gene contains this sequence.

Figure 4 illustrates the difference in hybridization pattern obtained using either an internal, or a terminal portion of the *Mu1* element as a hybridization probe on maize DNA. In this example, the genomic DNA is digested with *Eco*RI which does not cut within the *Mu1* element (50), thus each band hybridizing to the internal probe may represent one copy of the element. The more repetitive pattern observed when the terminal *Mu1* probe is used, indicates the presence of DNA fragments in the genome containing this sequence, either with or without the internal sequence also being present on the fragment. The less repetitive, central portion of the *Mu1* element would identify fewer DNA fragments when used as a probe to clone a Mutator-induced 'target' gene. Initial work suggested that this internal *Mu1* fragment would be a good probe, for cloned Mutator-induced mutants of the maize *Adh1*, *A*, *Bz*, and *Sh1* genes contain this portion of *Mu1* (5,20,30, Rowland and Strommer in ref. 11). Recently, however, a *Mu3* element has been cloned from a Mutator-induced mutant allele of *Adh1* (*Table 2*) and this element does not contain this internal *Mu1* sequence (K.Oishi, personal communication). In addition, no *Mu1*- or *Mu1.7*-containing clone has been correlated with a Mutator-induced allele of the maize *B* gene, *B-Peru-Mu5*, even after reducing the transposon copy number while following the unstable phenotype (V.Chandler, personal communication).

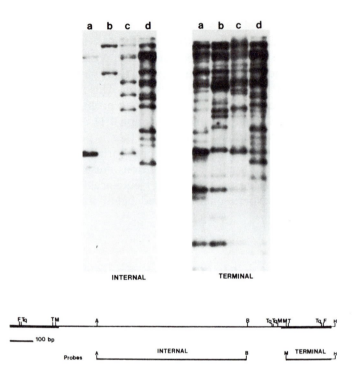

Figure 4. Comparison of two regions of the *Mu1* transposon when used as a radioactive probe on a genomic Southern blot of maize DNA: internal *Mu1* probe (leftmost four lanes) and terminal *Mu1* probe (rightmost four lanes). The internal and terminal regions of the element used as a probe are as indicated on the *Mu1* restriction enzyme map below (50): F, *Hin*fI; H, *Hin*dIII; Tq, *Taq*I; T, *Tth*III-I; M, *Mlu*I; A, *Ava*I; B, *Bst*NI. Maize DNA samples (ranging from 4 to 8 μg) were digested to completion with *Eco*RI (which does not cut within the element). This DNA was electrophoresed through a 0.5% agarose gel, transferred to a Genetran filter (Plasco) and hybridized with a ^{32}P-labelled DNA fragment from the internal region of the *Mu1* element. After autoradiography, this internal probe was removed and the DNA on the filter was re-hybridized with the terminal portion of *Mu1*. **Lanes a** and **b**: DNA extracted from non-Mutator maize plants. **Lanes c** and **d**: DNA extracted from plants exhibiting 'Mutator' activity (38). This figure was kindly supplied by Vicki Chandler, University of Oregon.

Just as in the example for the Mutator system above, the choice of a probe from other transposons is not an easy task. *Figure 5* illustrates a simplified restriction map of three cloned transposons, *Ac9*, *En-1* and *Tam3*. The shaded portion of the maps represent portions of these 'complete' or autonomous elements which have been used as hybridization probes in the gene-tagging experiments of *Table 1*. Note that here again, the portion of the element chosen as a molecular probe for an autonomous element is often from a less repetitive, central portion of the transposon. Choice of a probe is even more difficult when the target gene may contain a non-autonomous element. For example, the genetically defined *Ds* element of maize has been cloned from several different mutable genes (*Table 2*). Two of these elements, *Ds* and *Ds2*, cross-hybridize with the autonomous *Ac* element while another, *Ds1*, does not. This third element is homologous only at the inverted repeat of the termini to *Ac* (51).

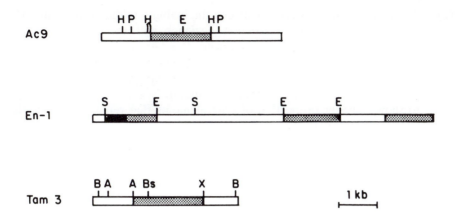

Figure 5. A partial restriction enzyme map of the *Ac9*, *En-1* and *Tam3* elements from refs 3, 7 and 6 respectively. An example of the portion(s) of the element used as the initial hybridization probe on the genomic library in a gene-tagging experiment is indicated by the filled rectangles. The 1.6 kb *Hind*III fragment of *Ac9* (pAcH1.6) has been used in several experiments (see *Table 1*). The *Ds5933* probe does not contain this region (10). The three filled rectangles of *En-1* represent the three probes used consecutively in cloning *c-m668655* (7). The solid rectangle of *En-1* represents the *Ban*II–*Xba*I fragment of the *Spm-I8* element used as a probe by K.Cone *et al*. (8). This portion of the element is less repetitive (8) and contains part of the 5′ end of gene 1 (18). Other probes used successfully from the *Spm-I8*, *En-1* and *Mu1* element are not shown. (H, *Hind*III; P, *Pvu*II; E, *Eco*RI; S, *Sal*I; A, *Ava*II; Bs, *Bst*EII; B, *Bal*I).

Both types of elements, *Ac*-hybridizing and non-*Ac*-hybridizing, have been identified in several different maize genes, thus either may be responsible for any newly found *Ds*-induced mutation. More information concerning the sequences necessary for *cis* or *trans*-activation functions is necessary to refine the probe selection process.

4.2 Identifying the correct(?) fragment before cloning

Using the transposon as a molecular probe, one may perform several experiments prior to cloning, to see if one particular hybridizing genomic fragment can be correlated with the mutant gene. The methods given below are not necessarily presented in order of importance or performed in the order given.

4.2.1 Use of DNA modification pattern

Some DNA restriction endonucleases fail to digest DNA when the recognition sequence is methylated (for review, see 52). Several researchers have identified a correlation between the ability of such enzymes to digest the transposon DNA sequence, and the transposon's activity. This includes three types of observation.

(i)   A correlation was observed between the ability of the *Pvu*II enzyme to digest the two internal *Pvu*II sites of the maize *Ac* element (*Figure 5*) and the ability of the element to autonomously transpose (53,54).

(ii)   A correlation was observed between the inability to digest the maize *Mu* elements with certain restriction enzymes and the loss of somatic reversion activity of a Mutator-induced allele, *bz2-mu1* (55).

(iii)   Most *Mu1* elements in an active 'Mutator' line of maize may be digested with the enzyme *Hin*fI which cuts twice within the element, while 'Mutator' lines which no longer induce mutations at a high frequency tend to show a higher proportion of *Mu1* elements that remain uncut by the enzyme (for review, see 55,29).

Such correlations led to the use of modification-sensitive enzymes to digest plant genomic DNA in hopes of seeing one particular transposon hybridizing sequence being cut with the enzyme while other fragments remained uncut; presumably the fragment being cut would contain an active element within the target gene.

This idea has been used successfully for the cloning of three maize genes containing an autonomous element (method 6 of *Table 1*). One way of determining if the approach is feasible for a newly found mutation is to digest the mutant plant material with an enzyme that does not cut within the element. A portion of this digest is then taken and re-digested with a modification sensitive enzyme that does cut within the element. Using the transposon-specific probe in a Southern experiment on both DNAs, the hybridization result is compared and one tries to determine whether a particular hybridizing fragment is more susceptible to the digestion of the modification-sensitive enzyme than the others. If so, this may be the fragment containing the desired 'target' gene.

A second general method is illustrated in *Figure 6*. This method requires having a segregating population of individual plants which may be separated by

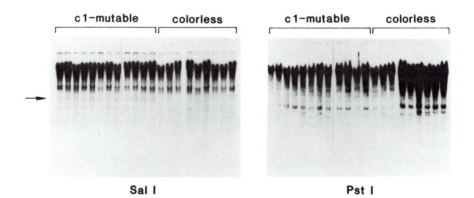

Figure 6. Co-segregation of a particular restriction fragment with the mutable phenotype of the *c1-m5::Spm* mutation. Genomic DNA was prepared from seedlings derived from kernels exhibiting a mutable or a stable recessive (colourless) phenotype of the maize *c1-m5::Spm* and *c1* alleles of the *C1* gene. This DNA was digested with *Sal*I or *Pst*I, run on an agarose gel, transferred to nitrocellulose and hybridized with a radioactive probe containing transposon sequence (8). The arrow denotes a 8.3 kb *Sal*I genomic restriction fragment that is present only in DNAs derived from kernels exhibiting the *c1*-mutable phenotype. This photo and experimental data are kindly provided by Karen Cone, Brookhaven National Laboratory.

208

phenotype as to either containing or not containing the mutant target gene. In this example the mutant allele, *cl-mutable*, is recognized by an unstable anthocyanin pigmentation of the maize kernel, as compared to the stable colourless phenotype of the *cl* allele. (The segregating population may be obtained by backcrossing a heterozygous plant, *cl-mutable/cl*, to a plant homozygous for the *cl* allele.) Genomic DNA is prepared from several individual plants of both phenotypes, digested with a modification-sensitive restriction enzyme (which cuts at least once within the element) and probed with the transposon sequence in a Southern hybridization experiment. One is then seeking a consistent correlation between the appearance of a smaller (i.e. cut) transposon-hybridizing sequence and the class of individuals containing the active, autonomous transposon in the target allele. In *Figure 6* this band is marked by an arrow and is present only in individuals representing the *cl-mutable* class of kernels. Note that such a correlation may be found using one enzyme (e.g. *Sal*I) but may not be observed with another (e.g. *Pst*I) when using the same plant DNA; therefore, it may be necessary to try many of the modification-sensitive enzymes. The choice of transposon probe is also important, since to be useful the hybridizing genomic DNA fragment must also contain 'target' gene sequence. For example, *Figure 5* shows the portion of the *Spm(En)* element used as a probe on *Sal*I-digested maize DNA in a gene-tagging experiment (8). Note that this region (solid bar) is flanked by *Sal*I sites within the transposon sequence. This fragment is only useful in the above approach if one of the *Sal*I sites is not cut in the active element, for if both sites are recognized and cut by the enzyme, the hybridizing fragment would not contain flanking target gene sequence.

4.2.2 *Use of other alleles*

Plant material from the immediate progenitor, or from revertant alleles of the desired gene, can be useful prior to the cloning step. For example, if a transposon probe hybridizes to only a few copies in the plant genome, then one may compare a Southern hybridization of progenitor, mutant and revertant plant material. This is done with the idea that the probe should identify some hybridization bands of a different size in the mutant plant material as compared to that of the progenitor and revertant. Similarly, as shown by Martin *et al.* in the cloning of the *pal-rec* allele of *Antirrhinum* (6), several independently derived germinal revertant plants may be probed to correlate the loss of a particular transposon-hybridizing band to the reversion phenomenon.

4.2.3 *Decreasing the element copy number*

If one can monitor the presence of the transposon within the mutant allele (e.g. recognize an unstable phenotype), then one may try to reduce the element copy number by outcrossing to plants containing fewer or modified elements (29, and E.Ralston and H.Dooner in ref. 11). This approach is outlined in *Figure 7* for the maize 'Mutator' system. A word of caution is in order when using this approach, for if the presence of the element at the target gene cannot be

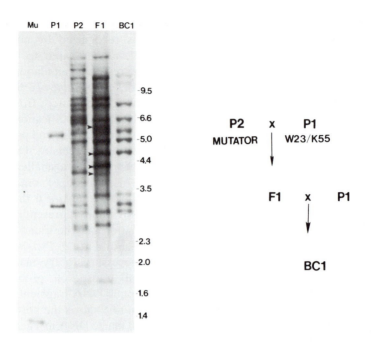

Figure 7. Genomic Southern blot showing *Mu1* copy number of individual maize plants and their progeny. The number of *Mu1* and *Mu1.7* elements in an individual maize plant may be estimated by using the internal *Mu1* fragment (see *Figure 4*) as a radioactive probe on a genomic Southern blot of DNA digested with either *Eco*RI or *Eco*RI and *Hin*dIII (neither enzyme cuts within the element). The lane marked *Mu* contains a *Hin*fI digest of the cloned *Mu1* element at a concentration of about 0.5 copies/diploid genome. **Lanes P1** and **P2** are representative of a non-Mutator (W23/K55 hybrid) and a 'Mutator' line respectively. P1 contains only two DNA fragments homologous to the *Mu1* internal probe while P2 has an estimated 24 *Mu1*-homologous fragments. A cross using P2 as female results in F1 progeny—a representative F1 individual containing 13 parental fragments and four new fragments (arrows) is shown in **Lane F1**. A backcross of the F1 by the non-Mutator line (P1) may result in some individuals containing fewer copies of the element. **Lane BC1**: a backcross plant that contains 10 parental fragments and no new fragments. This figure was kindly provided by Vicki Chandler, University of Oregon.

monitored, then the various crossing procedures required to reduce copy number may allow a second mutation of the target gene to arise unnoticed (e.g. deletion of part or all of the element or part of the target locus). The high 'mutability' of transposon sequences or of transposon-induced mutations is well-documented at both the genetic and molecular level (for review, see 1).

4.3 General comments

In cloning for gene-tagging purposes one must decide on three major items.

(i)   Whether to make a complete 'library' from the mutant plant material, or to purify a given size of DNA fragment (perhaps identified using steps of Section 4.2).

(ii)   Which probe(s) should be used for the transposon sequence in the initial library screen.

(iii) Which phage vector to use.

Obviously no one answer is correct; only after finding the desired 'target' gene can one be satisfied that the procedure used was sufficient. A few points may be made. It is better to clone DNA from a single mutant plant, rather than a pooled source of mutant plant material. The plant material used for cloning should be as close to the initial mutation event as realistically possible to try and avoid further DNA sequence rearrangements. To increase the probability of cloning the target gene, plant material showing a low level of somatic instability is preferable to that showing a high level. If the element at the locus is non-autonomous, then removal of the 'autonomous' element(s) by a crossing procedure should allow the cloning from a plant where the element is stable within the target locus. If possible, self- and out-cross the particular plant used for cloning, since this may allow genetic tests to confirm that the element was still at the target locus and provides progeny for future studies.

Once the recombinant phage library is prepared from mutant plant material, it may be best to screen the library with a transposon probe(s) that represents the complete element, internal as well as terminal sequences (7). A suggested method of doing this is to make several duplicate plaque lifts from the original plates and hybridize each one with a probe which is specific for a given region of the transposon. In this manner, one can score the initial 'positives' for how 'complete' the transposon sequence may be. This not only prevents elimination of good candidates, but may also help confirm hybridization signals. These initial candidates are then picked and replated for another round of screening. Depending upon the number of candidate clones chosen after this first screening, one may want to plate each candidate at both high and low phage density to ensure the positive phage is represented on the plate as well as aiming for a single plaque hybridization for the next round of screening and picking of positives. The positives should be purified with as few platings as possible, since rearrangements may occur due to the repetitive nature of many plant DNA sequences. The identification of many positive plaques on the first round of screening may be helpful if one later finds that no clone hybridizing to a more unique internal transposon probe leads to the identification of the desired gene. Amplification of the original library may reduce chances of finding the clone, since recombinant phage tend to multiply at unequal rates. Depending upon the number of positives in the initial screen, one may decide to rescreen with a more unique transposon fragment, such that all positives are not purified to a single plaque stage.

Once plaque-purified, phage DNA is prepared from the desired candidates, a portion is digested with a restriction enzyme, and fractionated on agarose gels. The task is then to eliminate clones containing the same genomic fragment thus making further screening more manageable. If the original cloning was performed using genomic DNA digested to completion using a given enzyme, then digestion of the recombinant phage DNA with this same enzyme allows classification of the DNA fragments according to size of the insertion. Similar insert size does not necessarily mean that the DNA fragments are the same, yet these phage DNAs may be grouped to run in parallel in the next experiment to determine whether a more frequent cutting enzyme also gives the same restriction fragment pattern for the clones.

If the original cloning was performed using partial enzyme digestion of the plant material by a frequently cutting enzyme such as *Mbo*I, then clones originating from the same genomic fragment are more difficult to eliminate. However, digestion of the phage DNAs with a frequently cutting enzyme, and then performing a Southern hybridization experiment using a probe for the complete transposon, will help to eliminate clones which do not contain expected transposon fragments (3). Elimination of clones is always risky (and plaque-purified clones should not be discarded) since one may be incorrect in assuming which transposon sequences will be present within the target gene. One must also be aware that a clone derived from a library of partial *Mbo*I-digested plant DNA may have an incomplete element due to the cloning procedure.

At this point, the number of clones which one must follow is usually quite reduced in number (e.g. from 25 initial clones hybridizing to the 1.6 kb *Hin*dIII fragment of *Ac* down to two clones in ref. 3; from 52 plaques showing homology to two different *En* probes, down to ten clones containing the expected *Eco*RI fragments of an autonomous *En* element in ref. 9). The next step is to subclone into a plasmid vector a non-repetitive plant DNA fragment adjacent to the transposon from each of the remaining candidate clones. Screening for this non-repetitive fragment may be done by nick-translating total plant DNA and hybridizing to various restriction fragments of the candidate clones (56). Fragments containing repetitive DNA sequences will hybridize to various degrees, while fragments representing a sequence of low copy number in the genome will not hybridize to the probe. These subcloned fragments (one from each candidate clone) are then used as a molecular probe in various experiments described below, to determine which (if any) may be part of or near the target gene.

5. IDENTIFICATION OF THE TARGET GENE

There is no one method to determine which of the recombinant phage contains DNA sequence within or very near the target gene. In general, the methods depend upon the availability of:

(i)   other alleles of the target gene;
(ii) segregating progeny;
(iii) information concerning the gene product (activity, location, induction, etc.).

Some of the same methods described in Section 4 may be used at this point, but the probe is now the subcloned DNA fragment that flanks the transposon sequence in the recombinant phage. These subclones will be referred to below as 'candidate clones' and should hybridize to only a few fragments in genomic plant DNA.

5.1 Use of other alleles

In 1982, Wienand *et al.* (33) described the use of progenitor, mutant and revertant plant material as a general method to identify plant structural genes among genomic DNA clones from transposon-induced mutations. This method involves the use of the candidate clones (see above) as a molecular probe to genomic DNA from the three plant materials in a Southern hybridization experiment. A transposon insertion within the mutant target gene may be detected by an alteration in the hybridization pattern, when the progenitor and mutant plant material are compared. This is dependent upon the particular restriction enzyme used, therefore several should be tried. A different hybridization pattern between progenitor and mutant plant material is not sufficient evidence to definitively assign a candidate clone to the target locus, since restriction site polymorphism is common in plants. However, if the mutant plant material has given rise to germinal revertants, then the restoration of the hybridization pattern to that resembling the progenitor is a good indication that the particular candidate clone is correct. As usual, there are words of caution, for the progenitor, mutant and revertant plant materials should not be separated by many generations of growth. Secondly, germinal reversion to the wild-type phenotype does not necessarily depend upon excision of the element, such that sometimes a hybridization pattern identical to the progenitor may not reappear. Finally, in most experiments the progenitor and/or revertant alleles are cloned to show that the transposon insertion is no longer present.

Many of the successful gene-tagging experiments have relied upon other mutable alleles to substantiate the cloning of the desired target gene. This is usually done by using candidate clone(s) as a probe in Southern experiments on genomic DNA from the other mutable alleles, and then using the 'best' candidate clone as a molecular probe to clone the other allele. For example, once a particular fragment had been 'assigned' to the *c* locus, this fragment was used to clone the homologous fragment from a library of the *c-m2* plant, a genetically defined *Ds*-induced mutation. This homologous DNA fragment was found to contain a *Ds* insertion as expected (7). Therefore although some transposons may be present in a high copy number and therefore difficult to use in a gene-tagging experiment as a probe, unstable alleles containing these elements are extremely useful as evidence that the correct candidate clone has been identified.

The use of a second mutable allele was also in a slightly different manner by O'Reilly *et al.* (5). Here genomic libraries were prepared from both an *En*-induced mutable allele of the maize *A* gene and a 'Mutator'-derived mutable allele of *A*. The libraries were screened with *En*- or *Mu1*-derived sequences respectively, and then the 'best' candidate clones from one library were used to detect the same sequence among the candidate clones from the other library.

Several researchers have used plant material from genetically defined deletion mutants of the target gene; the idea being that the correct candidate clone would no longer hybridize to this genomic DNA. This approach has proven useful, yet deletions often cover a large chromosomal segment and such plants are not always viable in the homozygous state.

5.2 Use of segregating progeny

The DNA fragment which contains the mutated target gene must always co-segregate with the mutant phenotype. One useful method to test this is that described previously for *Figure 6* in Section 4.2.1, using each candidate clone as a hybridization probe. Another method is illustrated in *Figure 8*. In this example a candidate clone has been shown to give a different hybridization pattern when used as a probe on homozygous wild-type (Wt.) and mutant (Mut.) plant material in a Southern experiment (lanes Wt. and Mut.). When these two plants are crossed the hybridization pattern of the heterozygote (lane Het.) will be representative of both parents since each contributes one of the alleles (G/g-m). When a homozygous mutant plant (g-m/g-m) is then crossed to or by a heterozygote (G/g-m), the resulting progeny will be of two types: g-m/G and g-m/g-m. When these two progeny are separated according to phenotype (wild-type and mutant respectively) then the segregation of the DNA fragments representative of the parental alleles should also segregate in the expected manner. That is, in the example of *Figure 8*, the progeny exhibiting a wild-type phenotype should show the two bands representative of the two parents and the progeny exhibiting the mutant phenotype should show the single band representative of the homozygous recessive mutant parent. It the candidate clone is unlinked to the target gene, then one would expect random segregation of the bands within a given phenotypic class. If the candidate clone is rather closely linked to the target gene (e.g. on the same chromosome but not the same entity) then only a few individuals may not give the expected pattern. Just as for any recombination experiment, the degree of certainty using co-segregation depends upon the number of individuals tested and a clear, correct classification

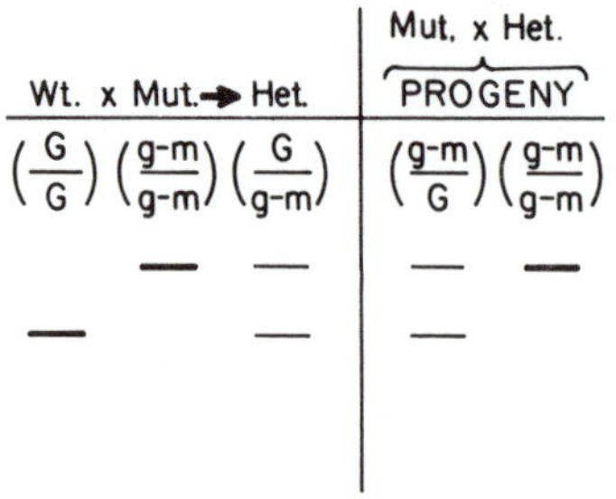

Figure 8. Hypothetical Southern hybridization experiment to determine if a particular candidate clone is representative of the gene *G*. Using a restriction enzyme that results in an altered hybridization pattern of wild-type (Wt.) versus mutant (Mut.) plant material, the candidate clone is used as a hybridization probe to screen many individual plant DNAs in a population segregating for the two phenotypes. This population may be obtained by crossing plants homozygous for the mutant allele, *g-m*, with heterozygous (Het.) plants, *G/g-m*.

of the individuals by phenotype. Discordance between a mutant kernel phenotype and the corresponding embryo genotype may be misleading. Also note that at this stage of screening, the mutant plant material does not have to be from the same allele from which the candidate clones were derived.

5.3 Mapping candidate clones

Several methods have been developed to molecularly map DNA fragments to chromosome (e.g. using monosomics as in ref. 57) or to chromosomal position (e.g. using RFLP's as in ref. 58 or recombinant inbreds as in ref. 59). If the chromosomal location of a specific 'target' gene has been determined and if these techniques are easily accessible to the researcher, then a prescreening of candidate clones by one of these approaches may lead to only a few candidate clones which must be analysed in a more tedious manner. These methods, however, give essentially the same information as that of examining segregating progeny.

5.4 Use of other information

In several successful gene-tagging experiments, researchers have used the candidate clones as a radioactive hybridization probe to mRNA in a 'Northern' experiment (see method 7, *Table 2*). The mRNA is prepared from various tissues of a wild-type plant and/or plants grown in different environmental conditions. One is interested in obtaining evidence that one particular candidate clone shows a hybridization pattern consistent with the phenotype. This approach, of course, depends upon the observed wild-type gene expression being due to regulation of transcription, rather than (or in addition to) other types of control.

The candidate clones may also be used to probe mRNA from plants which are homozygous for the transposon-induced mutation. Transcripts which hybridize both to a candidate clone and to a probe for the transposon may be identified (e.g. *Figure 9*). Using this as an initial step to screen candidate clones may be useful, since co-transcription of transposon and target allele is common (23,60,61). However, transposon insertion may eliminate target gene expression entirely, in which case searching for transcripts within the mutant plant material will not be rewarding.

One may also use the transposon probe against mRNA from the mutant plant material as an initial step prior to any genomic cloning. Identification of a transposon transcript specific to the mutant plant material in unusual amounts may suggest an alternative approach using a cDNA library as suggested by Freeling (62).

5.5 Transformation

Once identified, the best 'candidate clone' is used to isolate the homologous fragment of the wild-type allele so that identification of the gene transcript and characterization of its extent may proceed. If possible, this wild-type allele

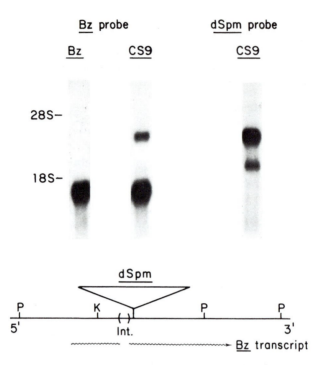

Figure 9. Example of co-transcription of element and mutated gene. mRNA prepared from maize plants containing the wild-type *Bz* gene or *dSpm*-induced mutable allele, *bz-m13CS9*, was hybridized in a Northern experiment (49) to a gene specific (*Bz*) or element specific (*dSpm* = *rSpm*) probe (61). The position of the *dSpm*(*rSpm*) element within the *Bz* transcription unit in the *CS9* allele as shown below is 38 bp after the intron (int)/exon junction (J.Schefelbein and D.B.Furtek, unpublished results). K. *Kpn*I; P, *Pst*I. This figure was kindly provided by Hwa Yeong Kim and Oliver Nelson, University of Wisconsin-Madison.

should be used in a plant transformation experiment (see Chapter 6, this volume) to definitively associate the DNA fragment with its expected gene function.

6. TRANSGENIC PLANTS AND GENE-TAGGING

At this point it should be clear that sorting through many transposon-hybridizing clones to identify the mutant target gene is very laborious and a direct result of the repetitive nature of the endogenous transposons cloned to date. The introduction of unique foreign DNA into a plant via transformation is a viable alternative, for one may more easily correlate a mutation event with a DNA insert if the insert is 'unique'. There are several requirements for this process, but the most basic is of course the ability to stably transform the plant species. The use of cultured cells for transformation is an advantage due to the large numbers of transformants which may be selected. Haploid plant material may be useful, since DNA insertion resulting in a recessive phenotype may be detected immediately. One might also consider using plant material that is heterozygous for a dominant 'target' gene. The efficiency of recovering transformants is extremely important when searching for a specific target.

Since most stable transformation procedures to date depend upon *Agro-bacterium*-mediated T-DNA integration (see Chapter 6) it is important to note that 80% of the transformants obtained under standard conditions transmit a T-DNA marker as a single locus (Budar *et al.*, as cited in ref. 63), thus T-DNA mediated insertions should provide the necessary insertional inactivation or alteration of gene expression. The lack of such reports in the literature to date is probably due to several reasons.

(i)   The use of tetraploid tobacco for a common model system for T-DNA transformation.

(ii)   The large amount of non-transcribed repetitive DNA in many plants.

(iii)   The presence of multigene families.

(iv)   A background level of mutations due to the use of protoplasts.

(v)   The lower frequency of dominant mutations expected from a DNA insertion event.

Recently, however, two groups have suggested the use of Ti plasmid-derived gene fusion vectors containing a promoter-less reporter gene (63,64). This essentially selects for DNA insertion into a transcribed gene, therefore these transformants are more likely to show an altered phenotype. The use of *Arabidopsis thaliana* for gene-tagging (65) should also prove helpful due to the small genome and low proportion of repetitive DNA in the species.

The use of plant transposons as the 'unique' foreign DNA is also enticing, since if the transposition is controllable one may be able to cope with a low transformation frequency. The maize *Ac* element has now been shown to transpose in tobacco (66), tomato (67), *Arabidopsis* and carrot (68), but negative results were obtained for transposition of *Mu1* in *Arabidopsis* (69).

7. SUMMARY

As suggested by the cautious tone of this article, the cloning of a gene via an endogenous transposon system is a risky endeavour—but it can work. The combination of genetics and molecular biology is necessary and each step of the process is dependent upon the knowledge gained in the previous step. Initial steps, involved in production and identification of a transposon-induced mutation, as well as final steps of target gene identification are dependent upon previously isolated mutants and studies of plant phenotypes and gene expression. There are several major problems in the method at present.

(i)   Few endogenous plant transposons have been cloned.

(ii)   The frequency of transposon-induced mutations is low.

(iii)   Recognition of a transposon-induced mutation is not simple.

(iv)   Many recombinant clones must be analysed.

Much of this may be circumvented in the future by learning more about the regulation of transposition, by the cloning of other transposons, or by the introduction of 'unique' DNA (perhaps transposon sequences) into the genome via transformation.

8. ACKNOWLEDGEMENTS

I would like to thank Marcus Rhoades and Ellen Dempsey for the *a-mrh* kernels used in *Figure 1*. I am also very grateful to Vicki Chandler, Karen Cone, Hwa Yeony Kim and Oliver Nelson for providing figures for the chapter. My special thanks to colleagues supplying unpublished information and allowing the reference of their articles from the *Maize Genetics Cooperation Newsletter*. Finally, I appreciate the patience displayed by my fellow researchers during the preparation of this article and the reading of the manuscript by Jennifer Sorrentino and James R.Wong.

9. REFERENCES

1. Nevers,P., Shepherd,N.S. and Saedler,H. (1986) In *Advances in Botanical Research.* Callow,J.A. (ed.), Academic Press, London, p. 103.
2. Bingham,P.M., Levis,R. and Rubin,G.M. (1981) *Cell*, **25**, 693.
3. Fedoroff,N.V., Furtek,D.B. and Nelson,O.E. (1984) *Proc. Natl. Acad. Sci. USA*, **81**, 3825.
4. Dooner,H.K., Weck,E., Adams,S., Ralston,E., Favreau,M. and English,J. (1985) *Mol. Gen. Genet.*, **200**, 240.
5. O'Reilly,C., Shepherd,N.S., Pereira,A., Schwarz-Sommer,Zs., Bertram,I., Robertson,D.S., Peterson,P.A. and Saedler,H. (1985) *EMBO J.*, **4**, 877.
6. Martin,C., Carpenter,R., Sommer,H., Saedler,H. and Coen,E.S. (1985) *EMBO J.*, **4**, 1625.
7. Paz-Ares,J., Wienand,U., Peterson,P.A. and Saedler,H. (1986) *EMBO J.*, **5**, 829.
8. Cone,K.C., Burr,F.A. and Burr,B. (1986) *Proc. Natl. Acad. Sci. USA*, **83**, 9631.
9. Wienand,U., Weydemann,U., Niesbach-Kloesgen,U., Peterson,P.A. and Saedler,H. (1986) *Mol. Gen. Genet.*, **203**, 202.
10. Theres,N.W. (1986) Inaugural-Dissertation, Universität zu Köln, 90 pp.
11. *Maize Genetics Cooperation Newsletter* (1986), **60** (articles cited from this reference by permission of the authors).
12. Dellaporta,S.L., Greenblatt,I., Kermicle,J., Hicks,J.B. and Wessler,S. (1988) In *Chromosome Structure and Function*. Gustafson,J.P. and Appels,R. (eds), Plenum Press, New York. p. 263.
13. Schwarz-Sommer,Zs., Shepherd,N., Tacke,E., Gierl,A., Rohde,W., Leclercq,L., Mattes,M., Berndtgen,R., Peterson,P.A. and Saedler,H. (1986) *EMBO J.*, **6**, 287.
14. Rhoades,M.M. and Dempsey,E. (1982) In *Maize Genetics Cooperation Newsletter*. Coe,E. (ed.), Vol. 56, p. 21.
15. Pohlman,R.F., Fedoroff,N.V. and Messing,J. (1984) *Cell*, **37**, 635.
16. Schiefelbein,J.W., Furtek,D.B., Raboy,V., Banks,J.A., Fedoroff,N.V. and Nelson,O.E.. (1985) In *Plant Genetics*. Freeling,M. (ed.), Alan R. Liss, Inc., New York, p. 445.
17. Merckelbach,A., Doering,H.-P. and Starlinger,P. (1986) *Maydica*, **31**, 109.
18. Pereira,A., Schwarz-Sommer,Zs., Gierl,A., Bertram,I., Peterson,P.A. and Saedler,H. (1985) *EMBO J.*, **4**, 17.
19. *Maize Genetics Cooperative Newsletter* (1987) **61** (articles cited from this reference by permission of the authors).
20. Taylor,L.P., Chandler,V.L. and Walbot,V. (1986) *Maydica*, **31**, 31.
21. Oishi,K.K. and Freeling,M. (1988) In *Plant Transposable Elements*. Nelson,O.E. (ed.), Plenum Press, New York, p. 289.
22. Blumberg vel Spalve,J. (1985) Inaugural-Dissertation, Universitaet zu Koeln, 96 pp.
23. Johns,M.A., Mottinger,J. and Freeling,M. (1985) *EMBO J.*, **4**, 1093.
24. Zack,C.D., Ferl,R.J. and Hannah,L.C. (1986) *Maydica*, **31**, 5.
25. Sommer,H., Carpenter,R., Harrison,B.J. and Saedler,H. (1985) *Mol. Gen. Genet.* **199**, 225.
26. Hudson,A., Carpenter,R. and Coen,E.S. (1987) *Mol. Gen. Genet.*, **207**, 54.
27. Peterson,P.A. (1978) In *Maize Breeding and Genetics*. Walden,D.B. (ed.), John Wiley & Sons, New York, p. 601.

28. McClintock,B. (1953) *Genetics*, **38**, 579.
29. Lillis,M. and Freeling,M. (1986) *Trends Genet.*, **2**, 183.
30. Bennetzen,J.L., Swanson,J., Taylor,W.C. and Freeling,M. (1984) *Proc. Natl. Acad. Sci. USA*, **81**, 4125.
31. Bonas,U., Sommer,H., Harrison,B.J. and Saedler,H. (1984) *Mol. Gen. Genet.*, **194**, 138.
32. Upadhyaya,K.C., Sommer,H., Krebbers,E. and Saedler,H. (1985) *Mol. Gen. Genet.*, **199**, 201.
33. Wienand,U., Sommer,H., Schwarz,Zs., Shepherd,N., Saedler,H., Kreuzaler,F., Ragg,H., Fautz,E., Hahlbrock,K., Harrison,B. and Peterson,P.A. (1982) *Mol. Gen. Genet.* **187**, 195.
34. Coen,E.S., Carpenter,R. and Martin,C. (1986) *Cell*, **47**, 285.
35. Motto,M., Marotta,R., Di Fonzo,N., Soave,C. and Salamini,F. (1986) *Genetics*, **112**, 121.
36. Nelson,O.E. (1968) *Genetics*, **60**, 507.
37. Walbot,V., Briggs,C.P. and Chandler,V. (1986) In *Genetics, Development, and Evolution*. Gustafson,J.P., Stebbins,G.L. and Ayala,F.J. (eds), Plenum Press, New York, p. 115.
38. Robertson,D.S. (1978) *Mutat. Res.*, **51**, 21.
39. Chandler,V., Rivin,C. and Walbot,V. (1986) *Genetics*, **114**, 1007.
40. Robertson,D.S. and Stinard,P.S. (1987) *Genetics*, **115**, 353.
41. Nelson,O.E. and Klein,A.S. (1984) *Genetics*, **106**, 769.
42. Robertson,D.S. (1985) *Mol. Gen. Genet.*, **200**, 9.
43. Stadler,L.J. (1939) In *The Spragg Memorial Lectures*, (Third Series), Michigan State College, p. 3.
44. Nowick,E.M. and Peterson,P.A. (1981) *Mol. Gen. Genet.*, **183**, 440.
45. van Schaik,N.W. and Brink,R.A. (1959) *Genetics*, **44**, 725.
46. Greenblatt,I.M. (1984) *Genetics*, **108**, 471.
47. Pryor,A.J. (1985) In *Plant Genetics*. Freeling,M. (ed.), Alan R. Liss, Inc., New York, p. 419.
48. Williams,E. and Brink,R.A. (1972) *Genetics*, **71**, 97.
49. Maniatis,T., Fritsch,E.F. and Sambrook,J. (1982) *Molecular Cloning A Laboratory Manual*. Cold Spring Harbor Laboratory, Cold Spring Harbor, New York.
50. Barker,R.F., Thompson,D.V., Talbot,D.R., Swanson,J. and Bennetzen,J.L. (1984) *Nucleic Acids Res.*, **12**, 5955.
51. Sutton,W.D., Gerlach,W.L., Schwartz,D. and Peacock,W.J. (1984) *Science*, **223**, 1265.
52. Kessler,C. and Hoeltke,H. (1986) *Gene*, **47**, 1.
53. Chomet,P.S., Wessler,S. and Dellaporta,S.L. (1987) *EMBO J.*, **6**, 295.
54. Schwarz,D. and Dennis,E. (1986) *Mol. Gen. Genet.*, **205**, 476.
55. Chandler,V.L. and Walbot,V. (1986) *Proc. Natl. Acad. Sci. USA*, **83**, 1767.
56. Gupta,M., Shepherd,N.S., Bertram,I. and Saedler,H. (1984) *EMBO J.*, **3**, 133.
57. Helentjaris,T., Weber,D.F. and Wright,S. (1986) *Proc. Natl. Acad. Sci. USA*, **83**, 6035.
58. Helentjaris,T., Slocum,M., Wright,S., Schaefer,A. and Nienhuis,J. (1986) *Theor. Appl. Genet.*, **72**, 761.
59. Burr,B., Burr,F.A., Thompson,K.H., Albertson,M.C. and Stuber,C.W. (1988) *Genetics*, **118**, 519.
60. Doering,H.-P., Freeling,M., Hake,S., Johns,M.A., Kunze,R., Merckelbach,A., Salamini,F. and Starlinger,P. (1984) *Mol. Gen. Genet.*, **193**, 199.
61. Kim,H.Y., Raboy,V., Schiefelbein,J., Furtek,D. and Nelson,O. (1986) In *Eukaryotic Transposable Elements*, Abstracts of 1986 Cold Spring Harbor Meeting. Cold Spring Harbor, New York.
62. Freeling,M. (1984) *Annu. Rev. Plant Physiol.*, **35**, 277.
63. Andre,D., Colau,D., Schell,J., Van Montagu,M. and Hernalsteens,J. (1986) *Mol. Gen. Genet.*, **204**, 512.
64. Teeri,T.H., Herrera-Estrella,L., Depicker,A., Van Montagu,M. and Palva,E. (1986) *EMBO J.*, **5**, 1755.
65. Feldman,K.A., Christianson,M.L. and Marks,M.D. (1988) In *J. Cell. Biochem.*, **12C**, 160.
66. Baker,B., Schell,J., Loerz,H. and Fedoroff,N. (1986) *Proc. Natl. Acad. Sci. USA*, **83**, 4844.
67. Yoder,J., Belzile,F., Alpert,K., Palys,J. and Michelmore,R. (1986) In *Tomato Biotechnology Symposium*. Nevins,D. and Jones,R. (eds), Allan R. Liss Inc., New York, in press.
68. Van Sluys,M.A., Tempe,J. and Fedoroff,N. (1987) *EMBO J.*, **6**, 3881.
69. Zhang,H. and Somerville,C.R. (1986) *Plant Sci. (Lett.)*, **48**, 165.
70. Schmidt,R.J., Burr,F.A. and Burr.B. (1987) *Science*, **238**, 960.
71. Chandler,V.L., Talbert,L.E., Mann,L. and Faber,C. (1988) In *Plant Transposable Elements*. Nelson,O.E. (ed.), Plenum Press, New York, p. 339.
72. Spell.M.L., Baran,G. and Wessler,S.R. (1988) *Mol. Gen. Genet.*, **211**, 364.

73. Bernatzky,R., Pichersky,E., Malik,V.S. and Tanksley,S.D. (1988) *Plant Mol. Biol. Rep.*, **10**, 423.
74. Herrmann,A., Schulz,W. and Hahlbrock,K. (1988) *Mol. Gen. Genet.*, **212**, 93.
75. Shirsat,A.H. (1988) *Mol. Gen. Genet.*, **212**, 129.
76. *Maize Genetics Cooperation Newsletter* (1988), **62** (articles cited from this reference by permission of the authors).
77. Shepherd,N.S., Sheridan,W.F., Mattes,M.G. and Deno,G. (1988) In *Plant Transposable Elements*. Nelson,O.E. (ed.), Plenum Press, New York, p. 137.
78. Coen,E.S. and Carpenter,R. (1988) *EMBO J.*, **7**, 877.

CHAPTER 9

Molecular plant virology

ROBERT H.A.COUTTS, PAULINE J.WISE and
SAMUEL W.MACDOWELL

1. INTRODUCTION

1.1 Use of cloned plant virus nucleic acids

Complete nucleotide sequences of a number of plant virus genomes are available and the determination of others is imminent. Such information facilitates high resolution transcript mapping and the identification of virus gene products *in vivo* using, in the latter case, antibodies raised to fusion proteins in bacterial expression vectors. The functions of gene products of DNA viruses are being investigated by site-directed mutagenesis *in vitro* and construction *in vitro* of recombinants between naturally occurring strains, or artificially induced mutants possessing different phenotypes. Likewise, methods are now available for some RNA viruses where, following construction of full length cDNA clones *in vitro*, mutagenesis or recombination at the DNA level and transcription back into infectious RNA is feasible (1,2).

An alternative method for studying gene expression is the integration of a virus gene or a cDNA copy into the plant genome either via an *Agrobacterium tumefaciens* Ti plasmid vector (Chapter 6) or by direct gene transfer (Chapter 7) and expression in plant cells from a suitable promoter. The effect of individual gene products on the host can then be monitored in the absence of virus replication. Whole virus genomes can also be introduced into plants using the Agrobacterium/Ti vector system. This process, termed 'agroinfection', results in the production of systemic infection and infectious virus (e.g. ref. 3) and has generated information on the sequences necessary for virus DNA replication.

Numerous aspects of virus pathogenicity may be investigated at the molecular level and future studies should include not only the analysis of gene products and their function, but also the role of host proteins in virus DNA and RNA replication, following the determination of virus genome sequences. Also dependent on a knowledge of genome sequences is the molecular analysis of a number of virus specific mechanisms such as:

(i)   vector transmission and specificity;
(ii)   host range;
(iii)   symptom expression;
(iv)   cell to cell transport;
(v)   host resistance;

(vi)   hypersensitive response;
(vii) resistance-breaking mechanisms of cross protection by related virus strains;
(viii) mechanisms of symptom modification by satellite RNAs.

Using information accumulated from the above studies, novel methods of producing virus-resistant plants are being investigated, following integration of virus-related sequences into plant genomes. For example, host production of:

(i)   suitable 'anti-sense' RNAs or antibodies to essential viral proteins which might interfere with virus replication or cell to cell spread;
(ii)   coat protein from related strains which might interfere with virus assembly or disassembly in an analogous fashion to *in vivo* cross-protection (e.g. ref. 4);
(iii) satellite RNAs which could reduce symptom expression (e.g. ref. 5).

These investigations are an alternative to the isolation of resistance genes and their introduction into crop plants by genetic manipulation. The availability of cloned and sequenced infectious viral nucleic acid has also attracted a great deal of interest in this context, especially concerning the potential contribution of DNA-containing viruses to the development of vectors for the transformation of plants (6). Additionally, a knowledge of plant virus molecular biology has already improved virus detection procedures, identification methods and disease diagnosis. To realize the goals described above, the molecular characterization of entire plant virus genomes must be achieved. This chapter, therefore, describes general methods for cloning and sequencing the genomes of single-stranded DNA- and RNA-containing plant viruses, which are applicable to other plant viruses. The biology and molecular biology of the DNA-containing plant viruses has been reviewed extensively (7–10). This chapter will describe only the single-stranded (ss) DNA viruses, geminiviruses. No further reference will be given in this chapter to the double-stranded (ds) DNA-containing caulimovirus group of plant viruses. The prototype virus of that group, cauliflower mosaic virus, and a number of other members have been analysed extensively at the molecular level and are described elsewhere (9,11). For general virological techniques the reader is referred to the volume in this series entitled *Virology—A Practical Approach* edited by B.Mahy.

1.2 Single-stranded DNA plant viruses: geminiviruses

Members of the geminivirus group are characterized by a genome of circular ssDNA encapsidated in twinned (or geminate) quasi-isometric particles from which the group derives its name (12). The majority of geminiviruses may be conveniently divided into two subgroups on the basis of host range and/or insect vector, namely: those that infect dicotyledonous plants and are transmitted by the same whitefly species, *Bemisia tabaci*; and those that infect mono-cotyledonous plants, each of which is transmitted by a different leafhopper vector. A third possible subgroup are transmitted by specific leafhopper vectors but, similar to the whitefly-transmitted geminiviruses, have host ranges confined to dicotyledonous plants (7). Cloned copies of the genomes of the whitefly-transmitted geminiviruses, cassava latent virus (CLV; 13), tomato golden mosaic virus (TGMV; 14) and bean golden mosaic virus (BGMV; 15) are infectious when mechanically inoculated to host plants and have been used to define a bipartite genome in each case (8). In contrast, only a single DNA component has so far been identified for the leafhopper-transmitted gemini-viruses, maize streak virus (MSV; 16) and wheat dwarf virus (WDV; 17). Demonstration of the infective unit of their genomes was hampered by an inability to mechanically transmit cloned copies of the genomes. These problems were overcome for both viruses when cloned copies of both MSV (18) and WDV (19) were delivered to susceptible host plants via 'Agroinfection' (3) and shown to be infectious.

Original observations (20,21) that crude extracts of nucleic acid from plants infected with two different geminiviruses contained a number of virus-specific DNA species, including a presumptive replicate intermediate (a ds supercoiled DNA form) prompted researchers to utilize extracts enriched in this species for studies on cloning and sequencing geminiviral DNA. Part of this chapter illustrates this procedure as applied to WDV, a unipartite, DNA-containing leafhopper-transmitted geminivirus of monocotyledons (17).

1.3 Single-stranded RNA plant viruses

RNA viruses constitute the largest group of plant viruses and fall into a number of different subgroups depending upon their morphology and genome organiz-ation. A review of the subject can be found in ref. 22. The work described in this chapter deals with tobacco necrosis virus (TNV) an icosahedral virus with a single component RNA genome (22). The virus is transmitted by zoospores of the fungus *Olpidium brassicae*. It can be mechanically transmitted to a wide range of plants, both dicotyledonous and monocotyledonous, usually causing necrotic lesions but rarely causing systemic infection. The viral RNA is not polyadenylated and a method is described in this chapter of generating random cDNA clones to the viral RNA for sequencing purposes.

2. ISOLATION, FRACTIONATION AND CHARACTERIZATION OF PLANT VIRUS NUCLEIC ACIDS

2.1 Isolation of geminiviruses

This section describes a protocol used for the purification of WDV. Refs 7, 8 and 10 list other publications which describe methods for the purification of a number of other geminiviruses and their genomic DNA. These methods and that outlined below are modifications of the general method described for the prototype geminivirus, MSV (23).

Wheat dwarf virus is isolated from systemically infected wheat tissue (*Triticum aestivum* L. cv. Diamant) following transmission by *Psammotettix alienus* (Dahlb.) by a modification to the method of Lindsten *et al.* (24). For efficient extraction of all geminiviruses, only fresh material should be used.

(i)   Homogenize 200 g of tissue in 400 ml of 10 mM Na_2HPO_4-NaH_2PO_4, pH 4.0 containing 10 mM Na_2EDTA and 0.1% (v/v) thioglycolic acid.

(ii)   Stir the homogenized fibres for 30 min at 4°C and strain through moistened cheesecloth.

(iii)   Centrifuge the filtrate for 10 min at 4°C in a MSE 6 × 250 ml rotor at 10 000 r.p.m. and pass the supernatant through fresh moistened cheese-cloth.

(iv)   Centrifuge the filtrate for 10 min at 16 000 r.p.m. in a MSE 8 × 50 ml rotor at 4°C and retain the supernatant on ice.

(v)   Centrifuge the supernatant for 2 h at 14 000 r.p.m. in a Beckman Ti 45 rotor at 4°C.

(vi)   Gently resuspend the pellets using a rubber policeman in 1 ml of 10 mM Na_2HPO_4-NaH_2PO_4, pH 7.0, prior to layering on 16 ml sucrose gradients [10–50% (w/v)] in 10 mM Na_2HPO_4-NaH_2PO_4, pH 7.0, in 17 ml Beckman polyallomer tubes. Preform the gradients 18 h before use and store at 4°C.

(vii)   Centrifuge the gradients for 16 h at 20 000 r.p.m. in a Beckman SW28.1 rotor at 5°C.

(viii) Fractionate the gradients by puncturing the bottom of the tubes and passing the fractions using a peristaltic pump through a Beckman UV spectrophotometer, for collection in a Gilson Microcol fractionator.

(ix)   Collect the UV-absorbing fractions; pool them into one tube and dilute with 10 mM Na_2HPO_4-NaH_2PO_4, pH 7.0, to approximately 5% (w/v) sucrose.

(x)   Centrifuge in a Beckman R65 rotor at 50 000 r.p.m. for 90 min, at 4°C.

(xi)   Gently resuspend the pellets using a rubber policeman in 100–200 µl of 10 mM Na_2HPO_4-NaH_2PO_4, pH 7.0, and store at 4°C.

(xii)   Assess the yield of virus spectrophotometrically, assuming $A_{260} = 7$ for a 1 mg/ml suspension (7).

2.2 Isolation of RNA-containing plant viruses

Due to the variations in isolation methods between viruses of different groups as well as within groups it is not possible to review all the methods here. Detailed protocols for the purification of a number of RNA viruses can be found in the Association of Applied Biologists' descriptions of plant viruses (25).

A method for the purification of TNV, the type member of its group, is given below. This virus, which causes a necrotic infection on French bean tissue (*Phaseolus vulgaris* cv Pinto), is isolated by a modification of the method of Lesnaw and Reichmann (26).

(i)   Homogenize 250 g of frozen tissue in 500 ml of acetate buffer (100 mM sodium acetate, pH 5.0, 10 mM Na_2EDTA) containing 1% (v/v) β-mercaptoethanol.

(ii)   Filter through muslin to remove large debris.

(iii)   Add an equal volume of chloroform/ether (1:1, v/v) and stir for 30 min at 4°C.

(iv)   Centrifuge for 10 min at 8000 r.p.m. in a MSE 6 × 300 ml rotor, at 4°C.

(v)   Remove the upper, aqueous layer to a clean flask.

(vi)   Add solid $(NH_4)_2SO_4$ to a concentration of 25% (w/v) and stir with aeration at 4°C for 2 h.

(vii) Pellet the precipitate as in step (iv) and discard the supernatant.

(viii) Resuspend the pellet in one-third of its original sap volume with acetate buffer [step (i)] and clarify by low speed centrifugation as in step (iv).

(ix)   Add solid $(NH_4)_2SO_4$ to 25% (w/v) to the supernatant and repeat the cycle of precipitation, resuspension and clarification until the volume is approximately 10 ml.

(x)   Layer the suspension onto two 36 ml sucrose gradients (10–40%, w/v) in acetate buffer [step (i)] pre-formed and stored at 4°C for 16 h before use. Centrifuge for 4 h at 25 000 r.p.m. in a Beckman SW28 rotor at 4°C.

(xi)   Fractionate the gradients by passing through a LKB Uvicord detector unit, collecting UV-absorbing fractions with a Gilson Microcol fractionator.

(xii) Collect the relevant peak fractions and pool together. Dilute with acetate buffer [step (i)] to give a final sucrose concentration of 5% (w/v).

(xiii) Add solid $(NH_4)_2SO_4$ to a concentration of 25% w/v and stir at 4°C for 16 h.

(xiv) Pellet the virus as in step (iv). Resuspend the pellet in 500 μl acetate buffer, clarify by centrifugation as in step (iv) and store at 4°C.

(xv)   Determine the yield of virus spectrophotometrically assuming $A_{260} = 5.5$ for a 1 mg/ml suspension.

2.3 Preparation and fractionation of nucleic acid from geminivirus-infected tissue

In order to produce sufficient amounts of DNA for restriction mapping and subsequent cloning of geminivirus DNA, it is necessary to generate preparations of nucleic acid enriched in ds forms of viral DNA, especially covalently closed circular (ccc) supercoiled DNA. The protocols described in the following sections produce such preparations and modifications of them have also been applied to produce extracts from a number of other plants infected with geminiviruses (e.g. 16). The procedure for the isolation of crude extracts is adapted from that Ikegami *et al.* (21).

2.3.1 *Preparation of unfractionated nucleic acid extracts*

The procedure for the preparation of WDV-specific DNA forms described below is generally applicable for the isolation of similar material from plants infected with other geminiviruses. Fresh tissue identical to that used in Section 2.1 is extracted as follows and can be pulverized in liquid nitrogen prior to extraction if necessary.

(i)   Homogenize the tissue with a pestle and mortar at a rate of 2 ml/g with 0.1 M Tris–HCl, pH 7.0, buffer containing 0.1 M NaCl, 10 mM

Na$_2$EDTA, 1% (w/v) SDS, 1% (v/v) β-mercaptoethanol, 0.325% (w/v) ascorbic acid and 100 μg/ml proteinase K (Sigma Chemical Co., previously auto-digested at 60°C for 1 h).

(ii)   Incubate the homogenate at room temperature (~22°C) for 30 min and filter through muslin.

(iii)   Extract the filtrate with an equal volume of a 4:1 mixture of phenol/chloroform and centrifuge for 10 min at 12 000 r.p.m. in glass Sorvall tubes in a MSE 8 × 50 ml rotor.

(iv)   Remove the aqueous phase, extract with an equal volume of chloroform and centrifuge as in step (iii).

(v)   Precipitate the nucleic acids from the aqueous phase with 2 vols of ice-cold ethanol at −20°C overnight.

(vi)   Recover the precipitate in glass Sorvall tubes by centrifugation for 10 min at 10 000 r.p.m. in a MSE 8 × 50 ml rotor, at 4°C.

(vii)   Dry the pellets *in vacuo* and resuspend in 1 × TAE buffer (40 mM Tris–HCl, pH 8.0, 20 mM acetic acid, 1 mM Na$_2$EDTA).

(viii)   Extract the nucleic acid solution with an equal volume of TAE-saturated phenol in Sorvall glass tubes by gentle mixing.

(ix)   Centrifuge as in step (vi).

(x)   Remove the aqueous phase and extract with an equal volume of a mixture of phenol, chloroform and isoamyl alcohol (25:24:1, v/v).

(xi)   Centrifuge as in step (vi).

(xii)   Remove the aqueous phase and extract with an equal volume of chloroform/isoamyl alcohol (24:1, v/v).

(xiii)   Centrifuge as in step (vi).

(xiv)   Precipitate the nucleic acids from the aqueous phase with 2 vols of ice-cold ethanol at −20°C overnight.

(xv)   Remove precipitate by microcentrifugation in sterile 1.5 ml Eppendorf tubes.

(xvi)   Dry the pellets *in vacuo* and resuspend in 1 × TAE buffer. Store at −20°C.

2.3.2 *Fractionation of nucleic acid extracts*

Nucleic acid extracts are fractionated in caesium chloride–ethidium bromide gradients according to the procedure of Sunter *et al.* (27).

(i)   Add solid CsCl to the nucleic acid extracts (Section 2.3.1) at a rate of 1 g/ml.

(ii)   Mix the contents until the CsCl dissolves and add 80 μl of ethidium bromide (EtBr, stock solution of 10 mg/ml in sterile distilled water) to every ml of the solution to achieve a final solution density of approximately 1.55 g/ml CsCl and an EtBr concentration of 600 μg/ml.

(iii)   Centrifuge the mixtures in sealed tubes in a Beckman SW50.1 rotor for 65 h at 32 000 r.p.m. at 20°C.

(iv)   Puncture the tubes at their base and collect 500 μl fractions using a Gilson Microcol fractionator.

(v)   Remove the EtBr by exhaustive extraction with 1-butanol (saturated with distilled water) in sterile 1.5 ml Eppendorf tubes.
(vi) Dialyse individual fractions against an excess of TE (10 mM Tris–HCl, pH 8.0, 1 mM Na_2EDTA) buffer at 4°C overnight to remove CsCl.
(vii) Precipitate the nucleic acids with 2 vols of ethanol at −20°C overnight.
(viii) Recover by microcentrifugation for 10 min at 4°C.
(ix) Dry the pellets *in vacuo* and resuspend in TE buffer.
(x)   Analyse fractions by agarose gel electrophoresis (Section 2.5).

In order to use DNA fractionated by the above procedures in further *in vitro* reactions, it is essential to remove all EtBr from the fractions prior to dialysis. The efficiency of this procedure is monitored by dye fluorescence after spotting aliquots of the fractions on strips of Parafilm and examination under a hand-held UV light source.

2.4 Isolation of nucleic acids

2.4.1 *Extraction of geminivirus DNA*

Geminivirus DNA can be prepared by the method of Adejare and Coutts (28). This method is primarily designed for isolation of DNA from virions, but is applicable to crude infected cell fractions.

(i)   Make samples 0.5% in SDS, and 0.5 mg/ml with respect to proteinase K [Section 2.3.1 (i)] in an Eppendorf tube.
(ii) Incubate at 65°C for 10 min.
(iii) Extract the mixture in the same tube with an equal volume of TAE-saturated phenol by gentle vortexing.
(iv) Microcentrifuge for 5 min.
(v)   Remove the aqueous phase to a clean Eppendorf tube and extract repeatedly with water-saturated di-ethyl ether.
(vi) Add to the aqueous layer $1/10$ vol. 3 M sodium acetate, pH 5.5, 2 vols of cold absolute ethanol and precipitate the DNA overnight at −20°C.
(vii) Collect the precipitate by microcentrifugation at 4°C for 10 min.
(viii) Discard the ethanol, resuspend the dried pellet in a minimal volume of TAE buffer and store at −20°C.

2.4.2 *Extraction of DNA other than geminiviral DNA*

All other DNA extractions (e.g. from recombinant phage) are performed according to a standard procedure described in *Table 1*.

2.4.3 *Extraction of plant virus RNA*

When extracting RNA it is important to avoid any contamination of buffers or solutions with RNase. All glassware should be baked and, where possible, solutions should be treated with diethyl pyrocarbonate (DEPC, Sigma Chemical Co.) at a concentration of 0.1% (v/v) for several hours before autoclaving.
   RNA can be prepared from TNV by the method of Coutts *et al.* (29).

Table 1. Extraction of DNA.

1. Mix equal volumes of solutions containing DNA and TAE-saturated phenol in sterile Eppendorf tubes by inversion until an emulsion forms.
2. Separate the aqueous phase by microcentrifugation at room temperature for 8 min and transfer it to a fresh Eppendorf tube.
3. Re-extract the aqueous phase first with a mixture of phenol/chloroform/isoamyl alcohol (25:24:1, by vol.) and then with chloroform/isoamyl alcohol (24:1, v/v) using two microcentrifugation steps as in step 2.
4. Remove the aqueous phase and precipitate the DNA by adding 2 vols of ethanol and leaving at $-20°C$ overnight. Adjust low ionic strength solutions to 0.25 M sodium acetate (from a 2.5 M sodium acetate, pH 5.2, stock solution) prior to ethanol addition.
5. Pellet the DNA by microcentrifugation for 10 min at 4°C.
6. Wash the pellet with 70% (v/v) ethanol and repeat the microcentrifugation as in step 5.
7. Remove the supernatant and dry the pellet *in vacuo*, prior to resuspension in buffer of choice (usually TE: 10 mM Tris–HCl, pH 8.0, 1 mM Na_2EDTA).

(i)   Resuspend TNV in TNE buffer (100 mM Tris–HCl, pH 7.5, 100 mM NaCl, 10 mM Na_2EDTA) and add SDS to 4% (w/v) and clarified bentonite to 0.2% (v/v).

(ii) Add an equal volume of TNE-saturated phenol.

(iii) Shake the mixture intermittently on ice for 10 min.

(iv) Centrifuge to separate the phases and recover aqueous phase.

(v)   Re-extract the phenol phase without added SDS, and combine the aqueous phases.

(vi) Extract the pooled aqueous phases repeatedly with ether to remove traces of phenol.

(vii) Add a 1/10 vol. of 3 M sodium acetate (pH 5.5) and 2.5 vols cold ethanol and allow the RNA to precipitate at $-20°C$ overnight.

(viii) Collect the precipitate by microcentrifugation at 4°C for 15 min.

(ix) Discard ethanol, resuspend the dried pellet in a minimal volume of TE and store in aliquots at $-70°C$.

To purify TNV RNA further, sucrose gradient centrifugation can be used.

(i)   Load the RNA onto 10–40% (w/v) sucrose gradients in 1 × TNE buffer and centrifuge for 3.5 h at 50 000 r.p.m. in a Beckman SW50.1 rotor at 4°C.

(ii) Fractionate each gradient into 200 μl aliquots. Remove a 5 μl aliquot from each fraction and display on a 1% agarose gel (Section 2.5) to identify the fractions containing full length genomic RNA.

(iii) Add to these fractions 200 μl of TNE and 1 ml of absolute ethanol. Precipitate the RNA overnight at $-20°C$.

(iv) Recover the precipitate by microcentrifugation. Dry and resuspend the RNA in TE and store in aliquots at $-70°C$.

RNA can be visualized on a number of denaturing gel systems such as methylmercury/agarose, formaldehyde/agarose or polyacrylamide/urea gels as described in ref. 30.

2.5 **Agarose gel electrophoresis and Southern blotting**

Agarose gel electrophoresis of viral DNA or RNA can be performed as follows.

(i)   Cast agarose gels in slabs for submarine horizontal electrophoresis. Several sizes of gels are used including 'mini-gels' (10.0 × 6.5 × 0.5 cm) and 'maxi-gels' (11.0 × 15.5 × 0.5 cm), which are used in 'mini-sub' DNA electrophoresis tanks (Biorad) or a larger apparatus of a similar design. Several comb sizes for casting gel slots are used varying in width from 0.2 to 0.8 cm. Agarose (Sigma Chemical Co., No. A6013) is used at concentrations ranging from 1 to 1.5% in TAE [Section 2.3.1 (vii)] or TBE (90 mM Tris–HCl, pH 8.2, 90 mM boric acid, 2 mM Na_2EDTA). Cast gels according to Maniatis *et al.* (30). Incorporate EtBr into the gels when required at a concentration of 0.5 μg/ml diluted from a 10 mg/ml stock solution in sterile distilled water.

(ii)   Mix the samples with a ¹⁄₁₀ vol. sterilized loading buffer [50% (v/v) glycerol, 10 × TAE or 10 × TBE 0.4% (w/v) bromophenol blue] and load into gel slots. Use sample volumes of 10–30 μl for 'maxi-gels' and 5–25 μl for 'mini-gels'. As size markers use restriction digests of phage DNA, for example *Hae*III digest of φX174 RF or *Hin*dIII digest of λ (both New England Biolabs).

(iii)   Electrophorese gels in 1 × TAE or TBE buffer at 200 mA constant current for 2–4 h for 'maxi-gels' and at 100 mA constant current for 30–60 min for 'mini-gels'. Terminate electrophoresis when the bromophenol blue dye front migrates approximately three-quarters the gel length.

(iv)   Stain gels electrophoresed without EtBr dye by immersion of the gels in the dye at a concentration of 0.5 μg/ml in sterile distilled water and destain with distilled water for 10 min.

(v)   Visualize nucleic acid bands on a UV Transilluminator. For a permanent record, take photographs with a suitable camera (e.g. Polaroid MP4) fitted with two Y (K2) yellow filters (Hoya) and one UV (O) filter (Hoya).

The nucleic acids displayed on agarose gels are often transferred to a solid support matrix for later hybridization probing. The protocol for maxi-gels described in *Table 2*, is a modification to that described by NEN in their handbook for GeneScreen plus hybridization membranes.

2.6 **Synthesis of random cDNA probes**

The protocol for the synthesis of random cDNA probes (*Table 3*), is generally applicable to all geminivirus ssDNAs studied thus far as well as to ss viral RNAs. The cDNA probes labelled with $\alpha^{32}P$ are prepared by modifications to the method described by Gould and Symons (31).

Between 10 and 30% of the radioactivity should be incorporated into cDNA. If the incorporation is poor it may be because the template DNA or RNA is degraded, or the reverse transcriptase preparation contains inhibitors of the

Table 2. Southern transfer of DNA to GeneScreen Plus membrane.

1. Cut the gel to the required size following electrophoresis and excise the slots with a sterile scalpel blade.
2. Incubate the gel with gentle agitation in 200 ml of 0.25 M HCl for 15 min at room temperature.
3. Pour off the HCl and incubate the gel with gentle agitation in 200 ml of 0.4 M NaOH, 0.6 M NaCl for 30 min at room temperature.
4. Pour off the alkaline salt solution and incubate the gel with gentle agitation in 200 ml of 0.5 Tris–HCl, pH 7.5, containing 1.5 NaCl for 30 min at room temperature.
5. Cut the GeneScreen Plus membrane to the exact size of the gel with sharp scissors and mark the concave side with a pencil asterisk. Wet the membrane thoroughly in sterile distilled water, before immersion in 100 ml of 10 × SSC[a] for 15 min.
6. Carefully place the gel onto a glass plate supported in a large glass casserole dish, covered with a sheet of Whatman No. 3 chromatography paper and surround with strips of Parafilm.
7. Half fill the dish with 10 × SSC, moisten the Whatman paper wick and carefully place the GeneScreen Plus membrane onto the gel with the asterisk mark against the gel. Excise a corner from the membrane for later orientation.
8. Gently smooth the gel down with a gloved hand and ensure that there are no trapped air-bubbles.
9. Stack four pieces of pre-cut Whatman No. 3 chromatography paper on top of the gel.
10. Stack on top of the paper a further 4 inches of folded Kleenex paper towels. Alternatively, Boots Disposable Nappy Pads provide excellent absorbency per unit cost.
11. Place a weight on top of the whole pile and surround the construction with 'cling-film' wrap. Leave undisturbed on a flat bench at room temperature overnight.
12. Remove all the towels and paper after transfer (both should be damp throughout). Remove the GeneScreen membrane into 100 ml of 0.4 M NaOH for 30 sec, shake off excess liquid and then immerse the membrane in 0.2 M Tris–HCl, pH 7.5, containing 2 × SSC[b] for 60 sec.
13. Blot the membrane dry, but do not dry completely. Store membrane in a slightly over-size polythene bag at 4°C for later hybridization.

[a] 10 × SSC is 1.5 M NaCl, 150 mM sodium citrate, pH 7.0, prepared according to Maniatis *et al.* (30).
[b] 2 × SSC is 0.3 M NaCl, 30 mM sodium citrate, pH 7.0, prepared according to Maniatis *et al.* (30).

reaction. Currently, a number of reverse transcriptase preparations derived from cloned material are commercially available. Whilst these are expensive they are preferred to others as these often contain inhibitors as mentioned above and lack RNase H activity. In the case of viral RNA, the inclusion of an RNase inhibitor (Human Placental Ribonuclease Inhibitor; Amersham) obviates this problem. The inhibitor is included in the reaction mixtures at concentrations varying from 1 to 2 U/μl.

2.7 Polyacrylamide gel electrophoresis of nucleic acid from geminivirus-infected tissue

Polyacrylamide gels are used to isolate the open-circular (oc) form of geminivirus dsDNA in a protocol described previously to separate the circular and linear forms of geminiviral DNA from host nucleic acids (32).

(i) Cast gels in siliconized glasss tubes (8.0 × 0.6 cm i.d.). The gel solution conntains 4% (w/v) acrylamide [prepared from a 40% (w/v) stock containing 39.6% (w/v) acrylamide (BDH Chemical Co.) and 0.4% *N,N'*-

Table 3. Construction of cDNA probes.

1. First prepare the primer for the reaction by digesting salmon sperm DNA (Sigma) with DNase I (Sigma) to generate a large number of random oligodeoxynucleotide fragments. Incubate 166 μl of salmon sperm DNA (6 mg/ml) with 20 μl of 10× primer reaction buffer[a], together with 94 μl of DNase I (1 mg/ml) in a sterile 1.5 ml microcentrifuge tube for 2 h at 37°C.
2. Punch a hole in the tube cap with a needle. Denature the DNase and primer DNA by boiling in a paraffin bath at 105°C for 10 min, and then cool on ice. Replace the punctured cap and use the primer immediately or store at −20°C.
3. Vacuum-dry 10 μCi [α-^{32}P]dCTP[b] in a sterile 1.5 ml microcentrifuge tube.
4. Add to the vacuum-dried [α-^{32}P]dCTP in the following order: 5 μl of 10 × cDNA synthesis buffer[c], 5 μl of 50 mM DTT, 2 μl of dATP, dGTP and dTTP each at 15 mM, and 25 μl of primer from step 2.
5. Add x μl of geminivirus DNA or viral RNA (up to 1 μg concentration).
6. Add an aliquot of sterile distilled water so that the final reaction volume will be 50 μl.
7. Mix the contents by vortexing and remove 2 × 1 μl aliquots onto Whatman 3 MM filter paper discs to assess the input and background levels of radioactive counts incorporated into acid-precipitable material.
8. Add 10–15 units of AMV reverse transcriptase (Pharmacia) to the tube and incubate at 37°C for 90 min.
9. Remove 1 μl aliquots and assay cDNA synthesis by determination of the radioactivity incorporated into acid-precipitable material.
10. Terminate the reaction by the addition of 1 μl of 0.5 M Na$_2$EDTA and store on ice.
11. Separate labelled cDNA from unincorporated dCTP by 'spin-column' chromatography (30) and store at 4°C.

[a] 10× primer reaction buffer is 0.1 M Tris–HCl, pH 7.4, 0.1 M MgCl$_2$.
[b] 2'-Deoxycytosine 5'-[α-^{32}P]triphosphate >410 Ci/mmol, ethanolic solution: Amersham.
[c] 10× cDNA synthesis buffer is 0.5 M Tris–HCl, pH 8.3, 60 mM MgCl$_2$, 400 mM KCl.
[d] One unit of reverse transcriptase catalyses the incorporation of 1 nmol of dTMP into acid insoluble product in 10 min at 37°C, using polyriboadenylate:deoxythymidylate as template primer.

methylene bisacrylamide (BDH Chemical Co.)], 0.05% (w/v) TEMED and 0.075% (w/v) ammonium persulphate in TAE [Section 2.3.1 (i)]. Allow acrylamide to polymerize under an overlay of sterile distilled water for at least 1 h at 22°C.

(ii) Mix crude nucleic acid samples with ¹⁄₁₀ vol. loading buffer (see Section 2.5) and layer onto gels in volumes as large as 50 μl.

(iii) Electrophorese the gels in 1 × TAE buffer at a constant current of 10 mA/tube for 2–3 h using a Shandon tube gel electrophoresis tank and a discontinuous buffer system.

(iv) After electrophoresis stain the gels with EtBr, and photograph under UV light (Section 2.5).

2.7.1 *Electroelution of nucleic acid*

Recover the oc form of geminivirus dsDNA from either agarose or polyacryl-amide by electroelution according to the procedure of Zassenhaus *et al.* (33) using electroelution cups (ISCO Co. Ltd) shown diagrammatically in *Figure 1*. The system consists of an electroelution cup and an electrophoresis tank with separate cathodic and anodic reservoirs, which are further subdivided by a semipermeable dialysis membrane creating four compartments A, B, C, and D.

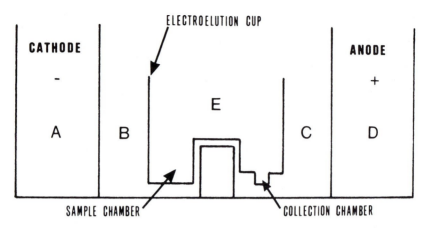

Figure 1. Simplified diagram of the ISCO electroelution apparatus. Semi-permeable boiled dialysis membranes are held in position at the bases of the open-ended sample and collection chambers (arrowed) by plastic compression rings prior to use.

(i)   Seal the dialysis membranes (held in position by plastic compression rings) at the bases of the sample and collection chambers of the electroelution cup.

(ii)   Fill compartments A, C, and D with sterilized high ionic strength buffer [3 M sodium acetate plus 1 × TEA (1 × TEA is 50 mM Tris–acetate, pH 7.8, 0.1 mM Na_2EDTA)] and compartment B with a lower ionic strength buffer (2 × TEA).

(iii)   Place the gel slice(s) containing the band(s) of interest in the sample chamber in the electroelution cup (E) and fill with 10 ml of a low ionic strength buffer (0.005 × TEA).

(iv)   Place the cups in the tank and electroelute the DNA into the collection chamber at 100 V for 1 h. Diffusion of the high ionic strength buffer in compartment C into the collection chamber forms a high salt barrier which prevents the electroeluting nucleic acid from either sticking to, or penetrating, the membrane.

(v)   Recover the electroeluate by transferring the buffer in the cup (but not the fluid in the collection chamber) to a 10 ml Sterilin polypropylene tube with a Pasteur pipette.

(vi)   Remove the fluid in the collection chamber with a Gilson automatic pippetor fitted with a disposable tip and transfer to Eppendorf tubes.

(vii)   Return the buffer to the cup and continue electroelution for a further 1 h.

(viii) Remove the fluid from the collection chamber and rinse the chamber with 100 μl of 0.15 M sodium acetate.

(ix)   Pool the collection chamber eluates and extract the nucleic acid (*Table 1*).

As an alternative several other methods for elution of nucleic acids are available and often preferable for RNA, particularly the methods of electro-elution in dialysis bags, 'freeze–squeeze' elution or elution from low melting-point agarose (30).

Table 4. Hybridization of DNA.

1. Pre-hybridize the membrane in a sealed plastic bag with pre-hybridization solution[a] (1 ml/ 10 cm[2] membrane) for at least 6 h at 42°C with constant agitation.
2. After incubation add denatured radioactive probe[b] to the pre-hybridization buffer, reseal the bag and incubate with constant agitation for 6–24 h at 42°C.
3. Remove the membrane after hybridization and wash with three changes of wash solution[c] at 70°C over a period of 60 min. Use 250 ml for each wash.
4. Blot the membrane dry, wrap in 'Saranwrap' and expose to Kodak X-Omat S film for several hours to several days at room temperature or at −70°C with the use of an intensifying screen.
5. Develop the autoradiogram either manually or using an automatic processor.
6. Keep the membrane moist during this process so that the radioactive probe can be stripped for later re-hybridization, if necessary.
7. For probe stripping two methods are available. Either:
   (i) Boil the membrane for 20–30 min in 10 mM Tris–HCl, pH 7.7, 1 mM Na$_2$EDTA, 1% SDS or
   (ii) Boil the membrane for 20–30 min in 0.1 × SSC[d], 1%SDS.
8. After blotting dry the membranes are now available for re-hybridization.

[a]Pre-hybridization solution contains 5 × SSPE (diluted from sterilized 20 × SSPE which is 2.4 M NaCl, 0.3 M tri-sodium citrate, 0.26 M NaH$_2$PO$_4$ and 20 mM Na$_2$EDTA), 1 × Denhardt's solution [diluted from filter-sterilized 50× Denhardt's solution, containing Ficoll 400, PVP, bovine serum albumin (BSA) each at 1% (w/v) concentration], 100 μg/ml sheared and denatured salmon sperm DNA and 50% (v/v) formamide (Rose Chemicals Ltd). De-ionize formamide before use by stirring 5 g of Amberlite MB-1 monobed resin with 100 ml of formamide for 30 min at room temperature. Filter de-ionized formamide through Whatman No. 3 paper and store at −20°C.
[b]Radioactive probes are either cDNA probes (see Section 2.6, *Table 3*) or nick-translated probes (see Section 2.7.2) and are denatured by heating at 90–105°C for 10 min in an Eppendorf tube and 'snap-cooling' in an ice/ethanol bath. Final concentration of the probe should be ~10 ng/ml (1–4 × 10^5 d.p.m./ml) for optimum signal to background ratios.
[c]Wash solution is 2 × SSC containing 1% (w/v) SDS.
[d]0.1 × SSC is 15 mM NaCl, 0.015 mM sodium citrate, pH 7.0.

2.7.2 *Preparation of nick-translated probes*

The electroeluted oc form of geminivirus dsDNA or plasmid DNAs are labelled with [α-^{32}P]dCTP (Amersham) by nick translation according to the procedure of Rigby *et al.* (34) as described in Amersham kit instructions. Fractionate the probes with 'Elu-tips' (Schleicher and Schuell) according to the manufacturer's instructions, to remove unincorporated nucleotides.

2.8 **Hybridization of Southern transfers**

Nick-translated probes (Section 2.7.2) or cDNA probes (Section 2.6, *Table 3*) are used in hybridization experiments to identify virus-specific nucleic acids. The protocol described in *Table 4* for hybridization is a modification of that described by NEN in their handbook for GeneScreen Plus hybridization transfer membranes.

2.9 **Nuclease digestions**

To characterize the various forms of DNA present in both unfractionated and fractionated nucleic acid extracts of geminivirus-infected tissue it is necessary to assess the sensitivity of the nucleic acid species to different nucleases. As a pre-requisite to cloning geminiviral dsDNA into a suitable vector a prior knowledge of the restriction endonuclease map of the DNA to be cloned is required. Enzymes which cut the vector DNA and the geminivirus DNA only once can then be identified and utilized. This section describes the restriction endo-nuclease analysis of fractionated geminivirus DNA. All nuclease digestions are carried out in either 0.5 or 1.5 ml Eppendorf tubes.

2.9.1 *Nuclease S1*

Perform digestions in 10 mM sodium acetate, pH 5.0, 5 mM $MgSO_4$, 0.5 mM $ZnSO_4$ (prepare as a 10× stock solution and store at −20°C) for 1–2 h at 37°C. Normally use 4.5–9 units of enzyme per reaction. Dilute nuclease S1 (type III, Sigma) to a concentration of 9 U/µl in a storage buffer [25% (v/v) glycerol, 10 mM sodium acetate, pH 5.0, 5 mM $MgSO_4$, 0.5 mM $ZnSO_4$] before use. After digestion extract DNA (*Table 1*) and resuspend in TE buffer prior to analysis by agarose gel electrophoresis (Section 2.5), or analyse directly following the addition of ¹⁄₁₀ vol. loading buffer (Section 2.5).

2.9.2 *DNase I and RNase A*

DNase I (type III) and RNase A (type XI-A) both obtained from Sigma, are prepared as 1 mg/ml solutions in sterile distilled water and stored at −20°C. In addition, RNase stock solutions are boiled for 10 min to destroy contaminating DNase.

Perform digestions in 10 mM Tris–HCl, pH 7.4, 10 mM $MgCl_2$ (prepare as a 10× stock solution and store at −20°C) for 1–2 h at 37°C. Normally use 1 µg of enzyme per reaction. Analyse the digestion products as described in Section 2.9.1.

2.9.3 *Exonuclease III*

Perform digestions in 20 mM Tris–HCl, pH 7.5, 7 mM $MgCl_2$, 100 mM KCl, 100 µg/ml gelatin, 1 mM DTT (prepare as a 10× stock solution and store at 4°C) for 1–2 h at 37°C. Dilute the enzyme 1:4 in 1 × reaction buffer and use 40–80 units of enzyme per reaction. Analyse as described in Section 2.9.1.

2.9.4 *Restriction endonucleases*

Perform restriction digestions in appropriate restriction buffers, the composition of which is described in Maniatis *et al.* (30). The conditions of digestion are as recommended for each particular enzyme in the 1988 New England Biolabs catalogue. For small-scale digests perform reactions in 10–20 µl in the appropriate buffer, with the enzyme at a concentration of 2–5 U/µg DNA, at DNA concentrations not exceeding 150 µg/ml. Add enzymes in double digests simultaneously, except where these require different incubation temperatures or NaCl concentrations. Check digestions by 'mini-gel' agarose electrophoresis (Section 2.5) if possible. Add calf intestinal phosphatase (CIP:BCL) to plasmid vector digests to prevent self-ligation. Reaction mixtures are incubated at the temperatures recommended by the manufacturers for periods of time varying from 1 to several hours, depending on the amount of DNA to be digested. Extract the DNA (Section 2.4.2, *Table 1*) resuspend in the desired buffer, and store for later analysis.

2.10 Examples of results

The autoradiogram in *Figure 2* shows a Southern transfer (from a 1% agarose gel) of WDV DNA prepared from purified virus as in Section 2.4.1. The cDNA probe used for hybridization (Section 2.8, *Table 4*) was constructed with WDV DNA as in Section 2.6, *Table 3*. Two encapsidated DNA species were noted, the larger and more abundant of which is thought to be virus genomic DNA; the smaller molecule is a sub-genomic DNA species (35). Similar sub-genomic DNAs have been identified previously in purified preparations of other geminiviruses (7,36). When a similar WDV-specific cDNA probe was prepared and used to probe a Southern blot of a preparation of nucleic acids from WDV-infected tissue produced as in Section 2.3.1, a number of other virus-specific DNA forms were

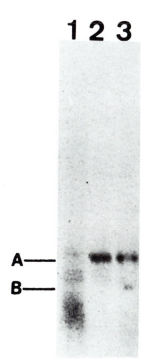

Figure 2. Three preparations of WDV virion DNA were electrophoresed on a 1% agarose gel and then Southern blotted onto GeneScreen Plus membrane. The autoradiogram shown was obtained following hybridization with a cDNA probe constructed with purified WDV DNA (Section 2.6, *Table 3*). The size markers indicated refer to the cloned genomic (A; 2660 nt, ref. 14) and sub-genomic (B; 1250 nt, ref. 34) DNAs of TGMV electrophoresed in separate lanes but not shown. The virus DNA shown in **Lane 1** was isolated from frozen leaf tissue while the preparations in **Lanes 2** and **3** were derived from fresh material.

visualized. By subjecting the nucleic acid extracts to treatment with a number of nucleases (Section 2.9) it is possible to tentatively identify the nature and form of the DNA species (20). For instance the autoradiogram shown in *Figure 3* is a Southern transfer of a 1.4% agarose gel and illustrates the effects of separate digestion of the nucleic acid preparations with exonuclease III, nuclease S1, RNase and DNase I. The bands shown probably represent genomic-length oc, linear and ccc dsDNA (bands C, D and E, respectively); genomic-length ssDNA (band G); dimeric oc and linear dsDNA (bands A and B, respectively) and subgenomic ccc dsDNA (band J) on the basis of their differential sensitivity to the nucleases investigated.

Single-stranded DNA is identifiable by its extreme sensitivity to nuclease S1 and resistance to exonuclease III, ccc dsDNA is identified by its relative resistance to both nuclease S1 and exonuclease III, while the oc and linear dsDNA are identified by their resistance to nuclease S1 and their sensitivity to exonuclease III. Linear and oc dsDNA forms are distinguished by their relative sensitivity to nuclease S1, which attacks ss regions in duplex molecules (e.g.

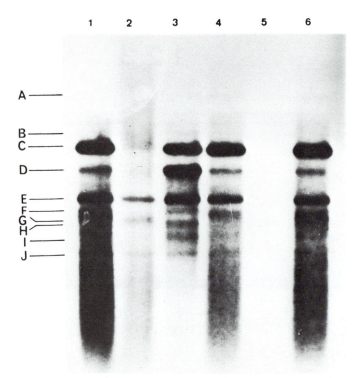

Figure 3. Nucleic acid extracts of WDV-infected leaf tissue were separated on a 1.4% agarose gel containing 0.5 mg/ml EtBr. The gel was blotted onto Gene-Screen Plus membrane and the blot was probed with a cDNA probe constructed with purified WDV DNA as in *Figure 2*. **Lanes 1** and **6** contain untreated nucleic acid extracts. The extracts shown in **Lanes 2, 3, 4** and **5** were treated previously with exonuclease III, nuclease S1, RNase A and DNase I, respectively as described in Section 2.9 prior to electrophoresis. The DNA bands illustrated, *viz.* A to J, are as described in Section 2.10.

nicks in dsDNA and unpaired regions in ccc dsDNA) and produces increased amounts of linear dsDNA relative to circular dsDNA of the same species. The other dsDNA species (bands F, H and I) are possibly oc and linear forms of subgenomic DNA. On the basis of the results shown in *Figure 3*, it would appear that in crude extracts there are at least two populations of differently sized subgenomic DNA species.

The oc form of geminiviral dsDNA (band C) can be fractionated away from the linear form of the DNA and host nucleic acid by electrophoresis of nucleic acid extracts from infected tissue (Section 2.3.1) through polyacrylamide gels (Section 2.7) and extraction of the DNA from the gel by electroelution (Section 2.7.1). Such a fractionation for WDV DNA is shown in *Figure 4*. Confirmation that the isolated DNA species were WDV specific is shown in *Figure 5* lane 6. The oc geminivirus dsDNA is now available for probe construction by nick translation (Section 2.7.2) or as a substrate for preliminary restriction endonuclease mapping (Section 2.9.4).

As an example of stripping and re-probing membranes, when the radioactive probe was stripped off the blot (*Table 4*) shown in *Figure 3* and the membrane checked by autoradiography, no positive signals were found. The blot was then re-hybridized with a nick-translated probe constructed with gel-purified oc WDV dsDNA. The profile of bands seen on the blot was identical to that in *Figure 3*.

Restriction endonuclease digestion is illustrated by the following study. The products of nuclease digestion of separate nucleic acid preparations from WDV-infected tissue were displayed on a 1.0% agarose gel, together with the products of endonuclease restriction of gel purified oc WDV DNA. Southern blots of the gel were probed with a cDNA probe prepared with purified WDV DNA. Four major virus-specific DNA species were noted (*Figure 5*, bands C, D, E and G). Four separate digestions with restriction endonucleases *Hind*III, *Sma*I, *Cla*I and *Eco*RI are shown. The results illustrate (lanes 7–10) that *Hind*III, *Sma*I and *Cla*I each produce a single DNA fragment (lanes 7–9) of similar electrophoretic mobility to viral linear dsDNA (lane 3, band D). Each of these enzymes apparently cleaved the oc WDV DNA at a single site, strongly suggesting the

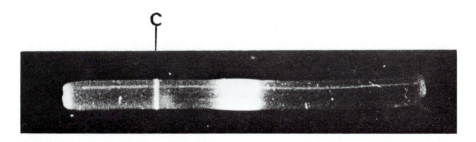

Figure 4. An EtBr stained polyacrylamide (4%) tube gel showing purified WDV oc dsRNA (Section 2.7). Confirmation that the band (c) was indeed the oc form of the DNA was obtained following separate nuclease treatment of the electroeluted DNA from the band (Section 2.7) and hybridization with a cDNA probe as in *Figures 2* and *3* (see *Figure 5*). Electrophoresis was from left to right.

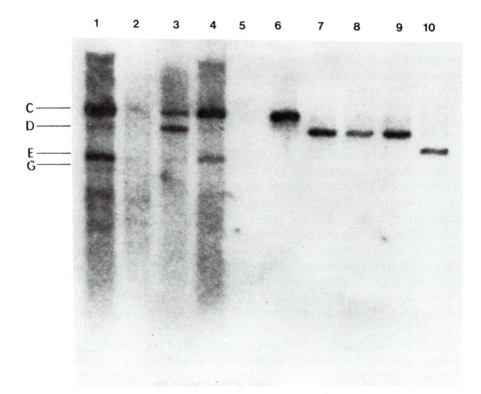

Figure 5. Unfractionated and purified nucleic acids from WDV-infected leaf tissue were electrophoresed on a 1% agarose gel, and Southern blotted onto GeneScreen Plus. The blot was probed as in *Figure 3* and **Lanes 1–5** contained the same samples as displayed in the same lanes in that figure. **Lane 6** contained electroeluted WDV oc dsDNA (see *Figure 4*) while **lanes 7, 8, 9** and **10** contained electroeluted WDV oc dsDNA following treatment with restriction endonucleases *Hind*III, *Sma*I, *Cla*I and *Eco*RI, respectively (Section 2.9.4). The DNA band identities are as in *Figure 3* and as described in Section 2.10.

presence of a single DNA component for the virus. However, the possibility still exists that one or more of these enzymes might cleave the DNA at two or more closely situated sites but the differentiation of such molecules is beyond the resolution of the gel system employed. *Eco*RI produces a fragment smaller than viral linear dsDNA (lane 10) indicating more than one site for this enzyme in oc WDV DNA. Smaller fragment(s) were not detected and this result might be explained by either poor transfer of small fragments of DNA to the GeneScreen Plus membrane or the presence of several closely clustered sites in the DNA. In the latter instance if the fragments were too small they might have migrated off the bottom of the gel. Whilst the oc form of geminiviral dsDNA is suitable for preliminary restriction mapping of the genome, because of a potential random loss of nucleotides from such DNA forms we prefer to use the supercoiled, ccc form of the DNA for cloning purposes.

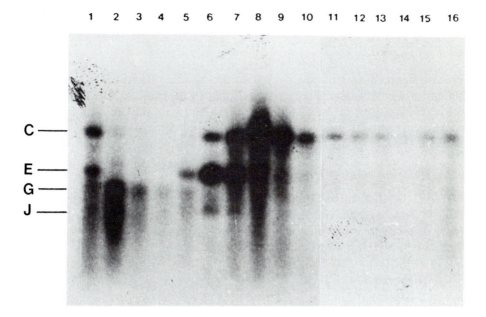

Figure 6. Unfractionated extract from WDV-infected leaf tissue was separated through a CsCl/EtBr gradient (Section 2.3.2). Aliquots of each fraction from the gradient was analysed on a 1% agarose gel, and blotted to GeneScreen Plus membrane. The blot was probed with a nick-translated probe (Section 2.7.1) constructed with WDV oc dsDNA extracted from gels (Section 2.7) as shown in *Figure 4*. **Lane 1** contained an unfractionated extract from WDV infected leaf tissue. **Lanes 2–16** contained fractions from the bottom to the top of the gradient respectively. The DNA band identities are as in *Figure 3* and as described in Section 2.10.

More stringent identification of the DNA forms present in WDV-infected tissue is facilitated by further fractionating the extracts, and subjecting the separate fractions to analysis by agarose gel electrophoresis, Southern blotting and probing. The results of a CsCl density gradient fractionation and subsequent analysis using nucleic acid extracts from WDV-infected leaf tissue are shown in *Figure 6*. Here positions of band E and band C (as shown in *Figure 4*) overlap in the gradient somewhat. The amounts of band E relative to band C are higher towards the bottom (lanes 5 and 6) and lower towards the top (lanes 9–16) of the gradient. This result confirms that band E has a greater density than band C (oc dsDNA) and is consistent with the former being ccc dsDNA. A second DNA species (band J) co-sediments with band E, has a higher electrophoretic mobility and is probably a subgenomic ccc dsDNA species. The DNA at the bottom of the gradient (band G) is probably ssDNA which has a greater density than dsDNA and binds more EtBr. Those fractions (fractions 4 and 5 in *Figure 6*) which appeared to consist predominantly of WDV ccc dsDNA were pooled for later cloning and subsequent sequencing. This procedure has been used successfully for the separation and eventual cloning of a number of genomic and subgenomic geminivirus DNAs (e.g. refs 27,36).

3. CLONING AND SEQUENCING OF PLANT VIRUS NUCLEIC ACIDS

3.1 **Media and buffers**

3.1.1 *Minimal agar*

Prepare minimal agar by mixing 500 ml of 3% (w/v) Bacto-agar (Difco) with 500 ml of 2 × M9 salts [prepared according to Maniatis *et al.* (30)], 1 ml of 0.1 M CaCl$_2$, 1 ml of 2 M MgSO$_4$, 1 ml 20% (w/v) glucose. Autoclave the components separately and combine with 0.5 ml of filter-sterilized 1% (w/v) thiamine–HCl (Sigma Chemical Co.) prior to pouring the plates.

3.1.2 *Luria (L) broth*

Prepare L broth by mixing together Bacto-tryptone (Difco), Bacto yeast extract (Difco), NaCl and glucose to concentrations of 1.0%, 0.5%, 1.0%, and 0.1% (w/v) respectively in distilled water. Adjust the pH to 7.5 with approximately 3 ml of 1 M NaOH per litre of broth and autoclave.

3.1.3 *2TY broth*

Prepare 2TY broth by mixing together Bacto-tryptone, Bacto yeast extract and NaCl to concentrations of 1.6%, 1.0% and 0.5% (w/v), respectively, in distilled water and autoclave.

3.1.4 *2TY bottom agar and 2TY H top agar*

Prepare 2TY bottom agar by dissolving 1.5% (w/v) Bacto-agar (Difco) in 2TY broth by autoclaving. Prepare 2TY H top agar by dissolving 0.6% (w/v) Bacto-agar in 2TY broth without yeast extract.

3.1.5 *SOB media*

Prepare SOB media (39) by mixing Bacto-tryptone Bacto yeast extract, NaCl and KCl to concentrations of 2% (w/v), 0.5% (w/v), 10 mM and 2.5 mM in water. Autoclave and add MgSO$_4$ and MgCl$_2$, to concentrations of 10 mM each from a filter-sterilized stock solution containing 1 M MgCl$_2$ and 1 M MgSO$_4$, before use.

3.1.6 *TFB Buffer*

Prepare a 0.5 M solution of MES (2[*N*-morpholino]ethane sulphonic acid, Sigma Chemical Co.) and adjust the pH to 6.2 with concentrated KOH, sterilize by filtration and store at −20°C. Use this solution to prepare a 10 ml K-MES solution and add solid KCl, MnCl.4H$_2$O, CaCl$_2$.2H$_2$O and hexamine cobalt (111)$^+$ trichloride (HA CoCl$_3$) to final concentrations of 100 mM, 45 mM, 10 mM and 3 mM, respectively. Filter solution and aliquot into sterile flasks before storing at 4°C.

R.H.A.Coutts, P.J.Wise and S.W.MacDowell

3.1.7 *DnD mix*

Dissolve 1.53 g of dithiothreitol (DTT, Sigma Chemical Co.) in 9 ml of dimethylsulphoxide (DMSO, Sigma Chemical Co.) with the addition of 200 μl of 0.5 M MES (pH 6.2) prepared as in Section 3.1.6. Final concentrations are 1 M, 90% (v/v) and 10 mM, respectively.

3.2 Cloning of ccc geminivirus dsDNA

3.2.1 *Restriction and ligation*

Produce a restriction endonuclease map of the oc form of the geminivirus DNA of interest. Select restriction endonucleases that cut the DNA once, especially those which also restrict the pEMBL9 (37) vector DNA once in the cloning region of the β-galactosidase gene to facilitate recombinant selection. This vector is chosen as an alternative to M13 (38) because of the ease of generating ss-template DNA for sequencing, and its high copy number. It is advisable to use restriction enzymes which generate DNA molecules with cohesive ended ('sticky-ended') rather than 'blunt-ended' molecules as the latter are more difficult to ligate together. Likewise it is not advisable to ligate together vector and insert DNA after restriction with enzymes which generate a hybrid site. Such sites are not recognized by the individual restriction enzymes in recombinants and obviate excision of the inserted DNA for sequencing. The ligation reaction is shown in *Table 5*. This protocol has been successfully utilized to clone a number of geminivirus DNAs.

Table 5. Endonuclease digestion and ligation.

1. Restrict geminivirus dsDNA with an appropriate restriction endonuclease (Section 2.9.4) in an Eppendorf tube and check the products of the reaction on a 'mini-gel' (Section 2.5) to ensure that the reaction is complete.
2. Simultaneously digest, in a separate Eppendorf tube, pEMBL9 DNA preferably with the same enzyme as in step 1. Include calf intestinal phosphatase (BCL) in the digestion (0.1 U/μl volume of digestion mixture) to prevent self-ligation of the vector DNA. Check the products of the reaction as in step 1.
3. If one enzyme generates a 'blunt-ended' molecule in either step 1 or step 2 fill in the DNA fragments with DNA polymerase I according to Maniatis *et al.* (30).
4. Mix the contents of the two Eppendorf tubes into a fresh tube and extract the DNA (Section 2.4.2, *Table 1*). The concentration of the insert DNA should be in 2- to 3-fold molar excess with respect to the vector DNA (normally 10 ng/μl).
5. Dry the DNA pellet from step 4 above and resuspend it directly in 1.5 μl of 10 × ligation buffer[a], 1 μl of T4 DNA ligase[b] and 12.5 μl sterile distilled water.
6. Incubate the mixture at room temperature for 5–6 h or at 6°C overnight.
7. Either extract the DNA from the mixture in step 6 (Section 2.4.2, *Table 1*) or use it directly for transformation of competent cells.

[a] 10× ligation buffer is 660 mM Tris–HCl, pH 7.6, 100 mM $MgCl_2$, 150 mM DTT, 1–5 mM spermidine, 10 mM ATP.
[b] T_4 DNA ligase (New England Biolabs) at a concentration of 10 U/μl is diluted 1:5 in 20 mM Na_2HPO_4-NaH_2PO_4, pH 7.6, 60 mM KCl, 5 mM DTT, 1 mM Na_2EDTA, 50% (v/v) glycerol and 2 U (1 μl) used for 'sticky-ended' ligations. For 'blunt-ended' ligations in an overnight incubation use 1 μl of undiluted T_4 ligase (10 U) in combination with 2.5 μl of 40% (w/v) PEG 6000 (BDH Chemical Co.) with the volume of sterile distilled water reduced accordingly.

241

3.2.2 Transformation

Either use ligation mixes directly or extract the DNA as in Section 2.4.2 and resuspend in a minimal volume of TE (<10 µl; *Table 1*) for transformation of competent JM101 cells. The protocol for the production of competent JM101 cells, shown below, is modified from that of Maniatis *et al.* (30).

(i)   Pick a single colony of *E.coli* JM101 cells from a minimal agar plate (Section 3.1) and inoculate into 1.5 ml of 2TY broth (Section 3.1.3) in a 10 ml Sterilin tube and incubate at 37°C with shaking for 16 h.

(ii)   Inoculate 7.5 ml of 2TY broth with 25 µl of the overnight culture from step (i) using a sterilized Gilson micropipette tip and incubate for approximately 4 h with vigorous shaking until $A_{550} = 0.3$.

(iii)   Cool the cells on ice for 10 min and centrifuge at 4000 *g* at 4°C for 5 min and discard the supernatant.

(iv)   Resuspend the cell pellet in 5 ml of cold 50 mM $CaCl_2$ with gentle mixing.

(v)   Keep the cells on ice for a further 20 min and re-centrifuge as in step (iii) prior to resuspension in 1 ml of cold 50 mM $CaCl_2$.

(vi)   Keep cells at 4°C for 12–24 h before use.

As an alternative method of producing competent cells a further protocol described by Hanahan (39) is detailed below.

(i)   Pick a single colony of *E.coli* JM101 cells from a minimal agar plate (Section 3.1.1) and inoculate into 1.5 ml of SOB medium (Section 3.1.5) in a 10 ml Sterilin tube and incubate at 37°C with shaking for 16 h.

(ii)   Inoculate 12 ml of SOB medium with 50 µl of the overnight culture from step (i) and incubate at 37°C with shaking for approximately 3 h until $A_{550} = 0.4$–0.5.

(iii)   Cool the cells on ice for 10 min and centrifuge at 4000 *g* for 5 min at 4°C and discard the supernatant.

(iv)   Resuspend the cell pellet in 4 ml of cold TFB buffer (Section 3.1.6).

(v)   Keep the cells on ice for a further 20 min and re-centrifuge as in step (iii).

(vi)   Resuspend cell pellet in 1 ml of cold TFB. Add 7 µl of DnD mix (Section 3.1.7) to every 200 µl of cells and mix immediately.

(vii)   Keep the cells on ice for 10 min then repeat the addition of DnD mix.

(viii) Leave the cells on ice for a further 10 min before transformation.

Transformation with pEMBL vectors is performed as described in *Table 6*. In all experiments two control transformations are performed, one using un-digested vector DNA, the other lacking any vector DNA. These experiments monitor the efficiency of transformation and antibiotic selection, respectively.

3.3 **Preparation of cloned geminiviral DNA for sequencing**

One strategy used to sequence geminivirus DNA consists of subcloning DNA fragments generated by restriction enzyme digestion of agarose gel-purified, full length inserts of geminivirus DNA. Normally the geminiviral DNA is excised from its cloning vector with the restriction enzyme used for cloning and

Table 6. Transformation with pEMBL vectors.

1. Resuspend a portion of the ligation mix, or ligated and extracted DNA, in a minimal volume of <10 μl TAE [Section 2.3.1 (vii)] containing 2–20 ng of DNA.
2. Add DNA from step 1 to 200 μl of competent JM101 cells prepared as in Section 3.2.2 and incubate on ice for 40 min.
3. 'Heat shock' the cells at 42°C for 3 min.
4. Add 800 μl of 2TY broth (Section 3.1.3) to the cells and incubate statically at 37°C for 15 min.
5. Spread 100–200 μl cells from step 4 onto 10 ml, 2TY bottom agar plates (Section 3.1.4) containing 100 μg/ml ampicillin (sodium salt; Sigma), 30 μl of X-Gal (5-bromo-4-chloro-3-indolyl-β-D-galactoside; Anglian Biotechnology Ltd; make up a 20 mg/ml solution in dimethyl formamide and store at −20°C) and 10–20 μl of IPTG (isopropyl-β-D-thiogalactopyranoside; Sigma; make up a 24 mg/ml solution in sterile distilled water and store at −20°C). Allow the plates to dry briefly.
6. Invert the plates and incubate at 37°C overnight.
7. Pick single white colonies onto replica agar plates as in step 5, using sterile toothpicks (30). Discard the blue colonies as these are assumed to harbour 'wild-type' pEMBL plasmid only.
8. Analyse a selection of replica-plated white colonies following small-scale isolation of plasmid DNA as described by Birnboim and Doly (40) or by colony hybridization as detailed in *Table 7*.
9. Dot-blot plasmid DNAs onto nitrocellulose filters (30, Schleicher and Schüll) and hybridize with a [α-^{32}P] nick-translated DNA probe (Section 2.7.2).
10. Select recombinant plasmids containing DNA inserts.
11. Analyse plasmids from step 10 by restriction enzyme digestion (Section 2.9.4) and 'mini-gel' agarose gel electrophoresis (Section 2.5).
12. Select plasmids containing DNA inserts of the same size and restriction pattern to oc geminivirus dsDNA (e.g. Section 2.10).
13. Initially select one plasmid from step 12 as a source of DNA for sequencing.

Table 7. Colony lifts onto GeneScreen Plus membranes and hybridization.

1. Cool the replica plate at 4°C for several hours. Cut the GeneScreen Plus membrane to Petri dish size. There is no need to autoclave the membrane.
2. Place the membrane onto the surface of the agar following the manufacturer's instructions concerning orientation. Use a scalpel blade to make three orientation marks around the edge of the membrane.
3. Carefully peel off the membrane and lay colony side up in a solution of 2.5 M NaCl and 0.5 M NaOH for 1 min.
4. Remove the membrane and lay it onto a neutralizing solution of 3 M sodium acetate, pH 5.5, for 1 min. A convenient method is to place 1–2 ml of solution onto a piece of 'cling-film' and lay the filter on the solution.
5. Blot the membrane dry on filter paper and pre-hybridize as described in *Table 4*.
6. Probe membrane with a reverse transcribed cDNA probe (*Table 3*).

This method can also be used for plaque hybridization of M13 clones, but then sterile membrane disks should be used since the plate will also act as the master plate.

electroeluted from agarose (Section 2.7.1). Subcloning is performed in M13 mp 18 or mp 19 bacteriophage vectors (38) and clones are sequenced by dideoxy chain termination (41). Subclones may also be generated by digestion of the DNA with *Bal*31 (Gibco BRL Ltd) as described in Maniatis *et al.* (30). Prepare template DNA for sequencing according to Sanger *et al.* (41). Larger amounts of the recombinant plasmid for these investigations are generated using a 'scaled-up' version of that described by Ish-Horowicz and Burke (42) including a LiCl precipitation step to remove the bulk of contaminating RNA (43). Restriction fragments are often randomly subcloned and the different subclones identified by 'T-tracking'. To identify further restriction enzymes that may be useful for subcloning, gel-purified geminiviral DNA is separately digested with a range of enzymes which generate termini compatible with those generated by cleavage in the M13 phage RF multiple cloning site (38). It is preferable to select enzymes which recognize tetranucleotide sequences in DNA. As more information about the sequence is obtained it becomes possible to gel-purify and subclone specific restriction fragments thereby accelerating the sequencing procedure. By cloning fragments in both M13 mp 18 and mp 19 vectors the sequences of the DNA can be determined in both orientations. The subcloning procedure is shown in *Table 8*. Alternatively viral nucleic acids cloned into pEMBL vectors can be sequenced directly following isolation of ssDNA as in Section 3.3.1.

3.3.1 *Preparation of ssDNA from pEMBL vectors*

The procedure below is essentially that described by Dente *et al.* (37) except that we use helper phage M13KO7 which is less efficiently replicated and packaged than the pEMBL plasmids during co-infection. M13KO7 carries a Kanamycin resistance marker which can be used to select for the growth of host cells superinfected with the phage.

(i)   Streak a colony carrying a pEMBL plasmid onto 2TY agar containing ampicillin at 100 µg/ml and incubate overnight at 37°C.

(ii) Inoculate a fresh grown colony from step (i) into 2 ml of pre-warmed 2TY broth containing ampicillin.

(iii) Grow with shaking at 37°C for 1 h (A_{550} = 0.1).

(iv) Inoculate the culture from (iii) with 1×10^{10} M13KO7 phage [approximate multiplicity of infection (m.o.i.) of 10].

(v)   Shake for a further 1 h then add kanamycin (Sigma Chemical Co.) to a concentration of 70 µg/ml.

(vi) Grow for a further 6 h before analysing the phage.

Table 8. Subcloning DNA fragments into M13 mp 18 or mp 19 vectors.

1. Digest gel-purified geminiviral DNA (Section 2.7) with the restriction endonuclease of choice (Section 2.9).
2. Ligate digested DNA from step 1 with similarly digested M13 phage RF DNA in a procedure identical to that shown in *Table 5*.
3. Use 2–20 ng (~3.35 µl) of the ligation mix from step 2 of ligated DNA to directly transform 200 µl competent JM101 cells (Section 3.2.2).
4. Incubate the cells on ice for 40 min.
5. Simultaneously prepare 3 ml of molten 2TY H top agar (Section 3.1.4) and hold at 42°C.
6. Add 30 µl of X-gal, 20 µl of IPTG (for both see *Table 6*), 200 µl of an overnight culture of JM101 cells (Section 3.2.2) and 50 µl of the transformation mixture from step 4 to 2TY H top agar from step 5.
7. 'Heat shock' the cells for 3 min.
8. Shake the 2TY H top agar and pour immediately onto pre-warmed (37°C) 2TY bottom agar plates (Section 3.1.4).
9. Allow the 2TY H top agar to set, invert the plates and incubate overnight at 37°C.

3.3.2 *Sequencing*

Preparation of template DNA for sequencing, preliminary characterization by T-tracking, full DNA sequencing, and reading sequencing autoradiograms is carried out as described by Davies (44) or as recommended by Amersham International (45).

3.4 **Examples of results**

The overall strategy for sequencing geminivirus DNA detailed herein has been used previously to sequence the genomic DNA of two confirmed geminiviruses TGMV (14) and WDV (18) and will undoubtedly be employed for the molecular analysis of other as yet uninvestigated geminivirus genomes. As a prequisite for sequencing cloned viral DNA it is essential to map a number of restriction enzyme sites which might be useful for cloning. For the reasons described in Section 3.2.1 the *Hin*dIII site in WDV DNA was selected for cloning into pEMBL9 to generate recombinant plasmid pWDI. Using electroeluted WDV DNA insert isolated from pWDI (Section 2.7) a number of single and multiple cutting enzymes were identified and these are shown in *Figure 7*. Using double digestions of this DNA a restriction map was constructed which was used for confirmatory purposes once sequence data was produced. The remaining gaps in the sequence were filled in by a semi-shotgun method using unmapped enzymes, for example *Sau*3A and *Taq*1.

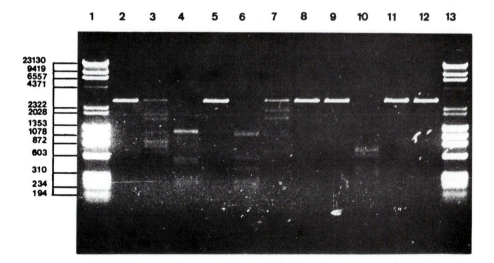

Figure 7. Restriction endonuclease analysis of cloned WDV dsDNA. Digestion products were electrophoresed on a 1% agarose gel in the presence of 0.5 μg/ml EtBr. **Lanes 1** and **13** contain mol. wt marker DNAs (nucleotide number shown) of λ phage DNA digested with restriction enzyme *Hin*dIII, plus φX174 phage RF DNA digested with restriction enzyme *Hae*III. **Lanes 2–12** contain gel purified cloned WDV DNA (excised from pWDI with restriction enzyme *Hin*dIII) digested with, respectively, **2**, no enzyme; **3**, *Hpa*II; **4**, *Sau*3A; **5**, *Acc*I; **6**, *Hae*III; **7**, *Pst*I; **8**, *Bam*HI; **9**, *Xho*I; **10**, *Taq*I; **11**, *Bgl*II and **12**, *Sal*I. There are no restriction sites in WDV dsDNA for the enzymes used in **lanes 5, 8, 9, 11** and **12**.

The strategy used for sequencing WDV DNA isolated from the recombinant plasmid pWDI is shown in *Figure 8* where the cloned DNA is represented as a thick black line with a number of selected restriction sites indicated. The arrows indicate the extent and direction of sequence obtained from each subclone. In order to differentiate the virion strand from the complementary strand, radiolabelled [α-^{32}P]cDNA *Hpa*I subclones (labelled A and B in *Figure 8*) were prepared (Section 2.6, *Table 3*) hybridized to virion ssDNA (Section 2.4.1). Only the cDNA to clone B hybridized illustrating that the sequence of this clone and all other clones in this orientation are complementary to the virion DNA.

Analysis of the WDV DNA sequence with the MAPSORT programme (46) revealed the presence of two *EcoRI* sites in the genome as was suspected from the original analysis of oc WDV from nucleic acid extracts of infected plants (Section 2.10). Both sites were within sequencing distance of the unique *Hind*III site initially selected for cloning. For any sequenced DNA it is essential to sequence across the cloning site used for the construction of the original

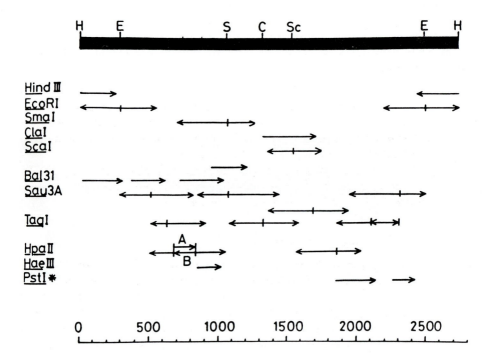

Figure 8. The DNA sequencing strategy used for cloned WDV DNA isolated from recombinant plasmid pWDI. Selected restriction sites are shown H = *Hind*III, E = *EcoRI*, S = *Sma*I, C = *Cla*I and Sc = *Sca*I. The arrows indicate the extent and direction of sequence obtained from each clone. The restriction enzyme used to generate the sequencing start-point is indicated at the left of each series of clones. *Bal*31 was also used as described in Section 3.3. The *Pst*I clones were generated by four base specific cleavages in *Pst*1 star activity (New England Biolabs 1988 catalogue). The scale indicates the distance in base pairs from the *Hind*III site. Labels **A** and **B** refer to clones used as templates for cDNA probes as described in Section 3.4.

recombinant clones. In the case of WDV DNA this was achieved by digesting gel purified oc WDV dsDNA with *Eco*RI and ligating the products to *Eco*RI digested M13 mp 9 DNA. The resultant recombinant clones were analysed by 'T-tracking', and full sequencing.

To confirm the full-length nature of any DNA plant virus clone it is essential to demonstrate that the cloned DNA is infectious for susceptible host plants and the procedure for performing these experiments is detailed in refs 47 and 48. Because of the difficulty of mechanically infecting monocotyledonous plant hosts with virion DNA or cloned DNA of geminiviruses the unequivocal demonstration that pWDI is an infectious full-length clone of WDV DNA could only be made using 'agroinfection', or more precisely agroinoculation, of wheat seedlings (49). In further investigations, particularly those with geminiviruses which are mechanically inoculable to dicotyledonous host plants, it is advisable to check the infectivity of the cloned DNA prior to sequencing to ensure that full-length molecules have indeed been produced.

3.5 **Production of cDNA clones of a RNA virus**

3.5.1 *First- and second-strand synthesis*

The protocol outlined here is similar to that of Watson and Jackson (50) in its production of first- and second-strand cDNAs.

(i)   If the RNA is not polyadenylated, prime the first-strand synthesis reaction with random DNA oligonucleotides and carry out as described in *Table 3*.

(ii)   Extract the RNA:DNA hybrid mixture with phenol/chloroform, then chloroform alone before the adding of ⅕ of its volume of ammonium acetate (8 M) and 2 vols of absolute ethanol.

(iii)   After precipitation at −20°C overnight, pellet the nucleic acid by microcentrifugation, dry and resuspend in TE before carrying out second-strand cDNA synthesis as described in *Table 9*.

3.5.2 *End filling with T4 DNA polymerase*

To produce 'blunt-ended' molecules of cDNA for ligation into vector DNA, T4 DNA polymerase is used to 'fill' the ends. In the presence of excess of all four deoxynucleotides this enzyme exhibits a 5′ to 3′ polymerase activity which is most active on a dsDNA molecule with a 5′ overhang, as well as a 3′ to 5′ exonuclease activity which is most active on a molecule with a 3′ overhang. A protocol for end-filling with T4 DNA polymerase is given in *Table 10*.

3.5.3 *'Blunt-end' ligation*

The DNA can be used at this stage for ligation (30) to pEMBL9 or pEMBL8, which has been cut with a restriction endonuclease which generates blunt ends, such as *Sma*I or *Hinc*II. Conditions for ligation are described in *Table 5*. When using 'blunt-end' ligated DNA for transformation, competent cells should be prepared by the method of Hanahan (39) described in Section 3.2.2, as this

Table 9. Second-strand cDNA synthesis.

1. Mix 25 μl of first-strand reaction product from Section 3.5.1, 10 μl of 5× second-strand buffer[a], 2.5 μl each of dATP, dGTP and dTTP at 1 mM, 2.5 μl of BSA (1 mg/ml) and 10 μCi of [α-^{32}P]dCTP, together.
2. Adjust the volume to 50 μl with sterile distilled water after addition of 1 U of RNase H (PL) and 15 U of DNA polymerase I (Amersham).
3. Remove a 1 μl aliquot and assay for radioactivity incorporated into acid-precipitable nucleic acid at zero time.
4. Incubate at 14°C for 1 h and at room temperature for a further 1 h.
5. Remove a 1 μl aliquot and assay for radioactive incorporation as in step 3. Calculate the increase from step 3.
6. Stop the reaction by adding 2.5 μl of Na$_2$EDTA (500 mM, pH 8.0) and 1.0 μl SDS (10% w/v).
7. Phenol/chloroform extract and then extract with chloroform/isoamyl alcohol (24:1, v/v). Add ⅕ vol. of ammonium acetate (3 M) and 2 vols of absolute ethanol. Precipitate at −20°C overnight.
8. Pellet the DNA by microcentrifugation for 15 min at 4°C, dry the pellet and resuspend in TE buffer.

[a]5× stock of second-strand buffer contains 250 mM Tris–HCl, pH 7.6, 40 mM MgCl$_2$, 67.5 mM KCl and 90 mM 2, β-mercaptoethanol.

Table 10. End filling with T4 DNA polymerase I.

1. Mix together 2 μl of 10 × T4 reaction buffer[a], 2 μl of BSA at 1 mg/ml and 2 μl of a solution containing dATP, dTTP, dCTP and dGTP at 1 mM each.
2. Add 10 μl of second-strand product (*Table 9*) and make up to 20 μl with sterile distilled water.
3. Add 5 U of T4 DNA polymerase I and incubate at 37°C for 30 min.
4. Phenol/chloroform extract the DNA and then extract with chloroform prior to the addition of ⅕ vol. of 3 M ammonium acetate and 2 vols of absolute ethanol. Precipitate the DNA overnight at −20°C.
5. Pellet, dry and resuspend the DNA by centrifugation, 5 min in a microcentrifuge, in TE buffer. Store at −20°C.

[a]10× stock of T4 reaction buffer contains 333 mM Tris–acetate, pH 7.0, 100 mM magnesium acetate and 660 mM potassium acetate.

protocol produces cells which have a higher transformation efficiency. Transformations are carried out as described in *Table 6*. Ligation of 'blunt-ended' DNA molecules can lead to a background of 'false positives' (i.e. colonies which appear white but lack a DNA insert) probably due to 'nibbling' of the ends of the vector molecule during restriction, particularly if digested with *Hinc*II. True recombinants are selected by colony hybridization to a cDNA probe. White colonies are toothpicked onto two replica agar plates and grown overnight at 37°C. One plate is kept at 4°C as a master plate, the other is used for a colony lift (Section 3.3, *Table 7*). Hybridization of a cDNA probe to the membrane is detected by autoradiography and used to determine which colonies on the master plate contain cDNA inserts. DNA is prepared from these clones for analysis by restriction endonuclease digestion or subcloning experiments. ssDNA can be prepared for determining the orientation of the inserted cDNA or for sequencing directly.

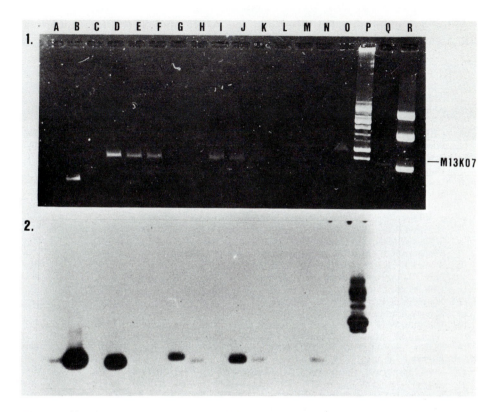

Figure 9. (**1**) 1% agarose gel containing ssDNA produced from super-infection of pEMBL clones with Fl phage. **Lane P** contains dsDNA prepared from pTN24, a pEMBL9 clone carrying a cDNA insert to TNV RNA. **Lane R** contains dsDNA of pEMBL9. Samples of ssDNA prepared from pEMBL clones carrying cDNA inserts are in **Lanes A–O**. (**2**) autoradiogram of a Southern blot of gel 1 after hybridization to a cDNA probe prepared by reverse transcription of TNV RNA (Section 2.6, *Table 3*).

3.5.4 *Examples of results*

The orientation of insertion of cloned cDNA is determined by running samples of ssDNA prepared from pEMBL clones on a 1% agarose gel (Section 2.5) and transferring to GeneScreen Plus. After hybridization to a strand-specific cDNA probe prepared to viral RNA by the protocol given in *Table 3*, the blot is washed and autoradiographed. A positive hybridization signal indicates that the cDNA clone is in the positive orientation with respect to the replication origin of Fl on pEMBL9 (*Figure 9*). Once cDNA clones to a RNA viral genome are produced the methods of subcloning and sequencing of DNA as applied to clones of a DNA viral genome can be used in an identical manner. Analysis of sequence data and alignment of clones is similarly carried out by computer analysis.

4. ACKNOWLEDGEMENTS

We should like to thank CAPES, Brazil for supporting S.W.MacDowell, the

AFRC for supporting P.J.Wise, and the AFRC and SERC for supporting R.A.Coutts.

5. REFERENCES

1. Ahlquist,P. and Janda,M. (1984) *Mol. Cell Biol.*, **4**, 2876.
2. Ahlquist,P. (1986) In *Methods in Enzymology*. Weissbach,A. and Weissbach,H. (eds), Academic Press, NY, Vol 118, p. 704.
3. Grimsley,N., Hohn,B., Hohn,T. and Walden,R. (1986) *Proc. Natl. Acad. Sci. USA*, **83**, 3282.
4. Powell Abel,P., Nelson,R.S., De,B., Hoffman,N., Rogers,S.G., Frayley,R.T. and Beachy, R.N. (1986) *Science*, **232**, 738.
5. Harrison,B.D., Mayo,M.A. and Baulcombe,D.C. (1987) *Nature*, **328**, 799.
6. Coutts,R.H.A., Buck,K.W., Roberts,C.J.F., Brough,C.L., Hayes,R.J., Macdonald,H., MacDowell,S.N., Petty,I.T.D., Slomka,M.J., Hamilton,W.D.O. and Bevon,M.W. (1987) In *Molecular Strategies for Crop Protection*. Arntzen,C.J. and Ryan,C. (eds), Alan R.Liss, NY, p. 307.
7. Harrison,B.D. (1985) *Annu. Rev. Phytopathol.*, **23**, 55.
8. Stanley,J. (1986) *Adv. Virus Res.*, **30**, 139.
9. Shepherd,R.J. (1979) *Annu. Rev. Plant Physiol.*, **30**, 405.
10. Stanley,J. and Davies,J.W. (1986) In *Molecular Plant Virology*. Davies,J.W. (ed.), CRC Press, Baton Rouge, FL, Vol. II, p. 191.
11. Hull,R. and Covey,S.N. (1983) *Sci. Prog.*, **68**, 403.
12. Matthews,R.E.F. (1982) *Intervirology*, **17**, 1.
13. Stanley,J. and Gay,M.R. (1983) *Nature*, **301**, 260.
14. Hamilton,W.D.O., Stein,V.E., Coutts,R.H.A. and Buck,K.W. (1984) *EMBO J.*, **3**, 2197.
15. Ikegami,M., Morinaga,T. and Miura,K. (1984) *Virus Res.*, **1**, 507.
16. Mullineaux,P.M., Donson,J., Morris-Krsinich,B.A.M., Boulton,M.I. and Davies,J.W. (1984) *EMBO J.*, **3**, 3063.
17. MacDowell,S.W., Macdonald,H., Hamilton,W.D.O., Coutts,R.H.A. and Buck,K.W. (1985) *EMBO J.*, **4**, 2173.
18. Grimsley,N., Hohn,T., Davies,J.W. and Hohn,B. (1987) *Nature*, **325**, 177.
19. Hayes,R.J., Macdonald,H., Coutts,R.H.A. and Buck,K.W. (1988) *J. Gen. Virol.*, **69**, 811.
20. Hamilton,W.D.O., Bisaro,D.M. and Buck,K.W. (1982) *Nucleic Acids Res.*, **10**, 4901.
21. Ikegami,M., Haber,S. and Goodman,R.M. (1981) *Proc. Natl. Acad. Sci. USA*, **78**, 4102.
22. Davies,J.W. and Hull,R. (1982) *J. Gen. Virol.*, **61**, 1.
23. Bock,K.R., Guthrie,E.J. and Woods,R.D. (1974) *Annu. Appl. Biol.*, **77**, 289.
24. Lindsten,K., Lindsten,B., Abdelmoeti,M. and Juntii,N. (1980) In *Proceedings of the 3rd Conference on Virus Diseases of Graminae*, Rothamsted, p. 27.
25. Association of Applied Biologists. In *Descriptions of Plant Viruses*. Spottiswoade Ballantyne Printers, Wellesbourne, UK.
26. Lesnaw,J.A. and Reichmann,M.E. (1969) *Virology*, **39**, 729.
27. Sunter,G., Buck,K.W. and Coutts,R.H.A. (1985) *Nucleic Acids Res.*, **13**, 4645.
28. Adejare,G.O. and Coutts,R.H.A. (1982) *Phytopath. Z.*, **103**, 198.
29. Coutts,R.H.A., Barnett,A. and Wood,K.R. (1975) *Nucleic Acids Res.*, **2**, 1111.
30. Maniatis,T., Fritsch,E.F. and Sambrook,J. (1982) *Molecular Cloning: A Laboratory Manual*, Cold Spring Harbor Laboratory, Cold Spring Harbor, New York.
31. Gould,A.R. and Symons,R.H. (1977) *Nucleic Acids Res.*, **4**, 3787.
32. Hamilton,W.D.O., Sanders,R.C., Coutts,R.H.A. and Buck,K.W. (1981) *FEMS Microbiol. Letts*, **11**, 263.
33. Zassenhaus,H.P., Butow,R.A. and Hannon,Y.P. (1982) *Anal. Biochem.*, **125**, 125.
34. Rigby,P.W.J., Dieckmann,M., Rhodes,C. and Berg,P. (1977) *J. Mol. Biol.*, **113**, 237.
35. Macdonald,H., Coutts,R.H.A. and Buck,K.W. (1988) *J. Gen. Virol.*, **69**, 1339.
36. MacDowell,S.W., Coutts,R.H.A. and Buck,K.W. (1986) *Nucleic Acids Res.*, **14**, 7967.
37. Dente,L., Cesarini,G. and Cortese,R. (1983) *Nucleic Acids Res.*, **11**, 1645.
38. Messing,J. and Vieira,J. (1982) *Gene*, **19**, 269.
39. Hanahan,D. (1985) In *DNA Cloning—A Practical Approach, Vol. I*. Glover,D.M. (ed.), IRL Press, Oxford, p. 109.
40. Birnboim,H.C. and Doly,J. (1979) *Nucleic Acids Res.*, **7**, 1513.
41. Sanger,F., Coulson,A.R., Barrel,B.G., Smith,A.J.H. and Roe,B.A. (1982) *J. Mol. Biol.*, **143**, 161.

42. Ish-Horowicz,R. and Burke,J.F. (1981) *Nucleic Acids Res.*, **9**, 1989.
43. Bisaro,D.M., Hamilton,W.D.O., Coutts,R.H.A. and Buck,K.W. (1982) *Nucleic Acids Res.*, **10**, 4913.
44. Davies,R.W. (1982) In *Gel Electrophoresis of Nucleic Acids—A Practical Approach.* Rickwood,D. and Hanes,B.D. (eds), IRL Press, Oxford, p. 117.
45. M13 Cloning and Sequencing Handbook, Amersham International.
46. Deveraux,J., Haerberli,P. and Smithies,O. (1984) *Nucleic Acids Res.*, **12**, 387.
47. Hamilton,W.D.O., Bisaro,D.M., Coutts,R.H.A. and Buck,K.W. (1983) *Nucleic Acids Res.*, **11**, 7387.
48. Stanley,J. (1983) *Nature*, **305**, 643.
49. Hayes,R.J., Macdonald,H., Coutts,R.H.A. and Buck,K.W. (1988) *J. gen. Virol.*, **69**, 891.
50. Watson,C.J. and Jackson,J.F. (1985) In *DNA Cloning—A Practical Approach*, *Vol. I.* Glover,D.M. (ed.), IRL Press, Oxford, p. 79.

CHAPTER 10

Molecular biology of *Chlamydomonas*

J.-D.ROCHAIX, S.MAYFIELD, M.GOLDSCHMIDT-CLERMONT
and J.ERICKSON

1. INTRODUCTION

Eukaryotic green algae provide powerful model systems for studying photo-synthesis. Work on *Chlorella*, *Scenedesmus* and *Chlamydomonas* has been of great importance for understanding the photosynthetic machinery. Among these algae *Chlamydomonas reinhardtii* offers special advantages because it is amenable to genetic analysis and because its photosynthetic function is dispensable when the cells are grown in the presence of a reduced carbon source such as acetate. This property has allowed for the isolation of numerous photosynthetic mutants which have been very valuable in examining photo-synthetic function (cf. ref. 1).

C.*reinhardtii* has also been used successfully for the analysis of the structure, function and assembly of the flagellar apparatus. The organism is also well suited for studies on cell–cell interactions (2). More recently, promising work has been initiated on the eyespot which appears to be an excellent model system for photoreceptors (3).

In this chapter we describe the basic methods used in the molecular analysis of *Chlamydomonas*. We shall limit ourselves to C.*reinhardtii* since our experience has been exclusively with this organism.

2. GROWTH OF CHLAMYDOMONAS REINHARDTII

C.*reinhardtii* can be grown on minimal medium with CO_2 as a carbon source (photoautotrophy), or with acetate as a source of reduced carbon in the dark (heterotrophy) or in the light (mixotrophy). The constituents and preparation of growth media are described in *Tables 1* and *2*. The algae are usually grown at 25°C under fluorescent lights (2000–4000 lux). Many photosynthetic mutants are light-sensitive, but often grow better in dim light (200–300 lux) than in complete darkness. It is convenient to set liquid cultures in cotton-plugged Erlenmeyer flasks, either on a rotary shaker (150 r.p.m.) or with a magnetic stirrer. Good aeration is important; atmospheric CO_2 is a limiting factor for cultures in minimal media. Faster growth is obtained by bubbling the cultures with a 5% mixture of CO_2 in air through a cotton-plugged pipette (inserted before sterilization through the cotton stopper of the flask). Both the growth rate (8–12 h/generation) and the saturation density depend on the growth medium. Saturation densities in minimal medium are around 5×10^6 cells/ml, but reach

Table 1. Stock solutions.

1. *4× Beijerinck salts*
    16 g NH_4Cl
    2 g $CaCl_2.2H_2O$
    4 g $MgSO_4·7H_2O$
    in 1 litre
2. *1 M (K)PO₄ pH 7*
    250 ml 1 M K_2HPO_4
    ~170 ml 1 M KH_2PO_4 (titrate to pH 7.0)
3. *2× PO₄ for HSM*
    0.08 M K_2HPO_4 14.34 g
    0.05 M KH_2PO_4 7.26 g
    adjust to pH 6.9 with KOH, to 1 litre with H_2O
4. *Trace elements* (from ref. 4)
1. Dissolve in 550 ml H_2O in the order indicated, then heat to 100°C
        11.4 g H_3BO_3
        22 g $ZnSO_4·7H_2O$
        5.06 g $MnCl_2·4H_2O$
        4.99 g $FeSO_4·7H_2O$
        1.61 g $CoCl_2·6H_2O$
        1.57 g $CuSO_4·5H_2O$
        1.1 g $(NH_4)_6Mo_7O_{24}·4H_2O$
2. Dissolve 50 g Na_2EDTA in 250 ml H_2O by heating, and add to the first solution at 100°C.
3. Heat the combined solutions to 100°C, cool to 80–90°C and adjust to pH 6.5–6.8 with 20% KOH (requires <100 ml). The pH meter should first be calibrated at 75°C; the temperature should remain above 70°C.
4. Adjust to 1 litre, and allow a rust-coloured precipitate to form, during 2 weeks at room temperature, in a 2 litre Erlenmeyer flask loosely stoppered with cotton. The solution will change from green to purple.
5. Filter several times through three layers of Whatman No. 1 under suction (Büchner funnel). Store the clear purple solution at −20°C.

Table 2. Media.

For 1 litre[a]	Tris–acetate phosphate (TAP)	Tris–minimal[a] (Tris Min)	High salt minimal[b] (HSM)
H_2O	975 ml	975 ml	925 ml
Tris	2.42 g	2.42 g	–
4× Beijerinck salts	25 ml	25 ml	25 ml
1 M (K)PO₄, pH 7	1 ml	1 ml	–
2× PO₄ for HSM	–	–	50 ml
Trace	1 ml	1 ml	1 ml
Glacial acetic acid	~1 ml[c]	–	–
Concentrated HCl	–	~1.5 ml[c]	–
References	5	4	6

[a] Supplement solid media with 20 g of agar (Difco) per litre.
[b] HSM is more stringent than Tris Min for photoautotrophic growth, probably because of small amounts of reduced carbon source in the Tris.
[c] Titrate to pH 7.0.

$2–4 \times 10^7$ cells/ml in mixotrophic conditions (acetate + light). The usual growth temperature is 25°C, but satisfactory growth is obtained between 18°C and 34°C. Cells can be counted under the light microscope in phase contrast using a haemocytometer [it is convenient to kill them with a trace of 2% iodine (I_2) in ethanol]. Good mixing of aliquots and dilutions is important because of the quick phototactic response of these motile algae.

3. CELL DISRUPTION

The first step in the preparation of cellular organelles (with the exception of flagellae), nucleic acids and proteins involves cell disruption. While organelle isolation and preparation of high molecular weight DNA requires that the cells be opened gently, the cells can be disrupted more thoroughly for RNA and protein isolation.

3.1 Gentle cell disruption

For gentle cell disruption it is helpful to use cell wall deficient strains such as cw15 (7) and cw2, which can easily be lysed by detergents or by passage through a Yeda Press. It is recommended to use the procedure described by Belknap and Togasaki (8) for organellar isolation.

(i)   Harvest the cells in exponential phase ($1-2 \times 10^6$ cells/ml) by centrifuging at 2500 g for 3 min.

(ii)   Wash the cells twice with 20 mM Hepes, pH 7.0 and resuspend in 20 ml (for 10^9 cells) breaking buffer [0.3 M sorbitol, 50 mM 2-(N-morpholino)-ethane sulphonic acid (MES), 10 mM Tris–HCl, pH 7.5, 2 mM EDTA, 1 mM $MgCl_2$, 1 mM $MnCl_2$ adjusted to pH 7.5 with NaOH, 1% bovine serum albumin (BSA, Sigma A 6003)] at room temperature.

(iii)   Quickly cool the cells in an ice bath. This step is important for obtaining an appropriate cell lysis later.

(iv)   Equilibrate the cells at 100 p.s.i. for 3 min in a pre-cooled Yeda Press. Rapidly release the suspension into a chilled tube.

It is possible to start from wild-type cells and to use the same procedure, if the cell wall is first removed with autolysin treatment. Autolysin is prepared as follows.

(i)   Plate 4×10^6 cells of mt^+ and mt^- separately on TAP (*Table 2*) plates.

(ii)   Incubate under strong illumination (3000–5000 lux) for 3–4 days until cell growth is confluent.

(iii)   Scrape the cells from each plate and resuspend in 10 ml of TAP medium lacking NH_4Cl and incubate the suspension with agitation under strong light. After 4 h small aliquots of mt^+ and mt^- cells are mixed and checked for mating. If the mating is adequate proceed to step (v).

(iv)   If the mating is poor, dilute the cells 3–4 times and incubate overnight with agitation.

(v)   Concentrate the cells to their original volume and mix both mating types. After 45 min sediment the cells at 13 000 g for 15 min. Save the supernatant containing the autolysin. Usually 1 ml of autolysin is used to remove the cell walls of 10^7-10^8 cells.

Removal of cell walls can be checked by the appearance of round cells or more quantitatively by measuring the released chlorophyll upon addition of a detergent to the autolysin-treated cells (9).

(i)   Add 2 vols of ice-cold 0.075% Triton X-100, 5 mM EDTA, pH 8.0, to the treated cell suspension, vortex briefly, incubate for 10 min in ice and centrifuge for 10 min in a microfuge.

(ii)   Measure the absorbance of the released chlorophyll in the supernatant at 435 nm. The crude autolysin extract can be stored at $-70°C$ for several months.

Cell wall deficient mutant cells can easily be lysed with 2% SDS, a treatment which is usually used during the isolation of high molecular weight DNA. It is recommended to subject wild-type cells to several freeze–thaw cycles prior to detergent treatment.

3.2 Mechanical cell disruption

Wild-type cells can be disrupted by several methods:

(i)   *Sonication*. Three to four 10–15 sec sonication pulses with a Branson-type sonicator are usually sufficient to achieve a complete breakage of cells. An advantage of this method is that it can be performed rapidly, an important point especially for RNA isolation.

(ii)   *French press*. Thorough breakage can be achieved by passing the cells once or twice through a French press at 4000 p.s.i.

4. PREPARATION OF DNA FROM CHLAMYDOMONAS REINHARDTII

A variety of methods have been developed for extracting DNA from wild-type strains of *C.reinhardtii* and from cell wall deficient mutants. Total DNA obtained from these algal cells can be purified by CsCl density gradient centrifugation. Nuclear DNA from *C.reinhardtii* is highly G/C rich with an average buoyant density of 1.724 g/ml, and can be readily separated from the relatively A/T rich chloroplast DNA, which bands at a density of about 1.696 g/ml. Several protocols for the isolation and purification of *Chlamydomonas* DNA are presented below and their relative merits, with respect to the quantity and quality of DNA obtained and the time needed for preparation, are discussed.

4.1 Large-scale preparation of nuclear and chloroplast DNA: Method I

This is the standard procedure for the preparation of nuclear and chloroplast DNA as described previously (10). A summary of the method is given below.

(i)   Grow 750 ml of a cell wall deficient strain of *C.reinhardtii* in TAP medium (see Section 2) at 25°C to stationary phase ($\sim 10^7$ cells/ml).

(ii)   Harvest the cells by centrifugation at 3000 g for 5 min at 4°C.

(iii)   Resuspend the pellet in 20 ml of TAP medium. Transfer into a 30 ml Corex tube and centrifuge at 3000 g for 5 min at 4°C.

(iv) Gently resuspend the cell pellet in 8 ml of A buffer (0.1 M NaCl, 50 mM EDTA, 20 mM Tris–HCl, pH 8.0) at 4°C, and transfer the suspension to a 50 ml Erlenmeyer flask.

(v) Add 0.5 ml of pronase solution (10 mg/ml in 0.01 M sodium citrate, pH 5.0, pre-digested for 2 h at 37°C and stored frozen) to the cells and mix by swirling.

(vi) Add 0.5 ml of 20% SDS to the cells, and place the flask in a 50°C water bath.

(vii) After 45 min incubation, add a further 0.5 ml of the pronase solution.

(viii) 45 min later, add the same amount of pronase and continue to incubate the cells at 50°C.

(ix) After a total of 2–2.5 h at 50°C, the cells begin turning from green to brown. When this happens, cool the cells on ice and add approximately 20 ml of distilled phenol (saturated with 0.1 M sodium borate) to the flask. Mix gently by shaking, and then leave for 20 min with occasional shaking.

(x) Centrifuge the phenol–cell mixture at 7000 g in a swinging bucket rotor for 15 min. After centrifugation, the top aqueous phase is a light pink, while the phenol phase is black. Unlysed cells and cellular debris appear as an olive green layer at the interphase of the tube. Starch may be seen as a white pellet in the bottom of the tube.

(xi) Gently remove the aqueous phase containing the DNA with a large bore pipette.

(xii) Purify the DNA contained in the aqueous phase further by either CsCl density gradient centrifugation in the presence (Section 4.1.1) or absence (Section 4.1.2) of ethidium bromide (EtBr).

4.1.1 *CsCl banding in the presence of ethidium bromide*

If the DNA is to be banded in the presence of EtBr, carry out the following protocol.

(i) Mix the aqueous phase from Section 4.1, step (xi) with 2 vols of absolute ethanol and collect the precipitated DNA by centrifugation at 7000 g for 10 min in a swinging bucket rotor.

(ii) Rinse the pellet in 70% ethanol, and vacuum-dry for several minutes.

(iii) Resuspend the pellet in 3.8 ml of 10 mM Tris–HCl, 1 mM EDTA, pH 8.0 (TE).

(iv) Add 4.15 g of CsCl and dissolve by inverting the tube.

(v) Add 0.5 ml of EtBr solution made in distilled water at 700 µg/ml, mix well, and transfer to a polyallomer centrifuge tube. Quick-seal tubes are the most convenient.

(vi) The recipe is adapted for the Ti50 Beckman rotor, in which DNA is banded by spinning at 35 000 r.p.m. (85 000 g), 20°C, for 60 h. The total volume can be adapted for tubes that fit other rotors, just keeping the ratios of CsCl, TE and EtBr constant. In the vertical rotor, such as the Beckman VTi65.2, centrifugation times are much shorter (12–16 h), and the nuclear and chloroplast DNAs are much better separated after one centrifugation.

(vii) After centrifugation, look at the tubes under UV light. There should be two fluorescent bands. The lower, much thicker band corresponds to nuclear DNA. The upper, fainter band is chloroplast DNA. These can be collected directly with a syringe, after the cap is removed from the tube, or the top of the quick-seal tubes is cut off. If there is a lot of DNA, it is often difficult to collect the chloroplast band separately. Better results may be obtained by dripping the gradient from the bottom, or pumping it out from the top using a denser solution.

(viii) Extract the banded DNA with salt-saturated isoamyl or isopropyl alcohol, add 3 vols of 70% ethanol, and centrifuge to precipitate the DNA.

(ix) DNA can then be dissolved in TE and stored at 4°C. For long-term storage, DNA is more stable in CsCl or in ethanol.

4.1.2 *CsCl banding in the absence of ethidium bromide*

If the DNA is to be banded in CsCl in the absence of EtBr, it must not be centrifuged after the addition of 2 vols of ethanol. Centrifugation at this point results in the co-precipitation of an unidentified substance which co-purifies with DNA in the CsCl gradient and interferes with restriction digestion of gradient purified DNA. To avoid this problem carry out the following protocol.

(i) Transfer the aqueous phase from Section 4.1, step (ix), into a small beaker, gently add 2 vols of absolute ethanol chilled to $-20°C$, and then spool the DNA onto a glass rod.

(ii) Rinse the spooled DNA by dipping the rod briefly into a test tube containing 70% ethanol.

(iii) Remove the rod, let the DNA dry in air at room temperature, and then place it into a small test tube containing enough TE [Section 4.1.1, step (iii)] to cover the DNA. This is usually a few millilitres. Leave the tube at 4°C overnight.

(iv) When the DNA is resuspended, treat it with pancreatic RNase at a final concentration of 30 µg/ml, for 45 min at 37°C.

(v) Stop the reaction by extracting the incubation mixture with an equal volume of phenol (saturated in 0.1 M sodium borate). Cover the tube with parafilm, mix gently by inverting the tube, and then centrifuge at 7000 g for 15 min.

(vi) Collect the aqueous phase, spool the DNA out of ethanol as described in steps (i) and (ii), rinse, dry, and resuspend in approximately 3.5 ml of TE.

(vii) Prepare the CsCl gradient by mixing 3.35 ml of DNA in TE with 4.27 g of solid CsCl. Cover with parafilm, and mix gently by inverting the tube until all the CsCl is dissolved. The solution will become very cold, and heating the tube to 37°C for a few minutes will speed up the time needed for dissolving the salt.

(viii) These volumes are again designed for the Ti50 rotor. In this rotor, spin the tubes at 35 000 r.p.m. (85 000 g) for at least 60 h, at 20°C.

(ix)   After centrifugation, the gradient may be dripped from the bottom, by puncturing the tube with a needle, or pumped out from the top by forcing a heavy solution into the bottom of the tube. The most convenient way to monitor the fractions is to have a UV monitor connected to the gradient outlet tube, and a recording chart. In this way, fractions containing chloroplast DNA can be immediately identified from fractions containing nuclear DNA, no DNA, and RNA (at the bottom of the tube, or pelleted).

(x)   The desired fractions are pooled, and precipitated with 3 vols of 70% alcohol.

4.2 Large-scale preparation of DNA: Method II

This procedure is taken from Cryer *et al.* (11) and works well for *Chlamydomonas* (A.Day, personal communication).

(i)   Grow 500 ml of cells to stationary phase in TAP (*Table 2*) medium. Check for contamination by examining the cells under the light microscope.

(ii)   Pellet the cells in large plastic bottles, at 6000 *g* for 10 min.

(iii)   Pour off the supernatant, and resuspend the cell pellet in 20 ml of distilled water. Transfer to a screw cap Oakridge tube or a 30 ml Corex tube.

(iv)   Add 2 ml of 0.5 M EDTA, pH 9.0, and 0.5 ml of concentrated β-mercaptoethanol. Mix gently.

(v)   If using a wild-type strain of *Chlamydomonas*, leave this at room temperature for 15 min, and then centrifuge at 12 000 *g* for 5 min. If the strain contains a cell wall mutation, centrifuge immediately.

(vi)   Pour off the supernatant, and freeze the cells in a dry-ice/ethanol bath. The cells can be stored at this point (frozen at −20°C), or thawed out immediately and the procedure continued. It is often useful to freeze and thaw cells several times if the strain has a wild-type cell wall.

(vii)   Add 5 ml of 0.2 M EDTA, pH 9.5–10, 0.15 M NaCl, 4% SDS.

(viii) Add pronase to a final concentration of 1 mg/ml. This is most readily accomplished by adding ¹⁄₁₀ vol. of a pronase stock solution at 10 mg/ml.

(ix)   Incubate the mixture at 50°C until the cells lyse (10–120 min). Lysis is detected by an increase in viscosity of the solution.

(x)   Add 8 ml of a 1:1 mixture of phenol and chloroform. Mix by gentle inversion of the tube, and centrifuge at 12 000 *g* for 10 min.

(xi)   Remove the top aqueous phase. Repeat the phenol extraction if the aqueous phase is not clean.

(xii)   Add 2 vols of absolute ethanol, mix well, and centrifuge at 12 000 *g* for 10 min.

(xiii) Pour off the supernatant, dry the pellet and resuspend in 1 ml of TE [Section 4.1.1, step (iii)], pH 8.0. Leave on ice or at 4°C overnight to resuspend.

(xiv) Add 50 µl of 20× SSC and pancreatic RNase to a final concentration of 50 µg/ml.

(xv) Incubate at 50°C for 10 min.

(xvi) Add 3.40–3.45 ml of distilled water and 50–100 µl of EtBr (10 mg/ml) so that the final volume added is 3.5 ml.

(xvii) Transfer 4.5 ml of this mixture to a tube containing 4.25 g of solid CsCl. Mix by inverting the tube until all the CsCl is dissolved.

(xviii) Centrifuge at 44 000 r.p.m. in a VTi 65.2 Beckman rotor, 20°C, 16 h.

(xix) Collect the fluorescent bands with a syringe and remove EtBr by extracting with iso-amyl alcohol as described in Section 4.1.

(xx) Precipitate the DNA with ethanol. The dried DNA pellet can be resuspended in 0.5 ml of TE. The yield is approximately 200–500 µg of DNA.

4.3 Rapid preparation of CsCl purified *Chlamydomonas* DNA

This method is described by Weeks *et al.* (12) and results in the isolation of fairly pure DNA in a very short period of working time. The procedure outlined below has been adapted for the VTi 65.2 Beckman rotor.

(i) Grow 200–400 ml of cells to stationary phase ($\sim 1 \times 10^7$ cells/ml) in TAP (*Table 2*) medium.

(ii) Collect the cells by centrifugation in 250 ml bottles, at 6000 g for 5 min.

(iii) Pour off the supernatant, drain the pellets, and resuspend the cells in 2.5 ml of distilled water.

(iv) Transfer 3.3 ml of cells (add more water to cells if you do not have enough to make this volume) to a 15 ml Corex tube.

(v) Add 6 ml of SDS solution (2% SDS, 0.4 M NaCl, 0.04 M EDTA, 0.1 M Tris–HCl, pH 8.0) and mix gently by inverting the tube.

(vi) Leave at 50°C for 15 min. The solution should be viscous.

(vii) Add 9.6 g of solid CsCl, and mix by inversion until dissolved. The solution will be less viscous at this point.

(viii) Add 0.6 ml of EtBr stock solution (10 mg/ml). Mix gently.

(ix) Centrifuge at 12 000 g for 10 min. This step clears cellular debris. A tight pellicle will be formed at the top of the liquid. Gently insert a Pasteur pipette past this pellicle and remove the cleared liquid from below, transferring it to two VTi65.2 quick seal tubes.

(x) Centrifuge at 44 000 r.p.m. (180 000 g) at 20°C for 16 h.

(xi) Collect the fluorescent bands. The nuclear band will be the lower thick band. The upper, fainter band is the chloroplast DNA.

(xii) Process the DNA in bands as described in Section 4.1.

The yield of total DNA per 40 ml of cells processed is approximately 20–30 µg. This DNA can be readily cut by restriction enzymes and can be ligated and cloned in *Escherichia coli*.

4.4 **Rapid mini-preps of** *Chlamydomonas* **DNA**

This method is a modification of the procedure by Davis *et al.* (13).

(i)   Collect 10 ml of cells at 3×10^6 cells per ml by centrifugation in a 15 ml Corex tube, at 3000 *g* for 5 min.

(ii)   Resuspend the pellet in 0.35 ml of 50 mM EDTA, 20 mM Tris–HCl, pH 8.0, 0.1 M NaCl.

(iii)   Transfer the resuspended cells to an Eppendorf tube (1.5 ml).

(iv)   Add 50 µl of pronase at 10 mg/ml, or of proteinase K at 2 mg/ml.

(v)   Add 25 µl of 20% SDS, and incubate for 2 h at 55°C.

(vi)   Add 2 µl of diethylpyrocarbonate, incubate for 15 min at 70°C in a hood.

(vii)   Cool the tube in ice briefly, then add 50 µl of 5 M potassium acetate.

(viii) Mix by shaking the tube thoroughly, leave on ice for 30 min or more.

(ix)   Centrifuge for 15 min in a microcentrifuge.

(x)   Transfer the supernatant into another Eppendorf tube.

(xi)   Extract the supernatant with an equal volume of phenol.

(xii)   Fill the tube to the top with room temperature ethanol.

(xiii) Centrifuge for 2 min at room temperature.

(xiv)   Rinse in 70% ethanol, and centrifuge for 1 min.

(xv)   Pipette off the supernatant and discard.

(xvi)   Dry the pellet and resuspend in 50 µl of TE 7.5, 1 µg/ml pancreatic RNase. 10–15 µl of this is adequate for one restriction analysis.

As an example *Figure 1* shows *Hind*III and *Bam*HI digests of *C.reinhardtii* DNA obtained by this method.

5. PREPARATION OF RNA

The first protocol (adapted from ref. 14) involves an overnight ultracentrifugation through a CsCl cushion (Section 5.1) which efficiently removes contaminating DNA and carbohydrates. The second procedure (Section 5.2) does not require ultracentrifugation, and is hence somewhat simpler.

Because of the ubiquity of RNases, it is important in the final steps of these procedures to use oven-baked glassware, plasticware treated with DEPC/H_2O (freshly prepared by vigorously shaking 2–3 drops diethylpyrocarbonate in 500 ml of H_2O) and sterile solutions. Avoid contact of solutions with fingertips or fingerprints (wear disposable gloves).

5.1 **Procedure I**

Materials and solutions are described in *Table 3*.

(i)   Collect the cells from 250 ml of culture in log phase ($1-2 \times 10^6$ cells/ml in TAP (*Table 2*) medium) by centrifugation (5000 *g* for 10 min).

(ii)   Resuspend the cells in TAP medium and transfer to a smaller tube.

(iii)   Collect the cells again by centrifugation.

(iv)   Lyse the cells in 3 ml of guanidinium–HCl (GuHCl)/sodium acetate (NaAc). (Use >10 times the volume of the pellet).

Figure 1. Restriction enzyme digests of *C.reinhardtii* DNA obtained from a minipreparation (Section 4.4). (**a**) *Hind*III digest; (**b**) *Bam*HI digest.

(v)   Vortex in the presence of glass beads or sonicate the suspension (glass beads or sonication can be omitted with cell wall deficient mutant cells).

(vi)   Remove the debris by centrifugation (10 000 *g* for 10 min in Sorvall HB-4 swinging bucket rotor). The clear supernatant can be stored at −20°C.

(vii)   Layer the clear supernatant over a CsCl/EDTA cushion in a tube for a swinging-bucket ultracentrifuge rotor. The volume of the cushion is about ⅓ of the tube volume (1.5 ml for the Beckman SW60 4.4 ml tube).

(viii) Centrifuge at 130 000 *g* for 16 h (35 000 r.p.m. in the SW60). The RNA pellets while other contaminants form bands in the CsCl or remain in the GuHCl/NaAc layer.

(ix)   Remove the GuHCl/NaAc layer by aspiration with a Pasteur pipette connected to a vacuum flask, and gently wash the sides of the tube with GuHCl/NaAc. This solution settles on the CsCl without coming in contact with the RNA pellet.

(x)   Remove the GuHCl/NaAc layer and part of the CsCl.

(xi)   Repeat this wash a second time with a stretched Pasteur pipette, gradually remove all the liquid, including droplets on the sides of the tubes.

(xii)   Fragment and resuspend the sticky transparent pellet (it looks like an eye contact-lens) using a siliconized Pasteur pipette, in 2 ml of GuHCl/NaAc. The pellet may not dissolve completely at this step.

Table 3. Materials and solutions for RNA preparations.

1. Glass beads (450–500 μm, Sigma). Wash with 0.1 M HCl, rinse several times with water, dry and bake at 180°C.
2. Use siliconized Pasteur pipettes. Wet the inside of the pipettes with a 2% solution of dimethyldichlorosilane in chloroform. Wash with water, dry and bake at 180°C.
3. GuHCl/NaAc. 6 M guanidium HCl plus 0.1 M sodium acetate, pH 5.2 (from 1 M–stock). Filter through 0.45 μm nitrocellulose.
4. CsCl/EDTA. 5.2 M CsCl prepared by mixing 10 g of CsCl + 8.5 ml of 50 mM Na_2EDTA, pH 7.5 Treat with 0.02% diethylpyrocarbonate (DEPC) then heat to 70°C for 15 min to inactivate DEPC.
5. 1 mM EDTA pH 7.5 (from 0.4 M stock of Na_2EDTA, pH 7.5). Treat with 0.02% DEPC and autoclave.
6. DEPC-H_2O. Treat with 0.02% DEPC for a few hours; then autoclave.
7. TEN–SDS   50 mM Tris–HCl, pH 8
               10 mM EDTA
               100 mM NaCl
               0.1% SDS

(xiii)   Add ¾ vol. of ethanol (1.5 ml) and leave at −20°C for a few hours.
(xiv)   Recover the RNA by centrifugation (10 000 g for 10 min).
(xv)   For large preparations, repeat steps (xii)–(xiv). Most of the pellet should dissolve this time (if necessary, heat to 65°C for 1–2 min).
(xvi)   Dissolve the pellet in 3 ml of 1 mM EDTA (heat to 65°C for 1–2 min if necessary).
(xvii)   Add NaCl to 0.2 M and 2.5 vols ethanol (7.5 ml), leave at −20°C for a few hours and recover the RNA by centrifugation (15 000 g for 10 min).
(xviii) Wash with 70% ethanol (centrifuge again for 5 min), then dry under vacuum.
(xix)   Dissolve the RNA in 0.5 ml of 1 mM EDTA (heat to 65°C, 1–2 min if necessary).
(xx)   Remove any insoluble material by centrifugation, and store frozen. The yield for 250 ml of cells at $2–3 \times 10^6$ ml is usually 0.5–1 mg RNA.

5.2 Procedure II

(i)   Collect the cells as in Section 5.1, steps (i)–(iii).
(ii)   Lyse the cells by sonication in 4 ml of GuHCl/NaAc.
(iii)   Remove the debris by centrifugation (10 000 g for 10 min) and keep the supernatant.
(iv)   Add ¾ vol. (3 ml) of ethanol to the supernatant and allow the nucleic acids to precipitate at −20°C for several hours.
(v)   Collect the precipitate by centrifugation (10 000 g for 10 min).
(vi)   Resuspend the pellet in 5 ml of TEN–SDS, extract with 2 vols (10 ml) of phenol/chloroform (1:1).
(vii)   Centrifuge at 15 000 g for 5 min, and take the upper (aqueous) phase.
(viii) Add one drop of glacial acetic acid to lower the pH, NaCl to 200 mM and ⁶⁄10 vol. of isopropanol. Leave at −20°C overnight.

(ix)   Collect the RNA by centrifugation.
(x)   Resuspend the pellet in 3 ml of DEPC-H_2O + 0.1% SDS. Add $MgCl_2$ to 1 mM, LiCl to 2 M (1 ml from 8 M stock) and allow the RNA to precipitate overnight at 4°C.
(xi)   Collect the precipitate by centrifugation at 15 000 g for 10 min.
(xii)   Dissolve the pellet in DEPC-H_2O and store at −20°C.

6. PROTEIN ANALYSIS

6.1 Sample preparation and electrophoresis

6.1.1 *Cell culture and harvesting*

(i)   Grow cells in TAP medium to a density of 2–5 × 10^6 cells/ml. No special precautions are necessary.
(ii)   Harvest the cells by centrifugation at 5000 g for 5 min.
(iii)   Resuspend the cells in $\frac{1}{10}$ vol. fresh media.
(iv)   Pellet again by centrifugation at 5000 g for 5 min.

6.1.2 *Sample resuspension, cell lysis and crude fractionation*

(i)   Resuspend the pelleted cells from Section 6.1.1 in a small volume of buffer A (*Table 4*) 1 ml/100 ml of original culture.
(ii)   Quick-freeze the cells in dry-ice/ethanol or liquid nitrogen. It is possible to store the cells frozen at this point. Cells may be stored at −20°C for several weeks or months with no apparent protein degradation or at −70°C indefinitely.
(iii)   Remove the cell suspension from the freezer and allow to thaw on ice for several minutes.
(iv)   Break the cells by sonication for 3 × 10 sec using a microtip. Keep the cell suspension on ice during all procedures. It is also possible to break the cells by passing them through a French press at 4000 p.s.i., but this procedure requires larger volumes of cell suspension for good lysis.
(v)   Check for cell lysis by examining under a light microscope. If the majority of the cells are not fragmented, repeat the sonication or French press procedure.
(vi)   The lysed cells are now separated into crude soluble and membrane fractions by centrifugation at 24 000 r.p.m. in a SW 60 (Beckman), or equivalent rotor, for 30 min at 4°C.
(vii)   Remove the supernatant, soluble fraction, to another tube and quickly freeze.
(viii) Resuspend the membrane pellet in a volume of buffer A equivalent to the amount of supernatant removed, and quickly freeze.

It is possible to perform all of the protocols which follow on lysed unfractionated cells, but we have found that this quick separation into membrane and soluble fractions allows for clearer results later.

264

6.1.3 *Preparation of 7.5–15% gradient polyacrylamide gels*

There are several different buffering systems which give equally good resolution of protein bands on polyacrylamide gels. We will discuss only one buffering system here, which is that of Neville as modified by Chua (15), and only 7.5–15% polyacrylamide gradient gels, which we have found useful for stained gels, radiolabelled protein gels and Western blots.

6.1.4 *Preparation of the gel plates*

It is important to have clean plates for good protein gels:

(i)   Scrub the gel plates well with wet alconox powder, rinse with large amounts of tap distilled water, then with 95% ethanol and dry with clean paper towel.

(ii) Clamp the desired width of spacers between the edges and bottom of plates. We standardly use 1.5 mm, and seal the edges and bottom with 1% molten agarose.

6.1.5 *Preparation of stock solutions and gel casting*

The grade of the chemicals used is listed with the manufacturer in *Table 4*. All chemicals are used as obtained from the manufacturer with the exception that the acrylamide is treated with activated charcoal, 2 g/100 ml of stock solution, for 5 min and then filtered through Whatman no. 1 paper and then Millipore (3.0 μ) before use.

(i)   Mix the stock solutions in the ratios shown in *Table 4*, but do not add the 10% APS.

(ii)   Add the APS to the light (7.5%) acrylamide, mix well, and put into the left chamber of a gradient maker.

(iii) Allow a small amount of solution to pass into the right chamber to remove any air bubble between the two chambers.

(iv)   Add the 10% APS to the heavy (15%) acrylamide and add it to the right chamber.

(v)   Open the passage between the two chambers and stir the contents of the right chamber, from which the solution will be pumped or poured into the gel former. In this way the gradient is formed from the bottom, with the lighter solution being layered on top. It should take approximately 15–25 min to pour a 40 ml gel.

(vi)   Stop the filling of the gel former when the solution is approximately 3–4 cm from the top.

(vii) Carefully layer on several millilitres of water-saturated butanol, which acts to stabilize the top of the gel during polymerization.

(viii) Allow the gel to set for 1 h, pour off the butanol and buffer which has collected at the top of the gel and wash with large amounts of distilled water.

Table 4. Polyacrylamide gel stock solutions.

1. *5× Resolving and lower buffer*
   Tris   514 g
   H_2O to   1800 ml
   HCl (conc.) ~28 ml, pH to 9.18 with HCl
   H_2O to   2000 ml, store at room temperature
2. *20× Upper buffer*
   Tris   49.5 g
   SDS   10 g
   Boric acid   14 g
   H_2O to   400 ml, pH to 8.64 with saturated boric acid solution
   H_2O to   500 ml, store at room temperature
3. *4× Stacking buffer*
   Tris   13 g
   H_2O to   450 ml
   H_2SO_4 ~2.2 ml, pH to 6.1 with H_2SO_4
   H_2O to   500 ml, store at room temperature
4. *30% Acrylamide, 0.8% bisacrylamide*
   30 g acrylamide (Kodak)
   0.8 g bisacrylamide (Kodak)-*N,N'*-methylene-bisacrylamide
   H_2O to 100 ml
   1 g activated charcoal (2.5 mm Merck)
   filter through No. 1 Whatman paper
   filter through 3.0 μ Millipore
   store at 4°C in the dark
5. *10% SDS*
   10 g sodium dodecylsulphate (research grade 99%)
   H_2O to 100 ml, store at room temperature
6. *TEMED*
   N,N,N',N'-tetramethylethylenediamine (Bio-Rad Labs)
7. *10% APS*
   1 g Ammonium-peroxodisulphate (Merck)
   H_2O to 1 ml store at 4°C, make fresh weekly
8. *60% Sucrose*
   60 g sucrose (Ultrapure, Schwarz/Mann)
   H_2O to 100 ml, autoclave, store at 4°C
9. *7.5–15% Gradient Gel* (40 ml)

	7.5%	15%
30% acrylamide, 0.2% bisacrylamide	5.0 ml	10 ml
5× resolving buffer	4.0 ml	4.0 ml
60% sucrose	1.7 ml	5.75 ml
10% SDS	0.2 ml	0.2 ml
H_2O	9.0 ml	
TEMED	9 μl	4.5 μl
10% APS	67 μl	67 μl

10. *Stacking gel (10.3 ml)*

30% acrylamide, 0.2% bisacrylamide	2.0 ml
4× stacking buffer	2.5 ml
H_2O	5.6 ml
10% SDS	100 μl
TEMED	10 μl
10% APS	100 μl

11. *Coomassie Blue R staining solution*

Coomassie Brilliant Blue R	1.25 g (Sigma)
Methanol	250 ml
Acetic acid	50 ml
H_2O	200 ml

   Store at room temperature

12. *Destaining solution*

Methanol	200 ml
Acetic acid (glacial)	50 ml
H_2O	250 ml

Protein resuspension buffers

13. *Buffer A*
   0.8 M Tris–HCl, pH 8.3
   0.4 M sucrose
   1% BME

14. *8 M Urea*
   40 g urea (Schwarz/Mann Ultrapure)
   H_2O to 125 ml, store at 4°C

15. *Loading buffer*
   2.5% SDS
   2.5% BME
   0.2 M Tris–HCl, pH 8.3
   0.1 M sucrose
   4 M urea
   0.1% bromophenol blue

Western buffers

16. *Transfer buffer*
   27 g sodium acetate, $3H_2O$
   1 g sodium azide
   H_2O to 4000 ml, pH 7.0 with a few drops acetic acid (glacial)

17. *Blocking buffer*
   2.5 g BSA
   18.5 g glycine
   H_2O to 250 ml

18. *Hybridization buffer*

14.8 g	NaCl
0.69 g	$NaH_2PO_4.H_2O$
4.02 g	Na_2HPO_4
2.0 g	NP-40 (or other non-ionic detergent)
0.2 g	sodium azide
2.0 g	BSA

   H_2O to 2000 ml, pH 7.4

19. *7.5 M guanidine 5% BMe*
   68 g guanidine–HCl (mol. wt 95.53) 99%
   5 ml BMe
   H_2O to 100 ml (work in fume hood)

20. *0.2 M glycine*
   6 g glycine
   H_2O to 350 ml, pH to 2.8 with HCl (conc.)
   H_2O to 400 ml

(ix)   Seal the gel in a damp plastic bag and store at 4°C. As long as the gel is not allowed to dry out it may be stored in this way for several days or weeks, and in fact we always store our gels for at least 24 h before use, as this improves the sharpness of the separated protein bands.

(x)   Approximately 1 h before use remove the gel from the cold and allow it to warm to room temperature.

(xi)   Mix the stacking gel solution in the ratios shown in *Table 4*, again adding the 10% APS last.

(xii)   Pour this solution on top of the polymerized separating gel and add a comb with an appropriate number of slots for your samples. This must be done quickly as this solution polymerizes within a few minutes.

(xiii) Allow the stacking gel to set for 15–30 min before use.

(xiv)   After the stacking gel has polymerized, remove the comb and the bottom spacer and clamp the cast gel into an appropriate electrophoresis tank.

(xv)   Fill the upper and lower reservoirs with 1× upper or lower buffer (same as resolving buffer).

(xvi)   Rinse the sample slots clear of any stray acrylamide with a syringe, and make sure the bottom of the gel is in good contact (no air bubbles) with the lower buffer.

6.1.6 *Sample preparation and electrophoresis*

(i)   Remove the crude protein samples from the freezer and allow to thaw on ice.

(ii)   Make a solution of 10% SDS, 10% β-mercaptoethanol (BME) and pipette 20 μl into a small Eppendorf tube.

(iii)   Add 20 μl of protein sample and mix by pipetting up and down several times.

(iv)   To this add 40 μl of 8 M urea, so that the final concentration of the mix is 4 M urea, 2.5% SDS and 2.5% BME.

(v)   Heat the sample to 70°C for 5 min.

(vi)   It is important to load both equal amounts of protein and equal volumes of sample to each sample lane, as both of these will affect the migration of the proteins. For membrane proteins it is usually sufficient to load equal amounts of chlorophyll, approximately 5–10 μg/lane, to obtain equal loading of membrane proteins. For water-soluble proteins either measure protein content (Lowry assay), or load a volume equal to the amount loaded for the membrane sample and stain the gel to estimate protein content. If two samples are different in concentration, dilute the concentrated sample to the same concentration as the weak sample with loading buffer prior to heating the sample.

(vii)   Carry out the electrophoresis at 4°C for maximal sharpness of the protein bands. Run the gels for 12–14 h, which should be sufficient to run the bromophenol marking dye to the bottom of the gel, at 8–10 mA constant current. The voltage will increase to several hundred volts during the running of the gel, so be sure that this does not become limiting on your power supply.

(viii) Remove the gel from between the plates by carefully prying them apart with a spatula.

(ix) Place the gel into Coomassie blue staining solution and gently shake for 1–2 h at room temperature.

(x) Pour off the staining solution and add destaining solution, changing it every 30 min.

(xi) After 2–3 h the protein bands should appear dark blue against an almost clear background.

6.2 *In vivo* labelling of proteins

(i) Grow cells in TAP medium to a cell density of $1-2 \times 10^6$ cells/ml.

(ii) Pellet 40 ml of cells at 5000 *g* for 5 min and resuspend in 40 ml of media lacking acetate.

(iii) Grow for an additional 1 h prior to labelling.

(iv) For labelling of cells add 50 μCi of [^{14}C]acetate (56 μCi/mmol, Amersham) to the 40 ml of cells in minimal media, and allow the label to incorporate for 10 min.

(v) Make the medium 50 mM with respect to acetate by the addition of 1 ml of 2 M non-radioactive acetate, and quickly harvest the cells by centrifugation for 2 min at 10 000 *g*.

(vi) Resuspend the pellet in buffer A and quickly freeze in dry-ice/ethanol.

It is possible to inhibit either chloroplastic or cytoplasmic protein synthesis prior to labelling by adding the appropriate drug (e.g. final concentration 10 μg/ml cycloheximide) 5 min prior to the addition of [^{14}C]acetate, in step (iii).

These *in vivo* labelled protein samples are now treated exactly like unlabelled proteins (Section 6.1.6) for subsequent crude fractionation and gel electrophoresis.

(i) Following electrophoresis, stain the gels for 1–2 h in Coomassie blue stain.

(ii) Destain for 2–3 h in destain, changing the destain solution every half hour.

(iii) Following destaining, soak the gel for 20–30 min in Enlighting (New England Nuclear), or similar autoradiographic enhancer.

(iv) Dry the gel onto 3 MM filter paper under vacuum.

(v) Expose to film at −70°C. We generally expose film for 48 h, but on poorly labelled samples the films can be left up for much longer exposures.

6.3 Transfer of proteins from gel to paper (Western blotting)

6.3.1 *Protein gel electrophoresis*

(i) Prepare the protein gel and the samples in the same way as for a stained gel (Section 6.1) adding the SDS/BME and urea.

(ii) Dilute the samples 10-fold with loading buffer.

(iii) Heat to 70°C for 5 min.

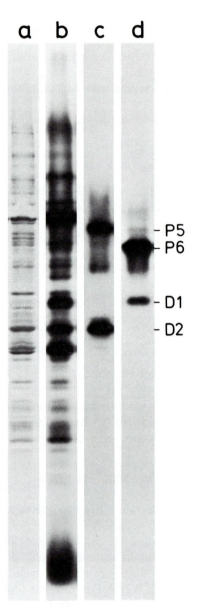

Figure 2. Membrane proteins from *C.reinhardtii*. Cells were labelled for 10 min with [^{14}C]acetate (Section 6.2). Membrane proteins were electrophoresed on a 7.5–15% SDS–polyacrylamide gel (Section 6.1). (**a**) Staining pattern obtained with Coomassie blue. (**b**) Autoradiogram of labelled proteins. (**c**) Western blot (Section 6.3) of membrane proteins with antibodies of polypeptides P5 and D2. (**d**) Western blot with antibodies of polypeptides P6 and D1.

(iv)   Load the same volume, which will be $^1/_{10}$ of the protein, as for the stained gels. Less protein is added to the gels used for blotting to ensure that the paper does not become saturated by any one protein, thus resulting in an aberrant signal on the autoradiograph.

6.3.2 *Electroblotting of the gel*

After electrophoresis carry out the following protocol.

(i)   Wash the gel for 2 × 10 min in 0.1% SDS to remove the Tris, which will compete with proteins for binding to the paper.

(ii)   Wash twice in transfer buffer, for 10 min each.

(iii)   Place the gel on a foam (or other) pad which has been wetted with transfer buffer.

(iv)   Carefully place a piece of cyanogen bromide (CNBr)-activated paper, or APT paper (BIO-RAD labs), on top of the gel (see *Table 5A*).

(v)   Place another wet pad on top of the CNBr paper, and secure the sandwich between two rigid plastic sheets.

(vi)   Place the entire sandwich into a transblot apparatus making sure that the paper is on the positive (anodal) side of the gel, as the proteins will migrate in this direction.

(vii)   Cover with transfer buffer, and blot for 1 h at 1 A. Do the blotting with a cooling core in place, or decrease the amperage and increase the time proportionally.

(viii) Carefully remove the sandwich from the blotter and lift the paper from gel directly into the blocking solution. Shake for 30 min at room temperature.

(ix)   Wash the filter for 30 min in hybridization buffer.

(x)   Place the filter in a sealable bag slightly larger than the filter. Add 20 ml, for a 15 × 17 cm gel, of hybridization buffer containing a 0.5–5000 dilution of antisera.

(xi)   Seal the bag and rock 6 h to overnight at room temperature.

(xii)   Wash the filter for 3 × 10 min in hybridization buffer.

(xiii) Repeat steps (x)–(xiii) replacing the antisera with ^{125}I-*Staphylococcus* protein A.

(xiv)   Expose to X-ray film with an intensifying screen; generally a 12–48 h exposure is sufficient.

7. TRANSFORMATION OF CHLAMYDOMONAS REINHARDTII

The establishment of a transformation system in *C.reinhardtii* is of great importance, because it would provide a powerful tool for understanding gene function and regulation, by the introduction of modified genes and/or regulatory elements into cells. It would also allow one to isolate genes by complementing defined mutations with wild-type genomic libraries. Indeed a large number of nuclear and chloroplast mutants have been isolated from *C.reinhardtii* (for an extensive discussion of *C.reinhardtii* genetics and for a list of available nuclear mutants see refs 1 and 16, respectively). Nuclear mutants include flagellar, ribosomal, drug resistant, auxotrophic and photosynthetic mutants, of which several have been mapped on the 18 nuclear linkage groups. Most characterized chloroplast mutants are ribosomal (including drug resistant) mutants and photosynthetic mutants. Transformation in *C.reinhardtii* is feasible, although the transformation yield is still rather low (10^{-6} transformants per treated cell).

Table 5A. CNBr-activation of paper.

WARNING: Cyanogen bromide is highly toxic. Perform all operations in a chemical fume hood.

1. Cut 17.5 × 14 cm sheets (12) of Whatman 540. Always wear gloves to handle paper.
2. Place the papers in a Pyrex baking dish and soak in 1 litre of 0.1 M $NaHCO_3$ for 15 min.
3. Pour off the $NaHCO_3$ and add 1 litre of distilled water for 30 min.
4. Dissolve 25 g of CNBr in 1 litre of distilled water in a 4 litre beaker. Stir on a magnetic stirrer in hood (20–30 min).
5. Pour off the water from the papers and place the Pyrex dish on a shaker in hood. Suspend an electrode in dish.
6. Add the CNBr solution to the papers, start the shaker and begin adding the NaOH. The reaction begins when a pH of 11 is reached.
7. Keep the pH between 10.5 and 12.5 by addition of NaOH for 8 min.
8. Pour the CNBr solution back into the 2 litre beaker and add 1 litre of 0.1 M $NaHCO_3$ to the papers. This stops the reaction. Shake for 5 min.
9. Pour the $NaHCO_3$ into the CNBr solution and wash the papers as follows:
    a. 2 × 5 min washes with 500 ml of 0.1 M $NaHCO_3$.
    b. 1 × 5 min wash with 1 litre of water.
    c. 1 × 5 min wash with 50% acetone.
    d. 1 × 5 min wash with acetone.
10. Hang the papers in hood to dry for 10 min. Store desiccated at 4°C.
11. Inactivate the CNBr solution by gradually adding concentrated sodium hypochlorite under constant stirring.

Table 5B. Preparation of CNBr paper.

Solutions for CNBr paper

1.	0.1 M $NaHCO_3$	3 litres
	Divide into two 1 litre and two 500 ml aliquots.	
2.	Distilled H_2O	3 litres
	Divide into three 1 litre aliquots. Use a 4 litre beaker for one of the containers.	
3.	2 M NaOH	500 ml
	Keep in a wash bottle.	
4.	50% acetone	250 ml
5.	Acetone	250 ml
6.	Sodium hypochlorite, conc. (~13%) ~1 litre.	

Equipment
1. Shaker, flat top.
2. pH meter.
3. Magnetic stirrer.

It is easier to use a cell wall-deficient strain, such as cw15, which behaves like a natural protoplast although it is less sensitive to osmoticum. Wild-type cells can also be used, provided their cell wall is first removed with autolysin (cf. Section 3.1).

7.1 Delivery of DNA to the cells

We have mostly used two methods which give comparable results.

7.1.1 *Poly-L-ornithine method (17)*

(i) Grow cells in TAP medium to a concentration of $<2 \times 10^6$ cells/ml.

(ii) Centrifuge the cells in a Sorvall HB-4 rotor at 4000 g for 5 min.

(iii) Wash the cells once with TAP–0.2 M mannitol.

(iv) Resuspend the cells ($2–5 \times 10^7$ cells) in 0.95 ml of TAP–0.2 M mannitol.

(v) Prepare the DNA mixture for transformation: 5 µl of DNA (1–2 mg/ml, containing the selective marker), 25 µl of carrier DNA (calf thymus or salmon sperm DNA at 1 mg/ml), 10 µl of poly-L-ornithine (1 mg/ml), 0.5 µl of 0.5 M $ZnSO_4$, 9.5 µl of H_2O. Prepare similarly a control sample which lacks the transforming DNA. Pre-incubate the mixtures for 1 h in ice. A faint precipitate should form.

(vi) Add the DNA mixture to the cells, together with 10 µl of 0.5 M $ZnSO_4$ and incubate 30 min at room temperature.

(vii) Dilute with 10 ml TAP–0.2 M mannitol and sediment the cells at 4000 g for 5 min in the Sorvall HB-4 rotor.

(viii) Resuspend the cells in 10 ml of TAP–0.2 M mannitol and repeat.

(ix) Incubate the cells with gentle agitation in dim light overnight.

(x) Sediment the cells and resuspend them in 0.6 ml of TAP–0.2 M mannitol.

(xi) Mix 0.2 ml of cells with 3 ml of TAP top agar (0.6% agar in TAP–0.2 M mannitol).

7.1.2 *Chloroquine method*

This method was first devised by S.Hong and S.J.Surzycki (personal communication). The rationale of using chloroquine is that by increasing the pH inside the cells, it may diminish nuclease activity during the initial stages of the transformation.

(i) Prepare the cells as in Section 7.1.1 and resuspend them in 2 ml of TAP, pH 9.3. High pH has no apparent deleterious effect on cell viability although it stops growth.

(ii) Add 5 µl of 2 mM chloroquine and 5 µl of transforming DNA (1 mg/ml).

(iii) Incubate the mixture for 3 h at room temperature in dim light. It is important to keep the temperature below 25°C, otherwise a large proportion of cells may die (S.Hong and S.J.Surzycki, personal communication).

(iv) Dilute the suspension with 10 ml of TAP–0.2 M mannitol, sediment cells, wash once and resuspend the cells in TAP, pH 7.0, 0.2 M mannitol, 10 µM chloroquine.

(v) Proceed with steps (ix) to (xi) of Section 7.1.1.

7.2 **Markers for selection**

7.2.1 *Arginine prototrophy*

The ARG7 locus was the first nuclear locus of *C.reinhardtii* that was characterized both at the genetic and biochemical level. This locus codes for argininosuccinate lyase, the last enzyme of the arginine biosynthetic pathway, which converts argininosuccinate into arginine and fumarate. It is homologous to the yeast ARG4 and *E.coli arg*H loci. Since the *C.reinhardtii* ARG7 locus has not yet been cloned, the ARG4 gene of yeast has been used to transform a cw15 ARG7 double mutant (17). Arginine prototrophs are obtained at a frequency ranging between 10^{-6} and 10^{-7}, only slightly above the reversion rate of the ARG7 mutations.

7.2.2 *Resistance to kanamycin*

Hasnain *et al.* (18) have obtained kanamycin resistant transformants at a frequency of 10^{-6} by using a plasmid containing the yeast 2 μ circle origin of replication and a neomycin-phosphotransferase (NPT II) gene under control of the SV40 early promoter. About two-thirds of the transformants were shown to contain free plasmids that could be recovered by transforming *E.coli* with the DNA of the transformants. Similar results were obtained by Cox *et al.* (19) who transformed *C.reinhardtii* to kanamycin resistance by using a plasmid containing a chloroplast ARS element and the NPT II gene with its own (bacterial) promoter. In our hands kanamycin resistance is difficult to use for selection because most of the kanamycin resistant clones we recovered after transformation with plasmids harbouring the NPT II gene did not contain this gene.

7.2.3 *Homologous selective markers*

Because the codon usage of nuclear genes from *C.reinhardtii* is very restricted and biased towards C and G in the wobble position of the codons, the possibility exists that heterologous selective marker genes are poorly expressed. We have therefore started to use genes for which mutations are known to exist and that could be used for selection in the transformation. The nuclear gene encoding the 33-kd polypeptide of the oxygen-evolving complex of photosystem II in *C.reinhardtii* has been cloned and characterized (S.Mayfield and J.D.Rochaix, unpublished results). One nuclear mutant, FuD44, which is unable to grow without acetate, has been shown to contain an insertion in the 33 kd protein gene. Experiments are in progress to use this mutant for transformation with the homologous wild-type allele by selecting for phototrophic growth.

7.3 **Autonomously replicating plasmids in *C.reinhardtii***

Several plasmids capable of autonomous replication in *C.reinhardtii* have been described (cf. *Table 6*). Three of these plasmids contain short chloroplast DNA fragments and one contains the origin of replication of the yeast 2 μ circle. Whether the latter induces autonomous replication in *C.reinhardtii* has not yet been proved. These plasmids do not significantly increase the transformation yield as compared to plasmids that integrate into the nuclear DNA. While the plasmids containing the chloroplast fragments were lost after 60–80 generations, the plasmid with the yeast 2 μ circle origin of replication could be maintained for at least 1 year. Recent experiments suggest that the ribosomal DNA of

Table 6. Autonomously replicating plasmids in *C.reinhardtii*.

Plasmid	Size insert (bp)	Selective marker	Reference
pCA1	7470 C	ARG4	21
pCA2	7210 C	ARG4	21
pCA3	7160 C	ARG4	21
pCA4	7310 C	ARG4	21
pSV2-neo-2 μ	7280 2 μ	NPT II	19

The plasmids are pBR322 derivatives containing inserts from chloroplast DNA of *C.reinhardtii* (C) and from the yeast 2 μ circle (2 μ). ARG4 and NPT II refer to the yeast gene coding for argininosuccinate lyase and for the gene of neomycin-phosphotransferase II (which confers resistance to kanamycin), respectively.

Physarum polycephalum which exists as 60 kb linear palindromic molecules (20), is capable of transforming, at least transiently, *C.reinhardtii*, even without selection. However, because of its large size this DNA is difficult to manipulate.

8. ACKNOWLEDGEMENTS

We thank S.J.Surzycki, S.Hong and A.Day for communicating unpublished methods and O.Jenni for preparing the figures. The work was supported by grant 3.328.0.86 from the Swiss National Foundation.

9. REFERENCES

1. Gillham,N.W. (1978) *Organelle Heredity*, Raven Press, New York.
2. Snell,W.J. (1985) *Annu. Rev. Plant Physiol.*, **36**, 287.
3. Forster,K.W., Saranak,J., Patel,N., Zarilli,G., Okabe,M., Kline,T. and Nakanishi,K. (1984) *Nature*, **311**, 756.
4. Surzycki,S.J. (1971) In *Methods in Enzymology*. San Pietro,A. (ed.), Academic Press, New York, Vol. 23, p. 17.
5. Gorman,D.S. and Levine,R.P. (1965) *Proc. Natl. Acad. Sci. USA*, **54**, 1665.
6. Sueoka, N., Chang,K.S. and Kates,J.R. (1967) *J. Mol. Biol.*, **25**, 47.
7. Davies,D.R. and Plaskitt (1971) *Genet. Res.*, **17**, 33.
8. Belknap,W.R. and Togasaki,R.K. (1981) *Proc. Natl. Acad. Sci. USA*, **78**, 2310.
9. Snell,W.J. (1982) *Exp. Cell Res.* **138**, 109.
10. Rochaix,J.D. (1980) In *Methods in Enzymology*. Grossman,L. and Moldave,K. (eds), Academic Press, New York, Vol. 65, Part I, p. 785.
11. Cryer,D.R., Eccleshall,R. and Marmur,J. (1975) *Methods Cell Biol.*, **12**, 39.
12. Weeks,D., Beerman,N. and Griffith,O.M. (1986) *Anal. Biochem.*, **152**, 376.
13. Davis,R., Thomas,M., Cameron,J., John,T., Scherer, S. and Padgett,R.A. (1980) In *Methods in Enzymology*. Grossman,L. and Moldave,K. (eds), Academic Press, New York, Vol. 65, p. 404.
14. Chirgwin,J.M., Przybyla,A.E., MacDonald,R.J. and Rutter,W.J. (1979) *Biochemistry*, **18**, 5294.
15. Piccioni,R., Bellemare,G. and Chua,N.H. (1982) In *Methods in Chloroplast Molecular Biology*. Edelman,M., Hallick,R.B. and Chua,N.H. (eds), Elsevier Biomedical Press, p. 985.
16. Harris,E.H. (1982) In *Genetic Maps*. O'Brien,S.J. (ed.). Laboratory of Viral Carcinogenesis, National Cancer Institute, NIH, Frederik, MD 21701, p. 168.
17. Rochaix,J.D. and van Dillewijn,J. (1982) *Nature*, **296**, 70.
18. Hasnain,S.E., Manavathu,E.K. and Leung,W.L. (1985) *Mol. Cell Biol.*, **5**, 3647.
19. Cox,J.C., Bingham,S.E. and Bishop,R.J. (1985) *Abstracts 1st International Congress Plant Molecular Biology*, p. 115.
20. Ferris,P. and Vogt,V.M. (1982) *J. Mol. Biol.*, **159**, 359.
21. Rochaix,J.D., van Dillewijn,J. and Rahire,M. (1984) *Cell*, **36**, 925.

CHAPTER 11

The molecular biology of cyanobacteria

VALDIS A.DZELZKALNS, MIKLÓS SZEKERES and
BERNARD J.MULLIGAN

1. INTRODUCTION

The cyanobacteria are a diverse group of photoautotrophic prokaryotes (1) with
an oxygen evolving photosynthetic apparatus very similar to that of higher
plants. Early work on the molecular biology of the cyanobacteria has been
comprehensively reviewed (2) and later research into most aspects of the
biology of these organisms is the subject of a detailed monograph (3). Recent
advances in cyanobacterial molecular biology and genetics, together with the
application of gene cloning and analysis techniques, provide a valuable range of
experimental tools for studying the unique physiological and developmental
properties of these organisms. The most spectacular advances, employing
'conventional' techniques of recombinant DNA technology, have been in
unravelling the patterns of gene expression involved in the biochemistry and
control of nitrogen fixation and heterocyst development. Another area of
impressive advance has been in the isolation and analysis of 'photosynthesis
genes', the sequences of which have provided valuable information about the
structure and evolution of the photosynthetic apparatus. The ability of certain
cyanobacteria to adjust photosynthetic pigment synthesis to changes in light
quality (complementary chromatic adaptation) and intensity and the ability of
others to develop an anoxygenic photosynthesis system provide other areas of
photosynthesis research in which the techniques of molecular biology should
prove valuable.

Rapid progress in the understanding of genetic transformation in cyanobacteria has provided a new approach to problems in the molecular biology of
these organisms. It is difficult to overestimate the potential of transformation,
both as an experimental tool and as an exciting area of expanding interest in
cyanobacteriology. The identification and isolation of specific DNA sequences
with the aid of transformation techniques has already been achieved.

Cyanobacteria are hosts for a specific group of viruses, known as cyanophages, of which virulent and temperate forms exist. The study of cyanophage
biology is, perhaps, one of the most underexploited areas in cyanobacterial
research. Valuable information about photoautotrophic metabolism has been
obtained from investigations on the lytic cycle; the molecular biology of the
infection process has also been studied in some detail. The main drawback to the
use of cyanophages in genetic studies, for example, in gene mapping by

277

transduction, lies in the lack of a good range of suitable genetic markers in the host cyanobacteria. This disadvantage aside, however, other aspects of molecular genetics may be studied with the aid of cyanophages. One of the best examples of this is the use of mutated (irradiated) cyanophage DNA as a substrate in studies of DNA repair in cyanobacteria.

This chapter covers three broad areas. The first is a selective summary of general aspects of cyanobacterial work including growth of laboratory cultures and isolation of nucleic acids. A brief summary of information currently available on *in vitro* transcription and homologous and heterologous cell-free translation in cyanobacterial systems will also be included. The second is a detailed description of the methods of cyanophage biology. Finally, protocols for the transformation of cyanobacteria are described, together with a discussion of the achievements and aspirations of this branch of cyanobacterial genetics. As far as possible, detailed methods will steer clear of repetition of well documented methods of (bacterial) molecular biology.

2. GROWTH AND MAINTENANCE OF LABORATORY CULTURES OF CYANOBACTERIA

The best studied cyanobacteria, in terms of molecular biology, are mainly of freshwater origin. As photoautotrophs, the cyanobacteria require light, carbon dioxide and an inorganic medium to satisfy their nutritional requirements.

A convenient source of light is provided by fluorescent tubes, warm white illumination having a somewhat more useful spectral range than cool white. Incandescent tungsten bulbs may also be used, but the heat produced may be inconvenient in some cases. Light intensity should be carefully considered for each organism under study. Some, notably phycoerythrin-containing strains, are often rather light sensitive and illumination should not be provided in great excess of that required for growth. As a practical example, stock agar slope cultures of *Anacystis nidulans* and the phycoerythrin-containing *Nostoc* MAC grow well at 30°C at 13 cm from a bank of two warm white fluorescent tubes, each of 20 W power rating. Such slopes are well grown after 3–5 days; they can then be stored on a laboratory bench for 5–6 weeks. Temperature is obviously an important consideration. Many strains do well at 25–30°C, but higher temperatures (e.g. up to 40°C for *A. nidulans*) can often be used for liquid batch cultures to obtain faster cell growth. Clear plastic tanks can be provided with a thermostatically controlled heater/stirrer to give a convenient incubator for smaller batch culture vessels. Growth of liquid batch cultures employs higher light intensities to overcome self shading of cells in thickening cultures. The self shading problem can also be alleviated by the use of suitable culture vessels (see Section 6.1.1). Rapidly growing cultures require higher carbon dioxide levels than in the ambient atmosphere. This can be provided by gassing with a mixture of 0.5–5% carbon dioxide in air. Air may be replaced with nitrogen for strains particularly sensitive to oxygen. Solid $NaHCO_3$ (0.5 g/l) added to the liquid medium prior to autoclaving helps to counteract any fall in pH brought about by gassing with carbon dioxide. Alternatively, buffered media containing, for example, Tris or Hepes (4,5) may be employed.

A wide range of inorganic media suitable for the culture of cyanobacteria have been devised. These may be solidified with Difco Bacto-agar to a final concentration of 1–1.5%. Double strength preparations of liquid medium and agar in double distilled water are autoclaved separately, cooled to 48°C, mixed in equal volumes and poured into sterile culture tubes or Petri dishes (6). Three media in common use are described in *Table 1* (7–9). Considerable modification of these media may be necessary in some instances. For example, nitrate/ammonium should be omitted for the maintenance and growth of certain strains under nitrogen fixing conditions (1). Medium BG-11 and Allen's medium are particularly useful for maintenance on solid media and liquid culture of a wide range of freshwater strains. Medium C contains high levels of phosphate and is particularly good for obtaining dense cultures of species such as *A.nidulans*. More detailed consideration of the growth requirements of cyanobacteria is available in the literature (1,7,10).

3. ISOLATION OF CYANOBACTERIAL NUCLEIC ACIDS

In general, published protocols for the isolation of nucleic acids from cyanobacteria are similar to those encountered in the bacterial DNA and RNA

Table 1. Components of three media used for cyanobacterial cultures.

Ingredient	BG-11 g/l of medium	Allen's	C
$NaNO_3$	1.5	1.5	–
KNO_3	–	–	1.0
$K_2HPO_4 \cdot 3H_2O$	0.04	0.04	1.0
$MgSO_4 \cdot 7H_2O$	0.075	0.075	0.25
$CaCl_2 \cdot 2H_2O$	0.036	0.036	–
$Ca(NO_3)_2 \cdot 4H_2O$	–	–	0.025
Citric acid	0.006	0.006	–
Ferric citrate	–	0.006	–
Ferric ammonium citrate	0.006	–	–
Sodium citrate·$2H_2O$	–	–	0.165
Ferric sulphate·$6H_2O$	–	–	0.004
EDTA (disodium)	0.001[a]	0.001	–
Sodium silicate·$9H_2O$	–	0.058	–
Na_2CO_3	0.02	0.02	–
Microelements A5	–	–	1 ml
Microelements A6	1 ml	1 ml	–

Microelements A5 (g/l of distilled H_2O): H_3BO_4, 2.86; $MnCl_2 \cdot 4H_2O$, 1.81; $ZnSO_4 \cdot 7H_2O$, 0.222; $Na_2MoO_4 \cdot 2H_2O$, 0.391; $CuSO_4 \cdot 5H_2O$, 0.079.

Microelements A6: as for A5 with 0.0494 g/l $Co(NO_3)_2 \cdot 6H_2O$.

pH after autoclaving and cooling should be 7.4–7.8.

[a] Disodium magnesium EDTA in (1,7).
For convenience, concentrated stocks of the individual salts can be prepared and stored at 4°C.

literature. Such differences as there are probably reflect the very wide range of thickness and composition of cell wall seen in these organisms. Achieving efficient cell breakage is the first step in isolating any nucleic acid. Many methods have been used in disrupting different species. The cyanobacterial genome is distributed between the main chromosome and various genetically cryptic plasmids (11). Interest in the latter has shifted toward their use as vectors for introducing foreign or mutated DNA into the cyanobacterial cell (see Section 7.2.1).

3.1 Isolation of DNA

Cyanobacteria are Gram-negative organisms, the cell wall being composed of a peptidoglycan layer, an outer membrane and external slime layers (12). Removal of the cell wall is achieved by digestion with lysozyme. Digestion is usually carried out in an osmotic stabilizer, such as mannitol or sucrose. The cell is thus converted to a protoplast which is then lysed by detergents such as SDS and sarcosyl. Efficient lysis may require prolonged (>1 h) lysozyme treatment. The advantage of the lysozyme–detergent method of breakage is, of course, that shear forces are kept to a minimum. Thus high molecular weight chromosomal DNA and intact, supercoiled plasmids may be readily isolated. Isolation of plasmid DNA will be discussed in Section 7.1.2 and is well documented in the literature (13–15).

Various methods have been employed to obtain total DNA (16–18). We have found the method of Tomioka *et al.* (16), a variant of the Marmur procedure (19), to be very reliable for DNA preparations from *A.nidulans* and *Fremyella diplosiphon*. DNA obtained by this procedure is suitable for the preparation of genome libraries in λ phage and plasmid vectors. The protocol is given in *Table 2*.

It is not necessary to employ CsCl gradient centrifugation for the preparation of 'cloning quality' DNA. However, treatment of cyanobacterial lysates with high concentrations of salt in the form of CsCl and subsequent ultracentrifugation in the presence of ethidium bromide (EtBr) (17) provides a good alternative to deproteinization with organic solvents (see *Table 2*). A 'miniprep' method, given in *Table 3*, has also been devised, which for *A.nidulans* R2 and *Synechocystis* 6803 at least, gives DNA of sufficient purity for digestion with restriction enzymes and for cyanobacterial transformation experiments (20,21). The miniprep also yields RNA (see below) and this in turn is of sufficient quality for Northern hybridizations and nuclease S1 mapping procedures.

3.2 Isolation of RNA

Current interest in cyanobacterial RNA lies, in the main, with mRNA and its analysis by DNA:RNA hybridization (Northern blotting) and nuclease S1 mapping techniques. Methods for the isolation of cyanobacterial rRNAs and tRNAs are well documented in the literature (see ref. 2). Cyanobacterial mRNA is actually a preparation of high molecular weight RNA, from which smaller molecules have been selectively removed. Recent evidence suggests that some

Table 2. Isolation of total DNA from cyanobacteria.

1. Grow the organism under the appropriate conditions (e.g. batch culture, gassing with 5% carbon dioxide in air) until mid- to late-log phase.
2. Harvest the cells by centrifugation at room temperature at 5000 g for 10 min.
3. Wash the cells by resuspending in 50 mM Tris–HCl, 1 mM EDTA (pH 8) and recentrifuging. Store the cell pellet frozen, preferably at −70°C.
4. Thaw 1 g of frozen cells in 15 ml of 50 mM Tris–HCl, 20 mM EDTA, 50 mM NaCl, 0.25 M sucrose (pH 8). Add 40 mg of lysozyme (Sigma) dissolved in 5 ml of the same buffer. Shake the cell suspension gently at 37°C for 60 min.
5. Add self-digested pronase (20 mg/ml stock in water; 3 h at 37°C) to 1 mg/ml and SDS to 1% (w/v). Incubate overnight with gentle shaking at 37°C.
6. Add solid $NaClO_4$ to 1 M and dissolve the salt by gentle swirling.
7. Add 20 ml of chloroform:isoamyl alcohol (24:1, v/v) and mix the phases by gentle inversion for 5–10 min.
8. Centrifuge the extract at 5000 g for 10 min to separate the phases.
9. Remove the upper, aqueous phase and repeat the chloroform:isoamyl alcohol extraction.
10. Add 2.5 vols of absolute ethanol to the aqueous phase, mix and remove the DNA precipitate with a glass rod or by centrifugation if necessary.
11. Dissolve the DNA in 20 ml of 50 mM Tris–HCl, 20 mM EDTA, 0.1 M NaCl (pH 8). Add heat treated (boiling water bath for 10 min) RNase to 50 µg/ml and incubate at 37°C for 30 min.
12. Extract the DNA preparation with 20 ml of phenol [saturated with 50 mM Tris–HCl, 20 mM EDTA, 0.1 M NaCl (pH 8)] and centrifuge to separate the phases as in steps 8–9.
13. Repeat the phenol extraction.
14. Add 2.5 vols of ethanol to the aqueous phase and recover the DNA precipate.
15. Wash the precipitate with 70% ethanol and dissolve the DNA in 10 mM Tris–HCl, 1 mM EDTA (pH 7.5). (A few hours may be needed to achieve complete solubilization.)
16. Dialyse the dissolved DNA against several changes of the same buffer.
17. At this stage, the DNA is of sufficient purity for restriction enzyme digestion. If further purification is required, CsCl density gradient centrifugation should be employed.

Table 3. Simplified preparation of total nucleic acids.

1. Harvest the cyanobacteria (40–80 ml of culture) in mid- to late-log phase of growth. Optionally, (though note the precaution in Section 7.1.1), wash the cells in fresh growth medium.
2. Resuspend the cell pellet in 3 ml of 50 mM Tris–HCl, 20 mM EDTA (pH 7.8).
3. Add the cells dropwise to liquid nitrogen and grind to a fine powder.
4. Thaw the cell powder at 37°C and add proteinase K to 10 µg/ml. Incubate the mixture at 37°C for 30 min (see note to *Table 4*).
5. Extract the mixture three times with phenol/chloroform (5:1, v/v).
6. Precipitate the nucleic acids from the aqueous phase with ethanol in the presence of 0.5 M ammonium acetate. Recover the precipitate by centrifugation.
7. Dissolve the nucleic acid pellet in sterile distilled water or 10 mM Tris–HCl, 1 mM EDTA (pH 7.4). The DNA and RNA in this preparation can now be used for blotting experiments, transformation, etc.

cyanobacteria may contain poly(A)-tailed RNA species, though the significance of this observation remains unclear (22).

Methods suitable for disruption of cells for RNA extraction, include mechanical milling of frozen cells with glass beads (23), freezing and thawing in the presence of detergent (24), lysis with a detergent cocktail (18), grinding with quartz sand (16) and extrusion through a French pressure cell (25), to list but a few. All these methods succeed in their general aim of obtaining a relatively undegraded preparation of nucleic acid in which a transcript of interest can be detected using a suitable hybridization probe. Exogenous ribonuclease activity is minimized by autoclaving solutions, where possible, and baking glassware and wearing gloves during the extraction procedure. Endogenous ribonuclease can be minimized by the inclusion of phenol, Macloid clay, guanidinium–HCl, or a combination of these components in the cell disruption buffer (26). The protocol given in *Table 4* has proved reliable for preparing RNA from *F. diplosiphon* and *A. nidulans* containing readily detectable levels of, for example, the *psbA* gene transcript (encoding the 32 kd herbicide-binding protein) and its, possibly, processed form (27–28).

Some thought should be given to the method employed for harvesting cells for RNA extractions, in particular to avoid untoward perturbations which may have transcriptional consequences. Cyanobacteria are cold sensitive and thus the harvesting of cells in a chilled centrifuge should perhaps be avoided. Rapid filtration onto Whatman 1 MM filter paper can be employed for certain filamentous strains (29), while brief centrifugation can be used for unicells; both methods should ideally be carried out at the culture temperature. *Rapid* chilling, for example by addition of ice to cultures *prior* to centrifugation in a chilled centrifuge, is also suitable (18). Further modifications may be necessary in

Table 4. Isolation of total RNA from cyanobacteria.

All glassware should be baked. Where possible, solutions should be sterilized by autoclaving. Gloves should be worn throughout the preparation.

1. Harvest log-phase cells by centrifugation at the desired temperature at 5000 *g* (or by filtration on Whatman 1 MM paper for some filamentous strains).
2. Resuspend the cells quickly in a small volume of 50 mM Tris–HCl, 20 mM EDTA (pH 7.8) and pipette the solution dropwise into liquid nitrogen.
3. Grind the frozen cells to a fine powder in a blender, or with a chilled mortar and pestle adding more liquid nitrogen as necessary.
4. Add the ground cells to an equal volume of pre-warmed (37°C) lysis buffer [50 mM Tris–HCl, 0.1 M NaCl, 20 mM EDTA, 2% w/v SDS, 60 mM β-mercaptoethanol, 10 μg/ml proteinase K, (pH 8)]. Incubate at 37°C for 30 min.
4. If the solution is viscous at this stage, pass it several times through a syringe.
5. Extract the mixture three (or more) times: with an equal volume of lysis buffer-saturated phenol/chloroform (5:1, v/v). A clear, colourless, aqueous upper layer should be obtained. No protein should be visible at the aqueous–organic interface during the final extraction.
5. Remove the aqueous layer and adjust to 0.5 M ammonium acetate.
6. Precipitate the RNA by adding 2.5 vols of absolute ethanol and leaving at −70°C for a few minutes, or −20°C for a few hours.
7. Recover the nucleic acid by centrifugation at 12 000 *g*. Drain the pellet well.
8. Dissolve the pellet in sterile distilled water or a suitable buffer. Store at −20°C.
9. RNA can now be freed of contaminating DNA by treatment with DNase I, pelleting through CsCl or precipitation with 2 M LiCl (see Section 3.2).

If the RNA pellet proves difficult to dissolve, try extra phenol/chloroform extractions at step 5. If endogenous/exogenous RNase activity is thought to be a problem, the incubation in step 4 should be omitted and phenol extraction should proceed immediately. Additionally, RNase inhibitors should be included in the lysis buffer.

certain cases. For example, in studies on *nif* gene expression, anaerobiosis was maintained in cultures of *Anabaena* 7120 by purging with argon, while harvesting by continuous centrifugation (23).

Contaminating DNA may be removed from RNA preparations by digestion with RNase-free DNase I (Boehringer) and the use of placental RNase inhibitor may give additional protection to the RNA during this procedure (18). Alternatively, high molecular weight RNA may be selectively precipitated by adding LiCl to a final concentration of 2 M, allowing the mixture to stand on ice for a few hours, then recovering the RNA precipitate by centrifugation. This process can be repeated several times to increase the degree of purification. The RNA is finally precipitated with ethanol in the presence of 0.5 M ammonium acetate. Care should be taken with the LiCl method, however, since it is our experience that some high molecular weight RNA remains in solution during the LiCl treatment. Some losses may be acceptable, however, if greater purity is required. Selective precipitation of high molecular weight RNA also occurs in 3 M sodium acetate (18). RNA preparations may also be efficiently freed of DNA and other impurities by pelleting through a cushion of 5.7 M CsCl (30).

High molecular weight rRNAs are always found in these cyanobacterial RNA preparations. These are the 23S and 16S species and the two cleavage products of the former; the respective molecular sizes of these molecules are 2.95, 1.5, 2.5 and 0.5 kb and these are useful internal size markers for electrophoresis (see *Figure 1*).

As described in *Table 3*, a simplified method for the preparation of total nucleic acid has been devised. We routinely use this method for isolating RNA from *A.nidulans*. More than enough RNA is obtained from 40 ml of culture, at an optical density of 0.4 at 600 nm, for Northern blot analysis of specific transcripts. The method has also been used to obtain RNA for nuclease S1 mapping of transcripts from hybrid genes expressed in the cyanobacterium (20).

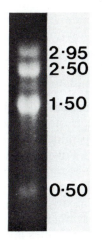

Figure 1. Electrophoresis of LiCl precipitated RNA from *F.diplosiphon* in 1.5% agarose. The gel was stained in 0.5 µg/ml EtBr and viewed under UV light.

However, the method has not been widely tested and may not be suitable for isolating low abundance transcripts or those with a rapid turnover.

To prevent the formation of base-paired secondary structures, RNA should, ideally, be denatured by glyoxylation prior to electrophoresis (31), or run in denaturing gels such as those containing formaldehyde (26,32). In certain cases, it may be possible to omit a denaturing step without incurring any obvious disadvantages in terms of anomalous migration of a particular transcript during electrophoresis (29).

4. APPROACHES TO THE CLONING OF CYANOBACTERIAL GENES

The identification and cloning of cyanobacterial genes requires, for the most part, the standard techniques of recombinant DNA technology. Libraries of cyanobacterial DNA can be constructed in bacteriophage and plasmid vectors by following standard protocols (26). To isolate genes from such libraries requires the availability of a suitable DNA probe. Cloned fragments of higher plant chloroplast DNA have been particularly valuable in this respect, especially in the identification of cyanobacterial 'photosynthesis' genes (28,33,34). There remains much scope for identification of specific sequences in this way. Heterologous probes of bacterial DNA have also proved useful. The nitrogen fixation genes of *Klebsiella*, for example, have allowed their cyanobacterial counterparts to be identified (35). Another, equally direct, approach has been to employ synthetic oligonucleotide probes. Sequences of suitable gene probes can be deduced from the primary structure of a polypeptide of interest. Several genes encoding the light harvesting phycobiliproteins have been isolated in this way (36,37). The development of shuttle vectors capable of replication in both *Escherichia coli* and a cyanobacterial host has provided another means of identifying cloned genes. The use of such vectors containing cloned fragments of wild-type DNA in genetic complementation by transformation of suitable mutants, can identify specific DNA sequences (38,39). Further discussion of this procedure can be found in Section 7.3.

5. EMERGING IN VITRO TECHNIQUES

5.1 Ribosomes and cell-free translation

Despite promising early work, there is a dearth of published information on cell-free protein synthesis by extracts of cyanobacteria. Cyanobacterial ribosomes are clearly '*E.coli*-like' in polypeptide-synthesizing activity, in both *A.nidulans* and *Anabaena variabilis* (40,41). Polyribosomes have been isolated from *Nostoc* MAC using a gentle lysozyme–detergent procedure; the RNA isolated from these polysomes stimulates polypeptide synthesis in an *E.coli* cell-free system, though the products remain poorly characterized (24). Total RNA from *F.diplosiphon* will direct the synthesis in an *E.coli* S30 cell-free translation system of several abundant polypeptides and other minor species (42). There would seem to be considerable scope for the use of *E.coli* translation systems in assaying for specific transcripts within a pool of total RNA but little of such work has been reported. The ribosomes of cyanobacteria and *E.coli* are also alike in their ability to synthesize 'magic-spot' nucleotides, the highly phosphorylated guanosine nucleotides such as ppGpp, in response to a nutritional or energy shift-down (5). The production of these compounds in cyanophage-infected *A.nidulans* has proved a useful marker of the complex metabolic changes occurring upon infection (43).

5.2 DNA-dependent RNA polymerase and cyanobacterial transcription

Interest in cyanobacterial DNA-dependent RNA polymerase will have been greatly stimulated by the recent convincing demonstration of specific transcriptional control of genes involved in nitrogen fixation and heterocyst development in *Anabaena* 7120. For example, transcription of the *glnA* (glutamine synthetase) gene is initiated from an '*E.coli*-like' promoter in cells grown in ammonium-containing medium, while a '*nif*-like' promoter is employed in nitrogen fixing cultures (23). *Anabaena* RNA polymerase purified from vegetative cells recognizes *E.coli*-like promoters *in vitro*, as well as different, *Anabaena*-specific vegetative promoters, as found in the *rbcL-rbcS* (large and small subunits of ribulose bisphosphate carboxylase) operon (44). The RNA polymerase core enzyme from *E.coli* can recognize *Anabaena* promoters in the presence of a component (sigma subunit?) of the cyanobacterial enzyme. Clearly, a full analysis of transcriptional control in cyanobacteria will require a detailed analysis of RNA polymerase activity. Earlier work indicated that stable enzyme preparations could be obtained from *A.nidulans* (45), while the RNA polymerase from *F.diplosiphon* was much more labile during purification (46). Both enzymes, however, showed an activity spectrum like that of the *E.coli* polymerase; both were sensitive to rifampicin and both recognized λ promoters *in vitro* (47).

A usefully detailed method for purification of a stable (~1 year at −20°C) RNA polymerase from *A.nidulans* has recently been published (48). The enzyme so obtained was free of DNase and transcribed the *Anacystis* 16S rRNA gene *in vitro* from the same, 'chloroplast-like' promoter as used *in vivo*. Whether this protocol produces a similar stable enzyme from other cyanobacteria such as *Fremyella* remains to be seen. But the *Anacystis* enzyme, at least, should prove of great interest in transcription experiments *in vitro*.

6. CYANOPHAGES

Cyanophages, cyanobacterial viruses discovered as recently as 1963, offer unique and powerful experimental advantages in the study of cyanobacterial photosynthesis. Such systems combine the advantages of bacteriophage methodology with those of the eukaryotic-type phototrophic metabolism of the host organism. Probably the most characteristic feature of the cyanophages is their extreme physiological adaptation to the light-dependent cyanobacterial functions. Since in most cases light is the ultimate source of all ATP-generating reactions and the cells lack significant amounts of metabolic reserves, undisturbed photosynthesis until very late in the infective process is a vital condition of a successful phage cycle. Consequently, a more or less intact photophos-phorylation can be studied with a completely different background of macro-molecular syntheses, serving the sole interest of effective virus production. The cyanophage literature is a valuable contribution to our understanding of cyanobacterial metabolic features such as redox control of the key metabolic pathways, ATP synthesis and its possible sources, maintenance and sizes of amino acid pools and the regulation of DNA repair mechanism(s). The recent discovery of a temperate *Anabaena* phage and the simplicity of genetic analysis suggest that cyanophages may play an important role in both *in vivo* and *in vitro* genetic manipulations with cyanobacteria in the near future (49–51).

All known cyanophages are tailed viruses with double-stranded DNA and show remarkable structural similarities to large bacteriophages. Some of the widely used representatives of the group are listed in *Table 5*. Complete bibliographies of cyanophage literature are published yearly and distributed upon request by R.S.Safferman and M.E.Morris, Environmental Monitoring and Support Laboratory, US Environmental Protection Agency, Cincinnati, OH 45268, USA.

6.1 Preparation of phage stocks

6.1.1 *Factors influencing virus production*

Cyanobacteria are fairly sensitive to the quality and intensity of illumination. The self shading/filtering effect of the cultures can greatly influence both of these characteristics. In a dense culture the transmitted light actually reaching the cells is weak and green. Choose culturing vessels which are easy to transilluminate. Light of both too high and too low an intensity can cause abortive infection. In general, laboratory conditions which are optimal for the growth of the host cyanobacterium are favourable for efficient viral infection.

6.1.2 *Preparation of crude lysate*

(i)   Inoculate 900 ml of Allen's liquid medium with a few millilitres of a starter cyanobacterium culture. Use a cylindrical glass culturing vessel with a diameter of 6–8 cm. Expose the cell suspension to fluorescent white light with an intensity of $3–5 \times 10^4$ mW/m^2 and gas the culture with 5% carbon dioxide in air.

(ii)   When the cell density reaches 5×10^7 cells/ml, infect the culture with virus adjusting to a multiplicity of infection (m.o.i.) of 5. Towards the end of the latent period the infected cells turn slightly yellowish and the onset of lysis is associated with massive foam production and a significant increase of transparency.

(iii)   When lysis is complete, remove the cellular debris by centrifugation (15 000 g for 10 min) and store the lysate at 4°C over chloroform. Under these conditions virus particles remain stable for several months.

Phage titres obtainable by direct lysis have a natural upper limit [$\sim 5 \times 10^9$ plaque-forming units (p.f.u./ml)] because increasing darkness in cultures with

Table 5. Some well characterized cyanophages.

Cyanophage	Type of infection	Mol. wt of DNA	Usual hosts[a]	Remarks
LPP-1	Virulent	27 Md	*Plectonema boryanum* PCC 73110	
N-1	Virulent	30 Md	*Anabaena variabilis* PCC 7120	
A-4(L)	Temperate	24 Md	*Anabaena variabilis* PCC 7120	Very short lytic cycle (~2.5 h)
AS-1 and AS-1M	Virulent	57 Md	*Anacystis nidulans* (*Synechococcus* spp.) PCC 6301 PCC 7942	
S-2(L)	Virulent	28 Md	*Synechococcus* sp. PCC 6907 CALU 698	2-Aminoadenine completely replaces adenine in phage DNA

[a]PCC: Pasteur Culture Collection.
CALU: Collection of Algae of Leningrad University.

cell densities over about 10^8/ml decreases the average burst size of cyanophages. In order to produce preparations of higher titres, virus particles need to be concentrated.

6.1.3 *Concentration and purification*

(i) *Differential centrifugation.* Phage particles of a cleared lysate can be sedimented by high speed centrifugation.

(1) Pellet the phage at 40 000 g at 4°C for 60 min.
(2) Discard the supernatant and layer a few millilitres of fresh liquid medium over the pellet.
(3) Leave it in the cold for a few hours.
(4) Resuspend the phage by vigorous vortexing.

(ii) *Concentration by DEAE–cellulose.* This method minimizes the chances of mechanical damage of the phage particles. Some cyanophages, however, may not bind readily to DEAE–cellulose.

(1) Prepare a thick (4 × 10 cm) column of equilibrated (10 mM Tris–HCl, pH 7.5, 0.1 mM EDTA) DEAE–cellulose (Whatman DE 52).
(2) Load 1 litre of cleared lysate on the column with a flow rate of 100 ml/h.
(3) Elute the phage with 30 ml of elution buffer (10 mM Tris–HCl, pH 8.0, 0.5 M NaCl, 0.1 mM EDTA). The final titre usually indicates a higher than 90% recovery of p.f.u.

Phage preparations can be purified to homogeneity, by centrifugation in sucrose or CsCl density gradients. Both methods may damage the virus particles causing substantial loss of infectivity. This problem can be avoided by the use of gel filtration.

(iii) *Fractionation of the phage concentrate by gel filtration*

(1) Pack a 2.5 × 50 cm Bio-Gel A-5 m (Bio-Rad) column and wash with buffer (10 mM Tris–HCl, pH 7.5, 1 mM $MgCl_2$).
(2) Load the carefully suspended phage concentrate and continue washing with the same buffer [step (1)] at a flow rate of 50 ml/h.
(3) Collect 3 ml fractions and measure A_{260}.

A typical elution profile of a cyanophage sample is shown in *Figure 2*. During the column purification, the dilution effect is negligible. Phage containing fractions have a milky appearance and a high (usually around 1.8) A_{260}/A_{280} ratio.

6.2 Infectivity assay

Much experimental work on the metabolism of phage infected cyanobacteria relies on the enumeration of infective virus particles released during the reproductive cycle. The agar layer method (*Table 6*) is the most direct way to determine the number of p.f.u. in a cyanophage sample.

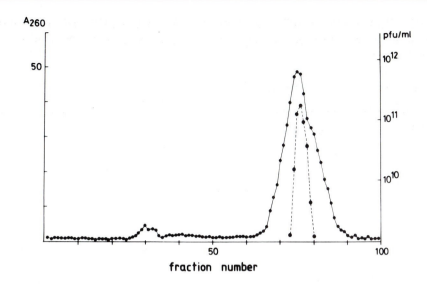

Figure 2. Purification of cyanophages. A fractionation profile on Bio-Gel A-5 m of concentrated cyanophage AS-1 lysate. A_{260}, continuous line; p.f.u./ml, dashed line.

Table 6. Agar layer method.

---

1. Prepare plates of Allen's medium supported with 1.5% agar.
2. Melt top agar (Allen's medium + 0.7% agar) and transfer 1 ml aliquots into test tubes held at 45°C.
3. Pellet a logarithmically growing cyanobacterium culture at room temperature and resuspend the cells in fresh medium, adjusting to a density of $\sim 10^9$ cells/ml.
4. Prepare serial dilutions of the phage sample and pipette 100 μl aliquots of these into the tubes containing top agar.
5. Mix 1 ml of the concentrated cell suspension in the phage containing top agar, vortex carefully and immediately spread the mixture over an agar plate.
6. Following solidification keep the plates under constant illumination ($3-5 \times 10^4$ mW/m^2) at room temperature. Colourless plaques on a fresh green background will appear in 3–4 days.

6.2.1 *Applicability of the plaque assay technique*

The agar layer method is suitable only for those indicator organisms which form a more or less even suspension in a liquid culture. In contrast to conventional bacteriophage plaque assays cyanobacteria grow very slowly within the top agar. Therefore, cyanophages are plated with a pre-grown lawn. For the same reason, in the case of temperate phages, the development of turbid plaque morphology requires prolonged incubation time. Lawns must be neither too dark nor too light to give reliable plaque counts.

6.3 **Characterization of the infective cycle**

Establishing the key parameters of the viral infective cycle is important in the characterization of a cyanophage as well as in the investigation of environmental and metabolic influences on the phage syntheses. The duration and productivity of the infective process can vary greatly under different laboratory conditions and published data are often difficult to reproduce. This is mostly due to ill-defined factors, such as spectral composition of light, geometry of culturing vessels, etc. Thus the following method helps the standardization of the infective process under our own laboratory conditions. We can determine the basic characteristics of phage infection in one experiment.

6.3.1 *One-step growth experiment*

(i)   Infect a 25 ml culture of logarithmically growing ($\sim 10^7$ cells/ml) cyano-bacteria with cyanophage at a m.o.i. of 0.01.

(ii)   Let adsorption proceed for 30 min then sediment the culture (8000 g for 5 min at room temperature).

(iii)   Discard the supernatant which contains non-adsorbed phage and re-suspend the cells in the same volume of fresh medium.

(iv)   Dilute the suspension 100-fold by transferring a sample into another culturing flask containing 25 ml of medium and incubate the culture under conditions ideal for virus multiplication.

(v)   Withdraw 100 μl aliquots at 30 min intervals and determine the number of infective centres (infected cells + free cyanophage) by the plating assay described in *Table 6*.

(i) *Latent period and average burst size. Figure 3* shows the one-step growth curve of cyanophage AS-1. A sudden increase in the number of infective centres

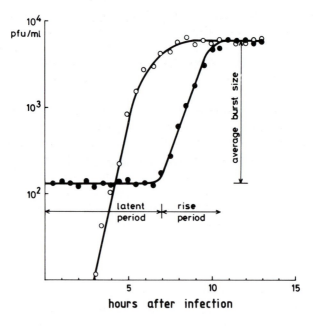

Figure 3. One-step growth curve of cyanophage AS-1. ●: data obtained by direct plating; ○: data obtained by premature lysis.

marks the end of the latent period. The ratio of the numbers of infective centres before and after the rise period gives the average burst size of the phage.

(ii) *Intracellular phage*. The number of infective centres does not provide information about the developmental stages of phage synthesis. In order to determine the number of completely assembled phage, including those inside the infected cells, the 100 µl aliquots collected throughout the one-step growth experiment should be subjected to premature lysis as follows:

(1) Add 10 µl of chloroform to each 100 µl sample in an Eppendorf tube.
(2) Homogenize carefully with 0.05 mm glass beads for 30 sec at 0°C.
(3) Add 200 µl of fresh medium.
(4) Spin the homogenate for 1 min in an Eppendorf centrifuge and assay the supernatant for p.f.u. of cyanophage (*Table 6*).

6.3.2 *General comments on the experimental system*

The following information may be useful in carrying out the experiment described above.

(i)   The infective cycle of a cyanophage is rather long and usually lasts for several hours. This is convenient for detailed study of the time course of infection.

(ii)   Adsorption can be very slow (e.g. 60 min to reach 90% adsorption) but the ratio of adsorption time to the duration of the lytic cycle, and the ratio of the latter to the generation time of healthy cells, are very similar to the corresponding time ratios seen in bacteriophage infection cycles.

(iii) The synchrony of the infection can be enhanced by introducing a low light intensity pre-incubation period following addition of the phage. Some cyanophages need light for efficient adsorption. Do not use an ice bath because chilling irreversibly destroys photosynthetic membrane structures.

(iv)   The cells of a cyanobacterial filament form an inseparable entity upon dilution and subsequent plating. Consequently, m.o.i. values in a one-step growth experiment using filamentous hosts should reflect the cyanophage/filament rather than the cyanophage/cell ratio.

(v)   Laboratory conditions for cyanophage infection are usually optimized to shorten the duration of the infective process as much as possible. When using high light intensity and temperatures, for example 30°C or over, in order to facilitate experimental work, bear in mind that such conditions rarely occur in nature.

6.4 **Radioactive labelling**

In most cyanobacteria the uptake of organic compounds is greatly limited. Thus direct addition of radioactive nucleic acid precursors or amino acids results in only a modest incorporation of the label, making pulse-labelling practically impossible. Inorganic label, however, is readily incorporated if supplied in the form of $^{32}PO_4^{3-}$, $^{35}SO_4^{2-}$ or $^{14}CO_3^{2-}$. Light greatly influences both uptake and metabolism of any radiolabelled substances.

In all cases of cyanophage infection studied to date the metabolic pools serving viral syntheses are mainly derived from the proteins and nucleic acids of the host cell. Efficient labelling of the virion can therefore be achieved by infecting *in vivo* labelled cyanobacterial hosts.

7. GENE TRANSFER IN CYANOBACTERIA

7.1 Cyanobacterial transformation

Certain cyanobacterial strains possess a natural ability to be transformed with homologous DNA. This property, though incompletely exploited at the present moment, offers great promise for the identification and manipulation of genes and gene products which are associated with photosynthesis and nitrogen fixation. In addition to their ability to be transformed, cyanobacteria can be easily mutagenized and quickly screened for a desired phenotype. These properties have yet to be fully achieved with plant genetic systems. Presented below are brief descriptions of cyanobacterial transformation protocols. For a recent review and a discussion of how these techniques can be applied to basic questions of plant molecular biology see ref. 52.

Initial characterization of transformation in the cyanobacterium, *A.nidulans* 602, suggested that few special treatments of the cells were required to achieve transformation (53). This observation has generally held true for other transformable cyanobacterial species. A simple protocol for the transformation of *A.nidulans* R2 and *Synechocystis* PCC 6803, used successfully in our laboratory, is given in *Table 7*. The comments presented below describe in greater detail the methodology involved, indicate where alternative procedures have been employed, and provide other data (e.g. procedures which do not work) that may be useful to those wishing to pursue cyanobacterial transformation.

7.1.1 *Growth and preparation of cells*

The period in the growth of a culture during which the cells are competent for plasmid transformation has been examined. Some cyanobacterial strains show a peak in transformation frequency at mid- to late-log phase of growth [such as for *A.nidulans* R2 (13,54), although see (55)], while others are apparently equally transformable throughout the active growth phase of a culture [e.g. *Agmenellum quadruplicatum* PR-6 (52)]. Treatments that have been found to either increase the frequency of transformation or which are able to induce transformation in other bacteria have either no effect or decrease the efficiency of transformation in *A.nidulans* R2. Treatments examined include: cold shock of the cells (13), CaCl$_2$ treatment (54) [this treatment, however, does render *Gloeocapsa alpicola* transformable (56)], PEG permeabilization (54), and the addition of ATP (57) (for *A.nidulans* PCC 6301).

Table 7. Transformation of *A.nidulans* R2 and *Synechocystis* PCC 6803 with DNA of chromosomal and plasmid origin (adapted from refs 20 and 21).

1. Grow the cyanobacterial strain in the appropriate medium [such as BG-11 for *A.nidulans* R2 (1) or a modified medium C of Kratz and Myers for *Synechocystis* PCC 6803 (21)], under the proper growth conditions (see, for example, ref. 21) to mid- to late-log phase of growth. This corresponds approximately to A_{730} values of 0.5 for *Anacystis* and 1.0 for *Synechocystis* PCC 6803.
2. Collect the cells by centrifugation at room temperature at 5000 g.
3. Resuspend the cells at one-half the original volume in sterile growth medium and recentrifuge as in step 2. It is important to maintain sterile conditions at all times.
4. Resuspend the cell pellet in growth medium at ⅟₁₀ the original volume (~2 × 10^8 cells/ml for *Anacystis* and *Synechocystis* PCC 6803).
5. Pipette 1 ml aliquots into sterile 1.5 ml clear microfuge tubes.
6. Add the DNA sample (0.5–2.0 µg) to the cells and incubate the mixture at 30°C for 4 h in light (3000 lux) with occasional manual agitation of the samples. Cells incubated without added DNA are useful controls.
7. At the end of the incubation period, plate the cells in top agar (growth medium plus 0.8% agar) onto solid medium and, after solidification of the agar, place the plates in light for 12–18 h.
8. Place the appropriate selective agent (e.g. antibiotic, herbicide, etc.) in a 1 ml vol. evenly underneath the agar. In this procedure, the agar is lifted from the plate with the aid of a sterile, slightly curved spatula and the 1 ml aliquot is uniformly applied to the bottom surface of the plate.
9. After absorption of the agent into the solid medium, return the plates to light. Transformant colonies are obtained in 5 days for *Anacystis* and 7 days for *Synechocystis* PCC 6803. The concentration of the selective agent should be empirically determined.
10. At 10–14 days, colonies are of sufficient size to be transferred to liquid selective medium and propagated for subsequent analysis.

Several cyanobacterial species are thought to secrete DNases and other substances that may inhibit transformation. It is, therefore, important to wash the cells several times in growth medium immediately prior to their transformation. We have also noted that preparation of the cells at room temperature increases the frequency of transformation. Occasionally, a particular culture of cyanobacteria may lose the ability to be transformed. In these instances several remedies may be taken to restore transformability to the cells:

(i) a fresh liquid culture may be inoculated from a colony grown on solid medium;
(ii) cells may be grown in liquid culture to stationary phase and an aliquot from this culture used to inoculate fresh medium for the preparation of transformation-competent cells.

7.1.2 *Preparation and uptake of DNA*

In general, two sources of donor DNA have been used to transform cyanobacteria: hybrid shuttle vectors capable of autonomous replication in both the cyanobacterial host and in *E.coli*; and DNA (either plasmid or genomic) capable of integration into host (cyanobacterial) DNA sequences.

The hybrid shuttle vectors consist of a minimum of three elements:

(i)   a fragment of cyanobacterial plasmid DNA that is capable of supporting autonomous replication in the cyanobacterium;
(ii)   a selectable marker (usually conferring antibiotic resistance);
(iii) an *E.coli* origin for plasmid DNA replication.

Integrative transformation is the insertion of a selectable marker into the chromosome of a recipient cell by way of homologous recombination between donor and host DNA. In the integrative plasmid vectors, a contiguous fragment of cyanobacterial DNA is either interrupted by the selectable marker or placed flanking the marker. A sampling of plasmid vectors available for autonomous replicative or for integrative transformation of various cyanobacterial species is presented in *Table 8*.

The use of shuttle vectors has been hampered by the lack of rec^- cyanobacterial strains. Consequently, transformants resulting from recombination between cyanobacterial-derived sequences found on the shuttle vector and the identical sequence located on a resident cyanobacterial plasmid can often be obtained (58,59). The availability of plasmid-less strains has, however, lessened this problem (13,59). Similarly, the re-introduction of a cyanobacterial gene cloned on a shuttle plasmid may (60) or may not (60,61) lead to recombination between the introduced and resident copy. The chances of maintaining stable merodiploids are increased by independently selecting for the introduced plasmid (60).

DNA prepared using so-called 'miniprep' protocols is of sufficient quality for use in transformation experiments; CsCl gradient centrifugation is not required in the preparation of plasmid and chromosomal DNA (21). The assimilation of DNA into *A.nidulans* R2 appears to be complete within 15 min [(13,54) but see (52,55)]. Quick uptake of chromosomal DNA by *Agmenellum* has also been observed with incubation times of only 30 min being needed to attain the

Table 8. Representative plasmid vectors for the transformation of cyanobacteria.

Species	Plasmid designation	Comments/features	Reference
Anacystis nidulans	pUC 303	Shuttle vector	59
	pSG 111	Shuttle vector	61
	pPLAN B2, etc.	Shuttle vectors	73
	pKW 1065, etc.	Integrative vectors	38
	pPL 1912	Vector for regulated gene expression	67
	pVAD 1	Chloroplast promoter fusion	20
Agmenellum quadruplicatum PR6	pAQE 2, etc.	Shuttle vectors	74
Anabaena	pRL 1, etc.	Vectors for conjugal transfer from *E.coli* to *Anabaena*	64
Synechocystis PCC 6803	pFCLV 7, etc.	Shuttle vectors	75
	p1BP1	Integrative vector	21

maximal number of transformants (62). The minimal amount of plasmid DNA necessary to saturate transformation again varies with species, but values of between 50 ng and 1.0 μg per 10^8 cells (in 1 ml) are frequently noted.

At present there is no consensus (even for *A.nidulans* R2) whether higher transformation frequencies are obtained by performing the transformation in darkness. Apparently, illumination during DNA uptake is not a requirement for transformation (13,55,57), darkness may or may not lead to more transformants (55,57), and, in certain cases, placing the cells in the dark during this incubation period may cause a decrease in the transformation frequency (57,38). The outcome varies with the isolate of the strain, the protocol employed, and even the particular plasmid (57) used.

7.1.3 *Selection for selectable marker*

After incubation of the cells with DNA is complete, the cells are plated either directly onto solid non-selective media or onto sterile nitrocellulose filters placed previously on the surface of the solid media (38). The cells are allowed to grow in the absence of selective pressure for 12–18 h. This period allows for recovery of the cells and for the expression of the selectable marker. The selective agent (e.g. antibiotic, herbicide, etc.) is placed evenly underneath the agar or, if the cells were plated onto a filter, the filter is lifted from the original plate and placed on the surface of a plate containing the selective agent. Cells transformed with genomic DNA are detected in 5 days in the case of *A.nidulans* R2 (54) or in 7 days for *Synechocystis* PCC 6803 (21). Transformation frequencies of 5×10^{-3} for *A.nidulans* R2 (54) and 5×10^{-4} per recipient cell for *Synechocystis* PCC 6803 (21,63) can be expected.

7.2 **Other gene transfer systems**

7.2.1 *Conjugal transfer*

Transformation in cyanobacteria has been limited to unicellular strains belonging to the classes of *Synechococcus* and *Synechocystis*. Recently, gene transfer to filamentous cyanobacteria (i.e. several strains of *Anabaena*) has been achieved (64). In this system a selectable plasmid containing an origin of DNA replication from *Nostoc* is transferred from *E.coli* to *Anabaena* by triparental conjugation. The transferred plasmid is capable of autonomous replication in *Anabaena* utilizing the *Nostoc* origin of replication. This advance is particularly significant considering that this is the first report of gene transfer to a nitrogen-fixing, heterocyst-forming, facultatively heterotrophic cyanobacterium.

7.2.2 *Permeaplast transformation*

Daniell *et al.* (65) have reported the transformation of *A.nidulans* PCC 6301 with pBR322. Transformants were obtained only after a minimum of an 18 h incubation period, with the transformation frequency reaching a maximum at 28 h. A 50-fold increase in transformation with pBR322 was observed when cells treated with lysozyme–EDTA (permeaplasts) were used as recipients. This is the first report of a plasmid containing solely *E.coli*-derived sequences being capable of replication in a cyanobacterium. Concomitant with transformation, the authors noted an amplification of chromosomal DNA and increased levels of ribulose bisphosphate carboxylase.

7.2.3 *UV-induced transformation*

Dzelzkalns and Bogorad (21) have reported that irradiation of *Synechocystis* PCC 6803 with low levels of UV light allows for stable integrative transformation of these cells by heterologous (non-cyanobacterial) DNA. In this system transformation does not rely on an autonomously replicating plasmid and insertion of DNA into the host genome occurs by non-homologous recombination. The UV treatment, it is proposed, either induces DNA repair/recombination processes that lead to the misincorporation of non-cyanobacterial DNA into host sequences or, alternatively, UV irradiation may diminish the levels of host restriction or exo-nucleases allowing the introduced DNA to survive in the cell long enough to be inserted into the genome. An advantage of this system is that it allows for direct, random integration of non-cyanobacterial DNA sequences into the cyanobacterial genome.

7.3 **Applications of cyanobacterial transformation**

7.3.1 *Gene expression*

Transformation allows the introduction of intact, modified, or foreign genes into recipient cells. The only requirement is that for a selectable or screenable phenotype.

Many of the applications of cyanobacterial transformation have naturally dealt with photosynthesis and the genes associated with this process. For example, by utilizing the transformation system of *A.nidulans* R2, Golden *et al.* (66) were able to unequivocally demonstrate that resistance to the herbicide diuron (DCMU) was conferred by the product of a *psbA* gene altered at a single codon. Additional applications to the study of photosynthesis are discussed below.

Transformation can, of course, be used to study the outcome of the process of transformation itself. A prerequisite for many experiments in cyanobacterial molecular biology is knowledge of the process of recombinant and, more specifically, how heteroduplex structures are resolved. Williams and Szalay (38) have demonstrated in *A.nidulans* R2 that structures predicted for two cross-over replacements can be obtained. In addition, single cross-over insertions are seen but at a lower frequency. The lower frequency may be a reflection of intra-chromosomal recombination between tandem repeats that are generated in this type of transformant. They also find transformants resulting from a non-reciprocal exchange with the chromosome followed by addition of the new construct to the chromosome.

Regulated and enhanced levels of gene expression are important to many experiments being undertaken with cyanobacteria. It is of both a practical and evolutionary interest that *E.coli* promoters are functional in cyanobacteria (58) as is the promoter of a chloroplast-encoded gene (20). With respect to inducible gene expression, Friedberg and Seijffers (67) have developed a system that will allow the regulated expression of genes introduced into *A.nidulans* R2. Utilizing a heat-labile repressor of phage λ, they have been able to facilitate the expression in *Anacystis* of an introduced chloramphenicol acetyl transferase gene by a simple shift in temperature.

A potentially valuable assay for gene expression during heterocyst development in *Anabaena* has been described by Schmetterer (68). The *lux* genes of *Vibrio harveyi* and *V.fischeri* have been introduced into *Anabaena* as reporter genes and the expression of these genes, as measured by light emission, can be monitored in transformed cells.

7.3.2 *Insertional inactivation*

The ability to transform cyanobacteria by homologous recombination allows one to selectively disrupt a given gene or to replace a gene with a modified copy of that gene. For example, the chloramphenicol resistance gene has been inserted into a presumptive *thi* gene from *A.nidulans* R2 (38). When *Anacystis* cells were transformed with this construct, the recovered chloramphenicol-resistant cells were found to be unhealthy.

In order to do these types of experiments with genes that are presumed to be required for photosynthetic function, it is necessary to have a cyanobacterial strain that, in addition to being transformable, is also capable of some type of heterotrophic growth. These criteria are met by *Synechocystis* PCC 6803 (1). Antibiotic resistance marker genes have been placed within selected genes cloned from this organism, returned to the cyanobacterium, and the resulting transformants tested for their ability to grow autotrophically. In this way Vermaas *et al.* (69) have shown that the *psbB* gene (chlorophyll-binding, photosystem II-associated protein) is essential for photoautotrophic growth. Similarly, genes that have been modified *in vitro* can be re-inserted into the genome and used to assign functional domains to the protein. This analysis, however, can be complicated if the target gene is present in multiple copies, as are the genes encoding the reaction centre polypeptides of photosystem II (28,33).

7.3.3 *Phenotypic complementation*

Phenotypic complementation of defined cyanobacterial mutations is a powerful approach in the identification of genes whose products are directly associated with a given process. In addition, this approach is of value in identifying genes or sequences which are required as regulatory elements or which are involved in assembly processes. One of the first cyanobacterial genes to be cloned was isolated by complementation of a transposon (Tn) 901-insertion mutation in *A.nidulans* R2 (70). The complementing gene was obtained by isolating the Tn-inactivated gene in *E.coli* and using this fragment as a probe to obtain the intact copy of the gene from a library of *Anacystis* DNA. The isolated wild-type copy of the gene is able to complement, in *Anacystis*, the Tn 901-induced mutation (namely *met⁻*). Genes of the nitrate reductase system of *A.nidulans* R2 have been obtained by use of a similar strategy (71). Recently, the isolation of mutants of *Synechocystis* PCC 6803 that are deficient in photosystem II activity has been reported (72). Complementation analysis suggests that the lesions are associated with several distinct genes and these complementing genes are being sought.

The related approach of using cloned cyanobacterial DNA to complement defined *E.coli* mutants has also been successfully employed. Williams and Szalay (38) were able to obtain a fragment of *A.nidulans* R2 DNA that complements *thi⁻ E.coli* cells and Porter *et al.* (39) were able to isolate several *Agmenellum* genes that restore prototrophy to various *E.coli* auxotrophic mutants.

8. REFERENCES

1. Rippka,R., Deruelles,J., Waterbury,J.B., Herdman,M. and Stanier,R.Y. (1979) *J. Gen. Microbiol.*, **111**, 1.
2. Doolittle,W.F. (1979) *Adv. Microbial Physiol.*, **20**, 1.
3. Carr,N.G. and Whitton,B.A. (eds) (1982) *The Biology of Cyanobacteria*. Blackwell Scientific Publications, Oxford.
4. Mann,N., Carr,N.G. and Midgley,J.E.M. (1975) *Biochim. Biophys. Acta*, **402**, 41.
5. Adams,D.G., Phillips,D.O., Nichols,J.M. and Carr,N.G. (1977) *FEBS Lett.*, **81**, 48.
6. Allen,M.M. (1968) *J. Phycol.*, **4**, 1.
7. Stanier,R.Y., Kunisawa,R., Mandel,M. and Cohen-Bazire,G. (1971) *Bacteriol. Rev.*, **35**, 171.
8. Allen,M.B. (1952) *Arch. Mikrobiol.*, **17**, 34.
9. Kratz,W. and Myers,J. (1955) *Am. J. Bot.*, **42**, 282.
10. Carr,N.G. (1969) In *Methods in Microbiology*. Norris,J.R. and Ribbons,D.W. (eds), Academic Press, Vol. 3B, p. 53.
11. Herdman,M. (1982) In *The Biology of Cyanobacteria*. Carr,N.G. and Whitton,B.A. (eds), Blackwell Scientific Publications, Oxford, p. 262.
12. Drews,G. and Weckesser,J. (1982) In *The Biology of Cyanobacteria*. Carr,N.G. and Whitton,B.A. (eds), Blackwell Scientific Publications, Oxford, p. 333.
13. Chauvat,F., Astier,C., Vedel,F. and Joset-Espardellier,F. (1983) *Mol. Gen. Genet.*, **191**, 39.
14. Reaston,J., Van den Hondel,C.A.M.J.J., Van der Ende,A., Van Arkel,G.A., Stewart, W.D.P. and Herdman,M. (1980) *FEMS Microbiol. Lett.*, **9**, 185.
15. Van den Hondel,C.A.M.J.J., Keegstra,W., Borrias,W.E. and Van Arkel,G.A. (1979) *Plasmid*, **2**, 323.
16. Tomioka,N., Shinozaki,K. and Sugiura,M. (1981) *Mol. Gen. Genet.*, **184**, 359.
17. Curtis,S.E. and Haselkorn,R. (1983) *Proc. Natl. Acad. Sci. USA*, **80**, 1835.
18. Lynn,M.E., Bantle,J.A. and Ownby,J.D. (1986) *J. Bacteriol.*, **167**, 940.
19. Marmur,J. (1961) *J. Mol. Biol.*, **3**, 208.
20. Dzelzkalns,V.A., Owens,G.C. and Bogorad,L. (1984) *Nucleic Acids Res.*, **12**, 8917.
21. Dzelzkalns,V.A. and Bogorad,L. (1986) *J. Bacteriol.*, **165**, 964.
22. Crouch,D.H., Ownby,J.D. and Carr,N.G. (1983) *J. Bacteriol.*, **156**, 979.
23. Tumer,N.E., Robinson,S.J. and Haselkorn,R. (1983) *Nature*, **306**, 337.
24. Gupta,M. and Carr,N.G. (1983) *J. Gen. Microbiol.*, **129**, 2359.
25. Nichols,J.M., Foulds,I.J., Crouch,D.H. and Carr,N.G. (1982) *J. Gen. Microbiol.*, **128**, 2739.
26. Maniatis,T., Fritsch,E.F. and Sambrook,J. (1982) *Molecular Cloning: A Laboratory Manual*, Cold Spring Harbor Laboratory Press, Cold Spring Harbor, New York.
27. Golden,S., Brusslan,J. and Haselkorn,R. (1986) *EMBO J.*, **5**, 2789.
28. Mulligan,B.J., Schultes,N., Chen,L. and Bogorad,L. (1984) *Proc. Natl. Acad. Sci. USA*, **81**, 2693.
29. Conley,P.B., Lemaux,P.G. and Grossman,A.R. (1985) *Science*, **230**, 550.
30. Glisin,V., Crkvenjakov,R. and Byus,C. (1974) *Biochemistry*, **13**, 2633.

31. Carmichael,G.G. and McMaster,G.K. (1980) In *Methods in Enzymology*. Grossman,L. and
    . Moldave,K. (eds), Academic Press, New York, Vol. 65, p. 380.
32. Lehrach,H., Diamond,D., Wozney,J.M. and Boedtker,H. (1977) *Biochemistry*, **16**, 4743.
33. Curtis,S.E. and Haselkorn,R. (1984) *Plant Mol. Biol.*, **3**, 249.
34. Shinozaki,K., Yamada,C., Takahata,N. and Sugiura,M. (1983) *Proc. Natl. Acad. Sci. USA*,
    80, 4050.
35. Mazur,B.J., Rice,D. and Haselkorn,R. (1980) *Proc. Natl. Acad. Sci. USA*, **77**, 186.
36. Mazel,D., Guglielmi,G., Houmard,J., Sidler,W., Bryant,D.A. and Tandeau de Marsac,N.
    (1986) *Nucleic Acids Res.*, **14**, 8279.
37. Pilot,T.A. and Fox,J.L. (1984) *Proc. Natl. Acad. Sci. USA*, **81**, 6983.
38. Williams,J.G.K. and Szalay,A.A. (1983) *Gene*, **24**, 37.
39. Porter,R.D., Buzby,J.S., Pilon,A., Fields,P.I., Dubbs,J.M. and Stevens,S.E.,Jr. (1986) *Gene*,
    41, 249.
40. Leach,C.K. and Carr,N.G. (1974) *J. Gen. Microbiol.*, **81**, 47.
41. Bazin,J.M. (1970) *Brit. Phycol. J.*, **5**, 155.
42. Larrinua,I.M. and Mulligan,B.J. unpublished observations.
43. Borbely,G., Kari,Cs., Gulyas,A. and Farkas,G.L. (1980) *J. Bacteriol.*, **144**, 859.
44. Tumer,N.E., Richaud,C., Borbely,G. and Haselkorn,R. (1985) *J. Cell Biochem.*, Suppl. *0* (9
    part C), 244.
45. Herzfeld,F. and Rath,N. (1974) *Biochim. Biophys. Acta*, **374**, 431.
46. Miller,S.S. and Bogorad,L. (1978) *Plant Physiol.*, **62**, 995.
47. Miller,S.S., Ausubel,F.M. and Bogorad,L. (1979) *J. Bacteriol.*, **140**, 246.
48. Kumano,M., Tomioka,N., Shinozaki,K. and Sugiura,M. (1986) *Mol. Gen. Genet.* **202**, 173.
49. Padan,E. and Shilo,M. (1973) *Bacteriol. Rev.* **37**, 343.
50. Gromov,B.V. (1983) *Ann. Microbiol.*, **134**, 43.
51. Sherman,L.A. and Pauw,P. (1976) *Virology*, **71**, 17.
52. Porter,R.D. (1986) *CRC Crit. Rev. Microbiol.*, **13**, 111.
53. Shestakov,S.V. and Khyen,N.T. (1970) *Mol. Gen. Genet.*, **107**, 372.
54. Dzelzkalns,V.A. and Bogorad,L. unpublished observations.
55. Golden,S.S. and Sherman,L.A. (1984) *J. Bacteriol.*, **158**, 36.
56. Devilly,C.I. and Houghton,J.A. (1977) *J. Gen. Microbiol.*, **98**, 277.
57. Daniell,H. and McFadden,B.A. (1986) *Mol. Gen. Genet.* **204**, 243.
58. Kuhlemeier,C.J., Borrias,W.E., Van den Hondel,C.A.M.J.J. and Van Arkel,G.A. (1981)
    Mol. Gen. Genet., **184**, 249.
59. Kuhlemeier,C.J., Thomas,A.A.M., Van der Ende,A., Van Leen,R.W., Borrias,W.E., Van
    den Hondel,C.A.M.J.J. and Van Arkel,G.A. (1983) *Plasmid*, **10**, 156.
60. Kuhlemeier,C.J., Hardon,E.M., Van Arkel,G.A. and Van de Vate,C. (1985) *Plasmid*, **14**,
    200.
61. Golden,S.S. and Sherman,L.A. (1983) *J. Bacteriol.*, **155**, 966.
62. Stevens,S.E.,Jr and Porter,R.D. (1980) *Proc. Natl. Acad. Sci. USA*, **77**, 6052.
63. Grigorieva,G. and Shestakov,S. (1982) *FEMS Microbiol. Lett.*, **13**, 367.
64. Wolk,C.P., Vonshak,A., Kehoe,P. and Elhai,J. (1984) *Proc. Natl. Acad. Sci. USA*, **81**, 1561.
65. Daniell,H., Sarojini,G. and McFadden,B.A. (1986) *Proc. Natl. Acad. Sci. USA*, **83**, 2546.
66. Golden,S.S. and Haselkorn,R. (1985) *Science*, **229**, 1104.
67. Friedberg,D. and Seijffers,J. (1986) *Mol. Gen. Genet.*, **203**, 505.
68. Schmetterer,G., Wolk,C.P. and Elhai,J. (1986) *J. Bacteriol.*, **167**, 411.
69. Vermaas,W.F.J., Williams,J.G.K., Rutherford,A.W., Mathis,P. and Arntzen,C.J. (1986)
    Proc. Natl. Acad. Sci. USA, **83**, 9474.
70. Tandeau de Marsac,N., Borrias,W.E., Kuhlemeier,C.J., Castets,A.M., Van Arkel,G.A. and
    Van den Hondel,C.A.M.J.J. (1982) *Gene*, **20**, 111.
71. Kuhlemeier,C.J., Logtenberg,T., Stoorvogel,W., Van Heugten,A.A., Borrias,W.E. and Van
    Arkel,G.A. (1984) *J. Bacteriol.*, **159**, 36.
72. Dzelzkalns,V.A. and Bogorad,L. (1987) In *Progress in Photosynthesis Research*. Biggins,J.
    (ed.), Martinius Nijhoff, Vol. IV, p. 841.
73. Gendel,S., Straus,N., Pulleyblank,D. and Williams,J. (1983) *J. Bacteriol.*, **156**, 148.
74. Buzby,J.S., Porter,R.D. and Stevens,S.E.,Jr (1983) *J. Bacteriol.*, **154**, 1446.
75. Chauvat,F., De Vries,L., Van der Ende,A. and Van Arkel,G.A. (1986) *Mol. Gen. Genet.*,
    204, 185.

APPENDIX

β-Glucuronidase (GUS)

A recently developed reporter gene that has rapidly become very popular among plant molecular biologists is the *E.coli* β-glucuronidase gene. This gene shares many useful characteristics with the most commonly used reporter gene outside the plant field, the *E.coli lacZ* gene, with the great advantage over *lacZ* that in many plant species no detectable background activity is present. It is important, however, to mention that significant background activities are present in some plant species such as rice and wheat.

Two methods to detect β-glucuronidase activity have been reported: a very simple and straightforward spectrophotometric assay that can be used in most cases, although it is only moderately sensitive, and a fluorometric assay that is highly sensitive but, as its name indicates, requires a spectrofluorimeter that may not be available in every laboratory.

Another advantage of this system is that it can be used to detect cell type specific expression using histochemical staining methods that are relatively simple and very sensitive.

Cellular extracts from different plant organs, callus tissue or protoplasts can be easily obtained by grinding the tissue with a glass rod in the presence of extraction buffer. Sonication is best when working with protoplasts. Extracts are stable for at least one month when stored at 4°C or −70°C, avoid storing at −20°C. To obtain more information about the GUS system consult ref. 1.

SPECTROPHOTOMETRIC ASSAY

Steps 2, 3, 4 and 6 should be carried out at 4°C.

(i)   Weigh 100 mg of tissue and transfer to Eppendorf tubes (it is possible to use less tissue and longer assay times or a fixed time point).

(ii)   Add 100 μl of extraction buffer (*Table 1*) and grind with a glass rod until homogenized.

(iii)   Spin down for 5 min in a microcentrifuge at 4°C.

(iv)   Transfer supernatant to clean Eppendorf tubes.

(v)   Determine total protein content using an aliquot of the extract (we recommend the microassay using the Bradford reagent from Biorad).

(vi)   Transfer to clean Eppendorf tubes aliquots of the extracts equivalent to 50–100 μg of total protein. Prepare several tubes for each extract if a kinetic analysis is intended.

(viii) Add prewarmed (37°C) reaction buffer (*Table 1*) to a final volume of 1 ml.

(ix)   Mix thoroughly by vortexing.

(x)   Incubate at 37°C from 15 min to overnight, depending on the level of enzyme present in the extract (intensity of yellow colour in the mix is a good indicator of the reaction). It is best to prepare several reaction mixes

Table 1. Reagents for β-glucuronidase assays.

Reagents
1. Extraction buffer
    50 mM $NaPO_4$
    10 mM β-Mercaptoethanol
    10 mM Na_2EDTA
    0.1% Sodium lauryl phosphate
    0.1% Triton X-100
2. Reaction buffer (spectrophotometric assay)
    1 mM *p*-Nitrophenyl β-D-glucuronide in extraction buffer
3. Stop buffer (spectrophotometric assay)
    2.5 M 2-Amino-2-methyl propanediol
4. Reaction buffer (fluorometric assay)
    1 mM 4-Methyl β-D-umbelliferyl glucuronide. Dissolve 2.2 mg of MUG directly in 10 ml of extraction buffer
5. Stop buffer (fluorometric assay)
    0.2 M Na_2CO_3

for each extract and stop them at different time points (usually 5 time points from 0–60 min are sufficient) to have a kinetic analysis and avoid saturation of the system.

(xi)   Stop the reaction by adding 0.4 ml of 2.5 M 2-amino-2-methyl propanediol. The zero time points must be stopped immediately after adding the reaction buffer.

(xii) Determine absorbance at 415 nm. Use as a blank for the spectrophotometer the zero time point corresponding to each extract.

In this assay it is always important to use the appropriate blank because of the possible interference of plant pigments or the presence of suspended particles in the extracts. When working with a new plant species, always include a non-transformed control to test for endogenous activity.

In a final volume of 1.4 ml an absorbance of 0.010 represents 1 nmol of product generated. One unit is the amount of enzyme that produces 1 nmol of product/min at 37°C and corresponds to 5 ng of β-glucuronidase.

FLUOROMETRIC ASSAY

Carry out steps (i)–(vi) as described for spectrophotometric assay.

(vii)   Transfer to clean Eppendorf tubes aliquots of the extracts equivalent to 1–50 μg of total protein.

(viii) Add prewarmed reaction buffer (*Table 1*) to a final volume of 0.5 ml.

(ix)   Mix thoroughly by vortexing.

(x)   Afer 1 min, transfer 100 μl of the reaction mix into an Eppendorf tube containing 0.9 ml of stop buffer (*Table 1*).

(xi)   Repeat step (iv) at regular time intervals. For high levels of GUS activity remove aliquots every 5 min (i.e. 35S promoter), or at 30 min intervals for lower levels of activity (i.e. nopaline synthase promoter).

(xii)   Determine the amount of reaction products (methylumbelliferone) with a spectrofluorimeter, using 365 nm as excitation wavelength and measuring emission at 455 nm. The spectrofluorimeter must be calibrated each time using dilutions of a standard solution of 1 mM methylumbelliferone (MU). If one establishes 100% relative fluorescence equal to the emission of 1 μM MU, the relative fluorescence of each sample can be directly read as nanomoles of MU.

This assay is about 100 times more sensitive than the spectrophotometric assay. To date the only commercially available fluorogenic substrate is 4-methyl umbelliferyl glucuronide, but probably several other alternative substrates will soon be developed. It should also be remembered that MU will fluoresce only above pH 9.

It is important to note that, due to the remarkable stability of GUS in living cells, the presence of the enzyme does not necessarily mean that the corresponding messenger is present. This is very important to keep in mind when studying regulatory sequences that direct gene expression temporally, in a developmentally regulated fashion or when transcription is switched off under certain environmental conditions.

REFERENCES

1. Jefferson,R.A. (1987) *Plant Mol. Biol. Rep.*, **5**, 387.

INDEX

A gene of maize, 189, 190, 193, 195–196
A2 gene of maize, 195–196
Actin, 126
Agarose bead protoplast culture, 167, 168, 180
Agarose gel electrophoresis of virus nucleic acids, 229, 230
Agroinfection, 221
Agrobacterium tumefaciens, 131, 221
   co-infection, 143
   co-cultivation with protoplasts, 171
   in planta transformation, 143, 144
   leaf disc transformation, 144, 145
   total DNA isolation, 140, 141
   transfer of gene constructs to, 132
Alleles,
   mutable, 193, 195, 196
   modifiers, 204
Alkaline phosphatase,
   calf intestinal, 45, 234, 241
Aminoglycoside antibiotics, 148, 152
Aniline blue, 124
Antibiotics,
   aminoglycoside, 148, 152
   elimination of *Agrobacterium*, 144, 172
   plant selectable markers, 148
Antibodies,
   blocking reagents, 111, 112
   fixatives, effect on antigenicity, 107
   fluorescent, 113
   labelling of sections, 110, 120
   monoclonal, 106
   polyclonal, 106
   production, 106
Antigens,
   heat stable, 110
   effect of fixatives, 107
Antisense RNA, 33, 222
Anoxygenic photosynthesis, 277
ATPase, 72, 96
Autofluorescence of plant tissue, 113
Autolysin of *Chlamydomonas*, 255
Autonomously replicating sequences (*ars*), 169
Autoradiography, 30–31, 32, 105
Auxin, 132
Avidin, 126
Axenic plants, 141, 175

β-Galactosidase assay, 155–156
β-Glucuronidase, 157, 301
Bacteriophage lambda, 42
   amplification, 48
   arm preparation, 45
   Charon 35, 42, 45
   concentration, 47
   in vitro packaging, 47
   plaque hybridization, 48
   recombinant phage, isolation, 48
   recombinant phage, large-scale isolation, 50
   recombinant phage, small-scale isolation, 50
Bacteriophage promoters,
   SP6, 17, 72
   T4, 17
   T7, 17, 72
Biotin labelling of DNA, 127–128
Biotinylated nucleotides, 126, 127
Binary vectors, 133, 136, 138–139
Bleomycin, 148
Border repeats of Ti-plasmid, 132, 136

Calcofluor, 113, 124
Callus cultures, 80
Cauliflower mosaic virus (CaMV), 169, 222
   35S promoter, 169
cDNA clones of mitochondrial RNA, 98
cDNA probes of RNA viruses, 247
cDNA probes of virus RNA, 229
CETAB miniprep of plant DNA, 183
Chlamydomonas reinhardtii, 253
   autolysin, 255
   cell disruption, gentle, 255
   cell disruption, mechanical, 256
   cell lysis for protein analysis, 264
   cell-wall deficient mutants, 255
   CsCl gradients with EtBr, 257
   CsCl gradients without EtBr, 258
   DNA preparation, 256, 259
   DNA isolation, rapid, 260
   DNA mini-preps, rapid, 261
   growth, 253
   mating, 255
   media, 254
   plasmids, 274
   protein labelling *in vivo*, 269
   proteins, 264
   proteins, membrane, 270
   proteins, PAGE, 265–269
   RNA isolation, 261–263
   transformation, 271–273
   transformation, chloroquinone, 273
   transformation, poly-L-ornithine, 273
   selectable markers, 273
   Western blotting, 269–271
Chloramphenicol acetyl transferase (CAT), 138, 152, 153, 169, 297

305